The methodology of
scientific research programmes

The methodology of scientific research programmes

Philosophical Papers

Volume 1

IMRE LAKATOS

EDITED BY

JOHN WORRALL AND GREGORY CURRIE

CAMBRIDGE UNIVERSITY PRESS

CAMBRIDGE

LONDON · NEW YORK · MELBOURNE

Published by the Syndics of the Cambridge University Press
The Pitt Building, Trumpington Street, Cambridge CB2 1RP
Bentley House, 200 Euston Road, London NW1 2DB
32 East 57th Street, New York, NY 10022, USA
296 Beaconsfield Parade, Middle Park, Melbourne 3206, Australia

First published 1978

Printed in Great Britain
at the University Press, Cambridge

Library of Congress Cataloguing in Publication Data
Lakatos, Imre.
The methodology of scientific research programmes
(His Philosophical papers; v. 1)
Bibliography: p.
Includes index.
1. Science – Methodology – Collected works.
2. Science – History – Collected works. 3. Science –
Philosphy – Collected works. I. Title.
Q175.L195 vol. 1   501s [501'.8]   77-71415
ISBN 0 521 21644 3

Contents

Editors' introduction

When Imre Lakatos died in 1974, many friends and colleagues expressed the hope that his unpublished papers would be made available. Some were also interested in seeing his contributions to journals and conference proceedings collected together in a book. At the request of the managing committee of the Imre Lakatos Appeal Fund we have prepared two volumes of selected papers which we hope will meet these demands.

None of the papers published here for the first time was regarded by Lakatos as entirely satisfactory. Some are early drafts, while others seem not to have been intended for publication. We have pursued a fairly liberal policy, including papers which, at least in their present form, Lakatos would not have allowed to go to print. As for previously published papers, we have included them all except for the two papers, 'The Role of Crucial Experiments in Science' and 'Criticism and the Methodology of Scientific Research Programmes', which would have introduced undue repetition, and except for *Proofs and Refutations*, which recently appeared in book form.

Volume 1 is a collection of Lakatos's best known articles developing the methodology of scientific research programmes, together with a hitherto unpublished essay on the effect of Newton's scientific achievement, and a new 'Postscript' to the already published paper on the Copernican Revolution.

Although Lakatos perhaps came to be better known for his work in the philosophy of the physical sciences, he regarded himself as primarily a philosopher of mathematics. Volume 2 contains papers on the philosophy of mathematics, as well as some critical essays on contemporary philosophers, and some short polemical pieces reflecting his concern with political and educational matters, which, among other things, give an impression of his forceful personality.

Information about the history of the material published here is included as introductory footnotes to each paper. These and other editorial footnotes are indicated by asterisks. (We have tried to minimise these editorial footnotes particularly in the case of previously published papers.)

Offprints of some of the published papers found in Lakatos's library contained handwritten corrections and we have incorporated these wherever possible. In preparing the previously unpublished papers for the press, we have taken the liberty of introducing some presentational alterations where the original text was incomplete, or seemed likely to be misleading, or where minor alterations seemed to produce major increases in readability. We felt justified in making these changes because Lakatos always took great care over the presentation of any of his material which was to be published and, prior to publication, he always had such material widely circulated among colleagues and friends for criticism and suggested improvements. These newly published papers would undoubtedly have undergone this treatment and the resulting changes have been much more far reaching than those we have dared to introduce. Wherever the device of enclosing our alterations within square brackets worked easily and smoothly we have adopted it. (However, square brackets within quotations from other authors enclose Lakatos's own insertions.)

Where Lakatos mentioned a paper reprinted in either of the present volumes, we have altered the style of reference. So, for example, 'Lakatos [1970a]' becomes 'this volume, chapter 1', and 'Lakatos [1968b]' becomes 'volume 2, chapter 8'.

Chapter 3 ('Popper on demarcation and induction') is reprinted by kind permission of Professor P. A. Schillp and the Open Court publishing company; chapter 4 ('Why did Copernicus's research programme supersede Ptolemy's?') is reprinted by kind permission of Professor Robert Westman and the Regents of California University Press.

A generous grant from the *Fritz Thyssen Stiftung* made possible the creation of an archive of Lakatos's papers – an essential preliminary to the publication of these volumes. We should like to thank Nicholas Krasso and Professors Kilmister and Yourgrau for helping us to supply some missing references, and Alex Bellamy and Allison Quick for compiling the indexes. We should also like to thank Sandra Mitchell for her help, especially for her research work in connection with chapter 5 of this volume. Several of our editorial problems were resolved during valuable discussions with John Watkins. We are especially grateful to Gillian Page for her kind cooperation in making Lakatos's papers available to us and for her consistently helpful advice.

The editing of these two volumes has been in many ways a sad and frustrating experience. 'If only we could talk this over with Imre', was a thought which often recurred. Nevertheless, as people whose own ideas were fundamentally affected by the force of his intellect and personality, we are very happy to have been involved in making Lakatos's work more widely available.

J.W.
G.C.

Introduction: Science and Pseudoscience*

Man's respect for knowledge is one of his most peculiar characteristics. Knowledge in Latin is *scientia*, and science came to be the name
of the most respectable kind of knowledge. But what distinguishes
knowledge from superstition, ideology or pseudoscience? The Catholic Church excommunicated Copernicans, the Communist Party
persecuted Mendelians on the ground that their doctrines were
pseudoscientific. The demarcation between science and pseudoscience is not merely a problem of armchair philosophy: it is of vital
social and political relevance.

Many philosophers have tried to solve the problem of demarcation
in the following terms: a statement constitutes knowledge if sufficiently
many people believe it sufficiently strongly. But the history of thought
shows us that many people were totally committed to absurd beliefs.
If the strength of beliefs were a hallmark of knowledge, we should
have to rank some tales about demons, angels, devils, and of heaven
and hell as knowledge. Scientists, on the other hand, are very sceptical
even of their best theories. Newton's is the most powerful theory
science has yet produced, but Newton himself never believed that
bodies attract each other at a distance. So no degree of commitment
to beliefs makes them knowledge. Indeed, the hallmark of scientific
behaviour is a certain scepticism even towards one's most cherished
theories. Blind commitment to a theory is not an intellectual virtue:
it is an intellectual crime.

Thus a statement may be pseudoscientific even if it is eminently
'plausible' and everybody believes in it, and it may be scientifically
valuable even if it is unbelievable and nobody believes in it. A theory
may even be of supreme scientific value even if no one understands
it, let alone believes it.

The cognitive value of a theory has nothing to do with its psychological influence on people's minds. Belief, commitment, understanding are states of the human mind. But the objective, scientific value
of a theory is independent of the human mind which creates it or
understands it. Its scientific value depends only on what objective
support these conjectures have in facts. As Hume said:

---

* This paper was written in early 1973 and was originally delivered as a radio lecture.
It was broadcast by the Open University on 30 June 1973. (*Eds.*)

If we take in our hand any volume; of divinity, or school metaphysics, for instance; let us ask, does it contain any abstract reasoning concerning quantity or number? No. Does it contain any experimental reasoning concerning matter of fact and existence? No. Commit it then to the flames. For it can contain nothing but sophistry and illusion.

But what is 'experimental' reasoning? If we look at the vast seventeenth-century literature on witchcraft, it is full of reports of careful observations and sworn evidence – even of experiments. Glanvill, the house philosopher of the early Royal Society, regarded witchcraft as the paradigm of experimental reasoning. We have to define experimental reasoning before we start Humean book burning.

In scientific reasoning, theories are confronted with facts; and one of the central conditions of scientific reasoning is that theories must be supported by facts. Now how exactly can facts support theory?

Several different answers have been proposed. Newton himself thought that he proved his laws from facts. He was proud of not uttering mere hypotheses: he only published theories proven from facts. In particular, he claimed that he deduced his laws from the 'phenomena' provided by Kepler. But his boast was nonsense, since according to Kepler, planets move in ellipses, but according to Newton's theory, planets would move in ellipses only if the planets did not disturb each other in their motion. But they do. This is why Newton had to devise a perturbation theory from which it follows that no planet moves in an ellipse.

One can today easily demonstrate that there can be no valid derivation of a law of nature from any finite number of facts; but we still keep reading about scientific theories being proved from facts. Why this stubborn resistance to elementary logic?

There is a very plausible explanation. Scientists want to make their theories respectable, deserving of the title 'science', that is, genuine knowledge. Now the most relevant knowledge in the seventeenth century, when science was born, concerned God, the Devil, Heaven and Hell. If one got one's conjectures about matters of divinity wrong, the consequence of one's mistake was eternal damnation. Theological knowledge cannot be fallible: it must be beyond doubt. Now the Enlightenment thought that we were fallible and ignorant about matters theological. There is no scientific theology and, therefore, no theological knowledge. Knowledge can only be about Nature, but this new type of knowledge had to be judged by the standards they took over straight from theology: it had to be proven beyond doubt. Science had to achieve the very certainty which had escaped theology. A scientist, worthy of the name, was not allowed to guess: he had to prove each sentence he uttered from facts. This was the criterion of scientific honesty. Theories unproven from facts were regarded as sinful pseudoscience, heresy in the scientific community.

It was only the downfall of Newtonian theory in this century which made scientists realize that their standards of honesty had been utopian. Before Einstein most scientists thought that Newton had deciphered God's ultimate laws by proving them from the facts. Ampère, in the early nineteenth century, felt he had to call his book on his speculations concerning electromagnetism: *Mathematical Theory of Electrodynamic Phenomena Unequivocally Deduced from Experiment*. But at the end of the volume he casually confesses that some of the experiments were never performed and even that the necessary instruments had not been constructed!

If all scientific theories are equally unprovable, what distinguishes scientific knowledge from ignorance, science from pseudoscience?

One answer to this question was provided in the twentieth century by 'inductive logicians'. Inductive logic set out to define the probabilities of different theories according to the available total evidence. If the mathematical probability of a theory is high, it qualifies as scientific; if it is low or even zero, it is not scientific. Thus the hallmark of scientific honesty would be never to say anything that is not at least highly probable. Probabilism has an attractive feature: instead of simply providing a black-and-white distinction between science and pseudoscience, it provides a continuous scale from poor theories with low probability to good theories with high probability. But, in 1934, Karl Popper, one of the most influential philosophers of our time, argued that the mathematical probability of all theories, scientific or pseudoscientific, given *any* amount of evidence is zero. If Popper is right, scientific theories are not only equally unprovable but also equally improbable. A new demarcation criterion was needed and Popper proposed a rather stunning one. A theory may be scientific even if there is not a shred of evidence in its favour, and it may be pseudoscientific even if all the available evidence is in its favour. That is, the scientific or non-scientific character of a theory can be determined independently of the facts. A theory is 'scientific' if one is prepared to specify in advance a crucial experiment (or observation) which can falsify it, and it is pseudoscientific if one refuses to specify such a 'potential falsifier'. But if so, we do not demarcate scientific theories from pseudoscientific ones, but rather scientific method from non-scientific method. Marxism, for a Popperian, is scientific if the Marxists are prepared to specify facts which, if observed, make them give up Marxism. If they refuse to do so, Marxism becomes a pseudoscience. It is always interesting to ask a Marxist, what conceivable event would make him abandon his Marxism. If he is committed to Marxism, he is bound to find it immoral to specify a state of affairs which can falsify it. Thus a proposition may petrify into pseudoscientific dogma or become genuine knowledge, depending on whether we are prepared to state observable conditions which would refute it.

Is, then, Popper's falsifiability criterion the solution to the problem of demarcating science from pseudoscience? No. For Popper's criterion ignores the remarkable tenacity of scientific theories. Scientists have thick skins. They do not abandon a theory merely because facts contradict it. They normally either invent some rescue hypothesis to explain what they then call a mere anomaly or, if they cannot explain the anomaly, they ignore it, and direct their attention to other problems. Note that scientists talk about anomalies, recalcitrant instances, not refutations. History of science, of course, is full of accounts of how crucial experiments allegedly killed theories. But such accounts are fabricated long after the theory had been abandoned. Had Popper ever asked a Newtonian scientist under what experimental conditions he would abandon Newtonian theory, some Newtonian scientists would have been exactly as nonplussed as are some Marxists.

What, then, is the hallmark of science? Do we have to capitulate and agree that a scientific revolution is just an irrational change in commitment, that it is a religious conversion? Tom Kuhn, a distinguished American philosopher of science, arrived at this conclusion after discovering the naïvety of Popper's falsificationism. But if Kuhn is right, then there is no explicit demarcation between science and pseudoscience, no distinction between scientific progress and intellectual decay, there is no objective standard of honesty. But what criteria can he then offer to demarcate scientific progress from intellectual degeneration?

In the last few years I have been advocating a methodology of scientific research programmes, which solves some of the problems which both Popper and Kuhn failed to solve.

First, I claim that the typical descriptive unit of great scientific achievements is not an isolated hypothesis but rather a research programme. Science is not simply trial and error, a series of conjectures and refutations. 'All swans are white' may be falsified by the discovery of one black swan. But such trivial trial and error does not rank as science. Newtonian science, for instance, is not simply a set of four conjectures – the three laws of mechanics and the law of gravitation. These four laws constitute only the 'hard core' of the Newtonian programme. But this hard core is tenaciously protected from refutation by a vast 'protective belt' of auxiliary hypotheses. And, even more importantly, the research programme also has a 'heuristic', that is, a powerful problem-solving machinery, which, with the help of sophisticated mathematical techniques, digests anomalies and even turns them into positive evidence. For instance, if a planet does not move exactly as it should, the Newtonian scientist checks his conjectures concerning atmospheric refraction, concerning propagation of light in magnetic storms, and hundreds of other conjectures which are all part of the programme. He may even invent a hitherto unknown planet and calculate its position, mass and velocity in order to explain the anomaly.

Now, Newton's theory of gravitation, Einstein's relativity theory,

quantum mechanics, Marxism, Freudianism, are all research programmes, each with a characteristic hard core stubbornly defended,
each with its more flexible protective belt and each with its
elaborate problem-solving machinery. Each of them, at any stage of
its development, has unsolved problems and undigested anomalies.
All theories, in this sense, are born refuted and die refuted. But are
they equally good? Until now I have been describing what research
programmes are like. But how can one distinguish a scientific or progressive programme from a pseudoscientific or degenerating one?

Contrary to Popper, the difference cannot be that some are still
unrefuted, while others are already refuted. When Newton published
his *Principia*, it was common knowledge that it could not properly
explain even the motion of the moon; in fact, lunar motion refuted
Newton. Kaufmann, a distinguished physicist, refuted Einstein's relativity theory in the very year it was published. But all the research
programmes I admire have one characteristic in common. They all
predict novel facts, facts which had been either undreamt of, or have
indeed been contradicted by previous or rival programmes. In 1686,
when Newton published his theory of gravitation, there were, for
instance, two current theories concerning comets. The more popular
one regarded comets as a signal from an angry God warning that He
will strike and bring disaster. A little known theory of Kepler's held
that comets were celestial bodies moving along straight lines. Now
according to Newtonian theory, some of them moved in hyperbolas
or parabolas never to return; others moved in ordinary ellipses. Halley,
working in Newton's programme, calculated on the basis of observing
a brief stretch of a comet's path that it would return in seventy-two
years' time; he calculated to the minute when it would be seen again
at a well-defined point of the sky. This was incredible. But seventy-two
years later, when both Newton and Halley were long dead, Halley's
comet returned exactly as Halley predicted. Similarly, Newtonian
scientists predicted the existence and exact motion of small planets
which had never been observed before. Or let us take Einstein's
programme. This programme made the stunning prediction that if
one measures the distance between two stars in the night and if one
measures the distance between them during the day (when they are
visible during an eclipse of the sun), the two measurements will be
different. Nobody had thought to make such an observation before
Einstein's programme. Thus, in a progressive research programme,
theory leads to the discovery of hitherto unknown novel facts. In
degenerating programmes, however, theories are fabricated only in
order to accommodate known facts. Has, for instance, Marxism ever
predicted a stunning novel fact successfully? Never! It has some
famous unsuccessful predictions. It predicted the absolute impoverish-
ment of the working class. It predicted that the first socialist revolution would take place in the industrially most developed society. It

predicted that socialist societies would be free of revolutions. It predicted that there will be no conflict of interests between socialist countries. Thus the early predictions of Marxism were bold and stunning but they failed. Marxists explained all their failures: they explained the rising living standards of the working class by devising a theory of imperialism; they even explained why the first socialist revolution occurred in industrially backward Russia. They 'explained' Berlin 1953, Budapest, 1956, Prague 1968. They 'explained' the Russian–Chinese conflict. But their auxiliary hypotheses were all cooked up after the event to protect Marxian theory from the facts. The Newtonian programme led to novel facts; the Marxian lagged behind the facts and has been running fast to catch up with them.

To sum up. The hallmark of empirical progress is not trivial verifications: Popper is right that there are millions of them. It is no success for Newtonian theory that stones, when dropped, fall towards the earth, no matter how often this is repeated. But so-called 'refutations' are not the hallmark of empirical failure, as Popper has preached, since all programmes grow in a permanent ocean of anomalies. What really count are dramatic, unexpected, stunning predictions: a few of them are enough to tilt the balance; where theory lags behind the facts, we are dealing with miserable degenerating research programmes.

Now, how do scientific revolutions come about? If we have two rival research programmes, and one is progressing while the other is degenerating, scientists tend to join the progressive programme. This is the rationale of scientific revolutions. But while it is a matter of intellectual honesty to keep the record public, it is not dishonest to stick to a degenerating programme and try to turn it into a progressive one.

As opposed to Popper the methodology of scientific research programmes does not offer instant rationality. One must treat budding programmes leniently: programmes may take decades before they get off the ground and become empirically progressive. Criticism is not a Popperian quick kill, by refutation. Important criticism is always constructive: there is no refutation without a better theory. Kuhn is wrong in thinking that scientific revolutions are sudden, irrational changes in vision. The history of science refutes both Popper and Kuhn: on close inspection both Popperian crucial experiments and Kuhnian revolutions turn out to be myths: what normally happens is that progressive research programmes replace degenerating ones.

The problem of demarcation between science and pseudoscience has grave implications also for the institutionalization of criticism. Copernicus's theory was banned by the Catholic Church in 1616 because it was said to be pseudoscientific. It was taken off the index in 1820 because by that time the Church deemed that facts had proved it and therefore it became scientific. The Central Committee of the Soviet Communist Party in 1949 declared Mendelian genetics pseudo-scientific and had its advocates, like Academician Vavilov, killed in concentration camps; after Vavilov's murder Mendelian genetics was rehabilitated; but the Party's right to decide what is science and publishable and what is pseudoscience and punishable was upheld. The new liberal Establishment of the West also exercises the right to deny freedom of speech to what it regards as pseudoscience, as we have seen in the case of the debate concerning race and intelligence. All these judgments were inevitably based on some sort of demarcation criterion. This is why the problem of demarcation between science and pseudoscience is not a pseudo-problem of armchair philosophers: it has grave ethical and political implications.

I

Falsification and the Methodology of Scientific Research Programmes*

1 SCIENCE: REASON OR RELIGION

For centuries knowledge meant proven knowledge – proven either by the power of the intellect or by the evidence of the senses. Wisdom and intellectual integrity demanded that one must desist from unproven utterances and minimize, even in thought, the gap between speculation and established knowledge. The proving power of the intellect or the senses was questioned by the sceptics more than two thousand years ago; but they were browbeaten into confusion by the glory of Newtonian physics. Einstein's results again turned the tables and now very few philosophers or scientists still think that scientific knowledge is, or can be, proven knowledge. But few realize that with this the whole classical structure of intellectual values falls in ruins and has to be replaced: one cannot simply water down the ideal of proven truth – as some logical empiricists do – to the ideal of 'probable truth'[1] or – as some sociologists of knowledge do – to 'truth by [changing] consensus'.[2]

Popper's distinction lies primarily in his having grasped the full implications of the collapse of the best-corroborated scientific theory of all times: Newtonian mechanics and the Newtonian theory of gravitation. In his view virtue lies not in caution in avoiding errors, but in ruthlessness in eliminating them. Boldness in conjectures on the one hand and austerity in refutations on the other: this is Popper's recipe. Intellectual honesty does not consist in trying to entrench

* This paper was written in 1968–9 and was first published as Lakatos [1970]. There Lakatos referred to the paper as an 'improved version' of his [1968b] and a 'crude version' of his forthcoming *The Changing Logic of Scientific Discovery*, a projected book which he was never able to start. He makes the following acknowledgments: 'Some parts of [my [1968b]] are here reproduced without change with the permission of the Editor of the *Proceedings of the Aristotelian Society*. In the preparation of the new version I received much help from Tad Beckman, Colin Howson, Clive Kilmister, Larry Laudan, Eliot Leader, Alan Musgrave, Michael Sukale, John Watkins and John Worrall.' (*Eds.*)

1 The main contemporary proponent of the ideal of 'probable truth' is Rudolf Carnap. For the historical background and a criticism of this position, cf. volume 2, chapter 8.
2 The main contemporary proponents of the ideal of 'truth by consensus' are Polanyi and Kuhn. For the historical background and a criticism of this position, cf. Musgrave [1969a] and Musgrave [1969b].

or establish one's position by proving (or 'probabilifying') it – intellectual honesty consists rather in specifying precisely the conditions under which one is willing to give up one's position. Committed Marxists and Freudians refuse to specify such conditions: this is the hallmark of their intellectual dishonesty. *Belief* may be a regrettably unavoidable biological weakness to be kept under the control of criticism: but *commitment* is for Popper an outright crime.

Kuhn thinks otherwise. He too rejects the idea that science grows by accumulation of eternal truths.[1] He too takes his main inspiration from Einstein's overthrow of Newtonian physics. His main problem too is *scientific revolution*. But while according to Popper science is 'revolution in permanence', and criticism the heart of the scientific enterprise, according to Kuhn revolution is exceptional and, indeed, extra-scientific, and criticism is, in 'normal' times, anathema. Indeed for Kuhn the transition from criticism to commitment marks the point where progress – and 'normal' science – begins. For him the idea that on 'refutation' one can demand the rejection, the elimination of a theory, is 'naive' falsificationism. Criticism of the dominant theory and proposals of new theories are only allowed in the rare moments of 'crisis'. This last Kuhnian thesis has been widely criticized[2] and I shall not discuss it. My concern is rather that Kuhn, having recognized the failure both of justificationism and falsificationism in providing rational accounts of scientific growth, seems now to fall back on irrationalism.

For Popper scientific change is rational or at least rationally reconstructible and falls in the realm of the *logic of discovery*. For Kuhn scientific change – from one 'paradigm' to another – is a mystical conversion which is not and cannot be governed by rules of reason and which falls totally within the realm of the (*social*) *psychology of discovery*. Scientific change is a kind of religious change.

The clash between Popper and Kuhn is not about a mere technical point in epistemology. It concerns our central intellectual values, and has implications not only for theoretical physics but also for the underdeveloped social sciences and even for moral and political philosophy. If even in science there is no other way of judging a theory but by assessing the number, faith and vocal energy of its supporters,

---

[1] Indeed he introduces his [1962] by arguing against the 'development-by-accumulation' idea of scientific growth. But his intellectual debt is to Koyré rather than to Popper. Koyré showed that positivism gives bad guidance to the historian of science, for the history of physics can only be understood in the context of a succession of 'metaphysical' research programmes. Thus scientific changes are connected with vast cataclysmic metaphysical revolutions. Kuhn develops this message of Burtt and Koyré and the vast success of his book was partly due to his hard-hitting, direct criticism of justificationist historiography – which created a sensation among ordinary scientists and historians of science whom Burtt's, Koyré's (or Popper's) message had not yet reached. But, unfortunately, his message had some authoritarian and irrationalist overtones.

[2] Cf. e.g. Watkins [1970] and Feyerabend [1970a].

then this must be even more so in the social sciences: truth lies in power. Thus Kuhn's position vindicates, no doubt, unintentionally, the basic political *credo* of contemporary religious maniacs ('student revolutionaries').

In this paper I shall first show that in Popper's logic of scientific discovery two different positions are conflated. Kuhn understands only one of these, 'naive falsificationism' (I prefer the term 'naive methodological falsificationism'); I think that his criticism of it is correct, and I shall even strengthen it. But Kuhn does not understand a more sophisticated position the rationality of which is not based on 'naive' falsificationism. I shall try to explain – and further strengthen – this stronger Popperian position which, I think, may escape Kuhn's strictures and present scientific revolutions not as constituting religious conversions but rather as rational progress.

2 FALLIBILISM VERSUS FALSIFICATIONISM

To see the conflicting theses more clearly, we have to reconstruct the situation as it was in philosophy of science after the breakdown of 'justificationism'.

According to the 'justificationists' scientific knowledge consisted of proven propositions. Having recognized that strictly logical deductions enable us only to infer (transmit truth) but not to prove (establish truth), they disagreed about the nature of those propositions (axioms) whose truth can be proved by extralogical means. *Classical intellectualists* (or 'rationalists' in the narrow sense of the term) admitted very varied – and powerful – sorts of extralogical 'proofs' by revelation, intellectual intuition, experience. These, with the help of logic, enabled them to prove every sort of scientific proposition. *Classical empiricists* accepted as axioms only a relatively small set of 'factual propositions' which expressed the 'hard facts'. Their truth-value was established by experience and they constituted the *empirical basis* of science. In order to prove scientific *theories* from nothing else but the narrow empirical basis, they needed a logic much more powerful than the deductive logic of the classical intellectualists: '*inductive logic*'. All justificationists, whether intellectualists or empiricists, agreed that a singular statement expressing a 'hard fact' may *disprove* a universal theory;[1] but few of them thought that a finite conjunction of factual

---

[1] Justificationists repeatedly stressed this asymmetry between singular factual statements and universal theories. Cf. e.g. Popkin's discussion of Pascal in Popkin [1968], p. 14 and Kant's statement to the same effect as quoted in the new *motto* of the third 1969 German edition of Popper's *Logik der Forschung*. (Popper's choice of this time-honoured cornerstone of elementary logic as a *motto* of the new edition of his classic shows his main concern: to fight *probabilism*, in which this asymmetry becomes irrelevant; for probabilists theories may become almost as well established as factual propositions.)

propositions might be sufficient to *prove* 'inductively' a universal theory.[1]

Justificationism, that is, the identification of knowledge with proven knowledge, was the dominant tradition in rational thought throughout the ages. Scepticism did not deny justificatonism: it only claimed that there was (and could be) no proven knowledge and *therefore* no knowledge whatsoever. For the sceptics 'knowledge' was nothing but animal belief. Thus justificationist scepticism ridiculed objective thought and opened the door to irrationalism, mysticism, superstition.

This situation explains the enormous effort invested by classical rationalists in trying to save the synthetic *a priori* principles of intellectualism and by classical empiricists in trying to save the certainty of an empirical basis and the validity of inductive inference. For all of them *scientific honesty demanded that one assert nothing that is unproven.* However, both were defeated: Kantians by non-Euclidean geometry and by non-Newtonian physics, and empiricists by the logical impossibility of establishing an empirical basis (as Kantians pointed out, facts cannot prove propositions) and of establishing an inductive logic (no logic can infallibly increase content). It turned out that *all theories are equally unprovable.*

Philosophers were slow to recognize this, for obvious reasons: classical justificationists feared that once they conceded that theoretical science is unprovable, they would have also to conclude that it is sophistry and illusion, a dishonest fraud. The philosophical importance of *probabilism* (or '*neojustificationism*') lies in the denial that such a conclusion is necessary.

Probabilism was elaborated by a group of Cambridge philosophers who thought that although scientific theories are equally unprovable, they have different degrees of probability (in the sense of the calculus of probability) relative to the available empirical evidence.[2] *Scientific honesty then requires less than had been thought: it consists in uttering only highly probable theories; or even in merely specifying, for each scientific theory, the evidence, and the probability of the theory in the light of this evidence.*

Of course, replacing proof by probability was a major retreat for justificationist thought. But even this retreat turned out to be insufficient. It was soon shown, mainly by Popper's persistent efforts, that under very general conditions all theories have zero probability, whatever the evidence; *all theories are not only equally unprovable but also equally improbable.*[3]

[1] Indeed, even some of these few shifted, following Mill, the rather obviously insoluble problem of inductive proof (of universal from particular propositions) to the slightly less obviously insoluble problem of proving *particular* factual propositions from other *particular* factual propositions.

[2] The founding fathers of probabilism were intellectualists; Carnap's later efforts to build up an empiricist brand of probabilism failed. Cf. volume 2, chapter 8, p. 164 and also p. 160, n. 2.

[3] For a detailed discussion, cf. volume 2, chapter 8, especially pp. 154 ff.

Many philosophers still argue that the failure to obtain at least a probabilistic solution of the problem of induction means that we 'throw over almost everything that is regarded as knowledge by science and common sense.'[1] It is against this background that one must appreciate the dramatic change brought about by falsificationism in evaluating theories, and in general, in the standards of intellectual honesty. Falsificationism was, in a sense, a new and considerable retreat for rational thought. But since it was a retreat from utopian standards, it cleared away much hypocrisy and muddled thought, and thus, in fact, it represented an advance.

(a) Dogmatic (or naturalistic) falsificationism. The empirical basis

First I shall discuss a most important brand of falsificationism: dogmatic (or 'naturalistic')[2] falsificationism. Dogmatic falsificationism admits the fallibility of *all* scientific theories without qualification, but it retains a sort of infallible empirical basis. It is strictly empiricist without being inductivist: it denies that the certainty of the empirical basis can be transmitted to theories. *Thus dogmatic falsificationism is the weakest brand of justificationism.*

It is extremely important to stress that admitting (fortified) empirical counterevidence as a final arbiter against a theory does not make one a dogmatic falsificationist. Any Kantian or inductivist will agree to such arbitration. But both the Kantian and the inductivist, while bowing to a negative crucial experiment, will also specify conditions of how to establish, entrench one unrefuted theory more than another. Kantians held that Euclidean geometry and Newtonian mechanics were established with certainty; inductivists held they had probability 1. For the dogmatic falsificationist, however, empirical *counter*evidence is the *one and only* arbiter which may judge a theory.

The hallmark of dogmatic falsificationism is then the recognition that all theories are equally conjectural. Science cannot *prove* any theory. But although science cannot *prove*, it can *disprove*: it 'can perform with complete logical certainty [the act of] repudiation of what is false',[3] that is, there is an absolutely firm empirical basis of facts which can be used to disprove theories. Falsificationists provide new – very modest – standards of scientific honesty: they are willing to regard a proposition as 'scientific' not only if it is a proven factual proposition, but even if it is nothing more than a falsifiable one, that is, if there are experimental and mathematical techniques avail-

---

[1] Russell [1943], p. 683. For a discussion of Russell's justificationism, cf. volume 2, chapter 1, especially pp. 11 ff.

[2] For the explanation of this term, cf. *below*, p. 14, n. 2.

[3] Medawar [1967], p. 144. Also cf. *below*, p. 93, n. 2.

able at the time which designate certain statements as potential falsifiers.[1]

Scientific honesty then consists of specifying, in advance, an experiment such that if the result contradicts the theory, the theory has to be given up.[2] The falsificationist demands that once a proposition is disproved, there must be no prevarication: the proposition must be unconditionally rejected. To (non-tautologous) unfalsifiable propositions the dogmatic falsificationist gives short shrift; he brands them 'metaphysical' and denies them scientific standing.

Dogmatic falsificationists draw a sharp demarcation between the theoretician and the experimenter: the theoretician proposes, the experimenter – in the name of Nature – disposes. As Weyl put it: 'I wish to record my unbounded admiration for the work of the experimenter in his struggle to wrest interpretable facts from an unyielding Nature who knows so well how to meet our theories with a decisive *No* – or with an inaudible *Yes.*'[3] Braithwaite gives a particularly lucid exposition of dogmatic falsificationism. He raises the problem of the objectivity of science: 'To what extent, then, should an established scientific deductive system be regarded as a free creation of the human mind, and to what extent should it be regarded as giving an objective account of the facts of nature?' His answer is:

The form of a statement of a scientific hypothesis and its use to express a general proposition, is a human device; what is due to Nature are the observable facts which refute or fail to refute the scientific hypothesis...[In science] we hand over to Nature the task of deciding whether any of the contingent lowest-level conclusions are false. This objective test of falsity it is which makes the deductive system, in whose construction we have very great freedom, a deductive system of scientific hypotheses. Man proposes a system of hypotheses: Nature disposes of its truth or falsity. Man invents a scientific system, and then discovers whether or not it accords with observed fact.[4]

According to the logic of dogmatic falsificationism, science grows by repeated overthrow of theories with the help of hard facts. For instance, according to this view, Descartes's vortex theory of gravity was refuted – and eliminated – by the *fact* that planets moved in ellipses rather than in

---

[1] This discussion already indicates the vital importance of a demarcation between provable factual and unprovable theoretical propositions for the dogmatic falsificationist.

[2] '*Criteria of refutation* have to be laid down beforehand: it must be agreed which observable situations, if actually observed, mean that the theory is refuted' (Popper [1963a], p. 38, n. 3).

[3] Quoted in Popper [1934], section 85, with Popper's comment: 'I fully agree.'

[4] Braithwaite [1953], pp. 367–8. For the 'incorrigibility' of Braithwaite's observed facts, cf. his [1938]. While in the quoted passage Braithwaite gives a forceful answer to the problem of scientific objectivity, in another passage he points out that 'except for the straightforward generalizations of observable facts...complete refutation is no more possible than is complete proof' ([1953], p. 19). Also cf. *below*, p. 29, n. 3.

Cartesian circles; Newton's theory, however, explained successfully the then available facts, both those which had been explained by Descartes's theory and those which refuted it. Therefore Newton's theory replaced Descartes's theory. Analogously, as seen by falsificationists, Newton's theory was, in turn, refuted – proved false – by the anomalous perihelion of Mercury, while Einstein's explained that too. Thus science proceeds by bold speculations, which are never proved or even made probable, but some of which are later eliminated by hard, conclusive refutations and then replaced by still bolder, new and, at least at the start, unrefuted speculations.

Dogmatic falsificationism, however, is untenable. It rests on two false assumptions and on a too narrow criterion of demarcation between scientific and non-scientific.

The *first assumption* is that there is a natural, *psychological* borderline between theoretical or speculative propositions on the one hand and factual or observational (or basic) propositions on the other. (This, of course, is part of the *'naturalistic approach'* to scientific method.[1])

The *second assumption* is that if a proposition satisfies the psychological criterion of being factual or observational (or basic) then it is true; one may say that it was *proved* from facts. (I shall call this the *doctrine of observational (or experimental) proof.*[2])

These two assumptions secure for the dogmatic falsificationist's deadly disproofs an empirical basis from which proven falsehood can be carried by deductive logic to the theory under test.

These assumptions are complemented by a *demarcation criterion*: only those theories are 'scientific' which forbid certain observable states of affairs and therefore are factually disprovable. *Or, a theory is 'scientific' if it has an empirical basis.*[3]

But both assumptions are false. Psychology testifies against the first, logic against the second, and, finally, methodological judgment testifies against the demarcation criterion. I shall discuss them in turn.

(1) A first glance at a few characteristic examples already undermines the *first assumption*. Galileo claimed that he could 'observe' mountains on the moon and spots on the sun and that these 'observations' refuted the time-honoured theory that celestial bodies are faultless crystal balls. But his 'observations' were not 'observational'

---

[1] Cf. Popper [1934], section 10.

[2] For these assumptions and their criticism, cf. Popper [1934], sections 4 and 10. It is because of this assumption that – following Popper – I call this brand of falsificationism 'naturalistic'. Popper's 'basic propositions' should not be confused with the basic propositions discussed in this section; cf. *below*, p. 22, n. 6.

It is important to point out that these two assumptions are also shared by many justificationists who are not falsificationists: they may add to experimental proofs 'intuitive proofs' – as did Kant – or 'inductive proofs' – as did Mill. Our falsificationist accepts experimental proofs *only*.

[3] The empirical basis of a theory is the set of its potential falsifiers: the set of those observational propositions which may disprove it.

in the sense of being observed by the – unaided – senses: their reliability depended on the reliability of his telescope – and of the optical
theory of the telescope – which was violently questioned by his contemporaries. It was not Galileo's – pure, untheoretical – *observations*
that confronted Aristotelian *theory* but rather Galileo's 'observations'
in the light of his optical theory that confronted the Aristotelians'
'observations' in the light of their theory of the heavens.[1] This leaves
us with two inconsistent theories, *prima facie* on a par. Some empiricists
may concede this point and agree that Galileo's 'observations' were
not genuine observations; but they still hold that there is a 'natural
demarcation' between statements impressed on an empty and passive
mind directly by the senses – only these constitute genuine 'immediate
knowledge' – and between statements which are suggested by impure,
theory-impregnated sensations. Indeed, *all* brands of justificationist
theories of knowledge which acknowledge the senses as a source
(whether as *one* source or as *the* source) of knowledge are bound to
contain a *psychology of observation.* Such psychologies specify the 'right',
'normal', 'healthy', 'unbiased', 'careful' or 'scientific' state of the
senses – or rather the state of mind as a whole – in which they observe
truth as it is. For instance, Aristotle – and the Stoics – thought that the
right mind was the medically healthy mind. Modern thinkers
recognized that there is more to the right mind than simple 'health'.
Descartes's right mind is one steeled in the fire of sceptical doubt which
leaves nothing but the final loneliness of the *cogito* in which the *ego*
can then be re-established and God's guiding hand found to recognize
truth. All schools of modern justificationism can be characterized by
the particular *psychotherapy* by which they propose to prepare the mind
to receive the grace of proven truth in the course of a mystical
communion. In particular, for classical empiricists the right mind is
a *tabula rasa*, emptied of all original content, freed from all prejudice
of theory. But it transpires from the work of Kant and Popper – and
from the work of psychologists influenced by them – that such
empiricist psychotherapy can never succeed. For there are and can be
no sensations unimpregnated by expectation and therefore *there is no
natural (i.e. psychological) demarcation between observational and theoretical
propositions.*[2]

(2) But even if there was such a natural demarcation, logic would
still destroy the *second assumption* of dogmatic falsificationism. For the

[1] Incidentally, Galileo also showed – with the help of his optics – that if the moon was
a faultless crystal ball, it would be invisible (Galileo [1632]).

[2] True, most psychologists who turned against the idea of justificationist sensationalism did so under the influence of pragmatist philosophers like William James who
denied the possibility of any sort of objective knowledge. But, even so, Kant's
influence through Oswald Külpe, Franz Brentano and Popper's influence through
Egon Brunswick and Donald Campbell played a role in the shaping of modern
psychology; and if psychology ever vanquishes psychologism, it will be due to an
increased understanding of the Kant–Popper mainline of objectivist philosophy.

truth-value of the 'observational' propositions cannot be indubitably decided: *no factual proposition can ever be proved from an experiment.* Propositions can only be derived from other propositions, they cannot be derived from facts: one cannot prove statements from experiences – 'no more than by thumping the table'.[1] This is one of the basic points of elementary logic, but one which is understood by relatively few people even today.[2]

If factual propositions are unprovable then they are fallible. If they are fallible then clashes between theories and factual propositions are not 'falsifications' but merely inconsistencies. Our imagination may play a greater role in the formulation of 'theories' than in the formulation of 'factual propositions',[3] but they are both fallible. Thus *we cannot prove theories and we cannot disprove them either.*[4] The demarcation between the soft, unproven 'theories' and the hard, proven 'empirical basis' is non-existent: *all* propositions of science are theoretical and, incurably, fallible.[5]

(3) Finally, even if there were a natural demarcation between observation statements and theories, and even if the truth-value of observation statements could be indubitably established, dogmatic falsificationism would still be useless for eliminating the most important class of what are commonly regarded as scientific theories. For even if experiments *could* prove experimental reports, their disproving power would still be miserably restricted: *exactly the most admired scientific theories simply fail to forbid any observable state of affairs.*

To support this last contention, I shall first tell a characteristic story and then propose a general argument.

The story is about an imaginary case of planetary misbehaviour. A physicist of the pre-Einsteinian era takes Newton's mechanics and his law of gravitation, (N), the accepted initial conditions, I, and calculates, with their help, the path of a newly discovered small planet, p. But the planet deviates from the calculated path. Does our Newtonian

---

[1] Cf. Popper [1934], section 29.

[2] It seems that the first philosopher to emphasize this was Fries in 1837 (cf. Popper [1934], section 29, n. 3). This is of course a special case of the general thesis that logical relations, like logical probability or consistency, refer to *propositions*. Thus, for instance, the proposition 'nature is consistent' is false (or, if you wish, meaningless), for nature is not a proposition (or a conjunction of propositions).

[3] Incidentally, even this is questionable. Cf. *below*, p. 42 ff.

[4] As Popper put it: 'No conclusive disproof of a theory can ever be produced'; those who wait for an infallible disproof before eliminating a theory will have to wait for ever and 'will never benefit from experience' ([1934], section 9).

[5] Both Kant and his English follower, Whewell, realized that all scientific propositions, whether *a priori* or *a posteriori*, are equally theoretical; but both held that they are equally provable. Kantians saw clearly that the propositions of science are theoretical in the sense that they are not written by sensations on the *tabula rasa* of an empty mind, nor deduced or induced from such propositions. A factual proposition is only a special kind of theoretical proposition. In this Popper sided with Kant against the empiricist version of dogmatism. But Popper went a step further: in his view the propositions of science are not only theoretical but they are all also *fallible*, conjectural for ever.

physicist consider that the deviation was forbidden by Newton's theory and therefore that, once established, it refutes the theory N? No. He suggests that there must be a hitherto unknown planet p' which perturbs the path of p. He calculates the mass orbit, etc., of this hypothetical planet and then asks an experimental astronomer to test his hypothesis. The planet p' is so small that even the biggest available telescopes cannot possibly observe it: the experimental astronomer applies for a research grant to build yet a bigger one.[1] In three years' time the new telescope is ready. Were the unknown planet p' to be discovered, it would be hailed as a new victory of Newtonian science. But it is not. Does our scientist abandon Newton's theory and his idea of the perturbing planet? No. He suggests that a cloud of cosmic dust hides the planet from us. He calculates the location and properties of this cloud and asks for a research grant to send up a satellite to test his calculations. Were the satellite's instruments (possibly new ones, based on a little-tested theory) to record the existence of the conjectural cloud, the result would be hailed as an outstanding victory for Newtonian science. But the cloud is not found. Does our scientist abandon Newton's theory, together with the idea of the perturbing planet and the idea of the cloud which hides it? No. He suggests that there is some magnetic field in that region of the universe which disturbed the instruments of the satellite. A new satellite is sent up. Were the magnetic field to be found, Newtonians would celebrate a sensational victory. But it is not. Is this regarded as a refutation of Newtonian science? No. Either yet another ingenious auxiliary hypothesis is proposed or...the whole story is buried in the dusty volumes of periodicals and the story never mentioned again.[2]

This story strongly suggests that even a most respected scientific theory, like Newton's dynamics and theory of gravitation, may fail to forbid any observable state of affairs.[3] Indeed, *some scientific theories forbid an event occurring in some specified finite spatio-temporal region (or briefly, a 'singular event') only on the condition that no other factor* (possibly hidden in some distant and unspecified spatio-temporal corner of the universe) *has any influence on it.* But then *such theories never alone contradict a 'basic' statement:* they contradict at most a conjunction of

---

[1] If the tiny conjectural planet were out of the reach even of the biggest *possible* optical telescopes, he might try some quite novel instrument (like a radiotelescope) in order to enable him to 'observe it', that is, to ask Nature about it, even if only indirectly. (The new 'observational' theory may itself not be properly articulated, let alone severely tested, but he would care no more than Galileo did.)

[2] At least not until a new research programme supersedes Newton's programme which happens to explain this previously recalcitrant phenomenon. In this case, the phenomenon will be unearthed and enthroned as a 'crucial experiment'; cf. *below*, p. 68 ff.

[3] Popper asks: 'What kind of clinical responses would refute to the satisfaction of the analyst not merely a particular diagnosis but psychoanalysis itself?' ([1963], p. 38, n. 3.) But what kind of observation would refute to the satisfaction of the Newtonian not merely a particular version but Newtonian theory itself?

a basic statement describing a spatio-temporally singular event and of a universal non-existence statement saying that no other relevant cause is at work anywhere in the universe. And the dogmatic falsificationist cannot possibly claim that such universal non-existence statements belong to the empirical basis: that they can be observed and proved by experience.

Another way of putting this is to say that some scientific theories are normally interpreted as containing a *ceteris paribus* clause:[1] in such cases it is always a specific theory *together* with this clause which may be refuted. But such a refutation is inconsequential for the *specific* theory under test because by replacing the *ceteris paribus* clause by a different one the *specific* theory can always be retained whatever the tests say.

If so, the 'inexorable' disproof procedure of dogmatic falsificationism breaks down in these cases *even if* there were a firmly established empirical basis to serve as a launching pad for the arrow of the *modus tollens*: the prime target remains hopelessly elusive.[2] And as it happens, it is exactly the most important, 'mature' theories in the history of science which are *prima facie* undisprovable in this way.[3] Moreover, by the standards of dogmatic falsificationism all probabilistic theories also come under this head: for no finite sample can ever *disprove* a universal probabilistic theory;[4] probabilistic theories, like theories with a *ceteris paribus* clause, have no empirical basis. But then the dogmatic falsificationist relegates the most important scientific theories *on his own admission* to metaphysics where rational discussion – consisting, by his standards, of proofs and disproofs – has no place, since a metaphysical theory is neither provable nor disprovable. The demarcation criterion of dogmatic falsificationism is thus still strongly antitheoretical.

(Moreover, *one can easily argue that ceteris paribus clauses are not exceptions, but the rule in science.* Science, after all, must be demarcated from a curiosity shop where funny local – or cosmic – oddities are collected and displayed. The assertion that 'all Britons died from lung cancer between 1950 and 1960' is logically possible, and might even have been true. But if it has been only an occurrence of an event with minute probability, it would have only curiosity value for the crankish fact-collector, it would have a macabre entertainment value, but no scientific value. A proposition might be said to be scientific only if it

---

[1] This '*ceteris paribus*' clause need not normally be interpreted as a separate premise. For a discussion, cf. *below*, p. 98.

[2] Incidentally, we might persuade the dogmatic falsificationist that his demarcation criterion was a very naive mistake. If he gives it up but retains his two basic assumptions, he will have to ban theories from science and regard the growth of science as an accumulation of proven basic statements. This indeed is the final stage of classical empiricism after the evaporation of the hope that facts can prove or at least disprove theories.

[3] This is no coincidence; cf. *below*, p. 88 ff.

[4] Cf. Popper [1934], chapter VIII.

aims at expressing a causal connection: such connection between being a Briton and dying of lung cancer may not even be intended. Similarly, 'all swans are white', if true, would be a mere curiosity unless it asserted that swanness *causes* whiteness. But then a black swan would not refute this proposition, since it may only indicate *other causes* operating simultaneously. Thus 'all swans are white' is either an oddity and easily disprovable or a scientific proposition with a *ceteris paribus* clause and therefore undisprovable. *Tenacity of a theory against empirical evidence would then be an argument for rather than against regarding it as 'scientific'. 'Irrefutability' would become a hallmark of science.*[1]).

To sum up: classical justificationists only admitted proven theories; neoclassical justificationists probable ones; dogmatic falsificationists realized that in either case no theories are admissible. They decided to admit theories if they are disprovable – disprovable by a finite number of observations. But even if there were such disprovable theories – those which can be contradicted by a finite number of observable facts – they are still logically too near to the empirical basis. For instance, on the terms of the dogmatic falsificationist, a theory like 'All planets move in ellipses' may be disproved by five observations; therefore the dogmatic falsificationist will regard it as scientific. A theory like 'All planets move in circles' may be disproved by four observations; therefore the dogmatic falsificationist will regard it as still more scientific. The acme of scientificness will be a theory like 'All swans are white' which is disprovable by one single observation. On the other hand, he will reject all probabilistic theories together with Newton's, Maxwell's, Einstein's theories, as unscientific, for no finite number of observations can ever disprove them.

If we accept the demarcation criterion of dogmatic falsificationism, *and* also the idea that facts can prove 'factual' propositions, we have to declare that the most important, if not all, theories ever proposed in the history of science are metaphysical, that most, if not all, of the accepted progress is pseudo-progress, that most, if not all, of the work done is irrational. If, however, still accepting the demarcation criterion of dogmatic falsificationism, we deny that facts can prove propositions, then we certainly end up in complete scepticism: then all science is undoubtedly irrational metaphysics and should be rejected. *Scientific theories are not only equally unprovable, and equally improbable, but they are also equally undisprovable.* But the recognition that not only the theoretical but *all* the propositions in science are fallible, means the total collapse of *all* forms of dogmatic justificationism as theories of scientific rationality.

[1] For a *much* stronger case, cf. *below*, section 3.

(b) Methodological falsificationism. The 'empirical basis'

The collapse of dogmatic falsificationism under the weight of falli-bilistic arguments brings us back to square one. If *all* scientific statements are fallible theories, one can criticize them only for inconsistency. But then, in what sense, if any, is science empirical? If scientific theories are neither provable, nor probabilifiable, nor dis-provable, then the sceptics seem to be finally right: science is no more than vain speculation and there no such thing as progress in scientific knowledge. Can we still oppose scepticism? *Can we save scientific criticism from fallibilism?* Is it possible to have a fallibilistic theory of scientific progress? In particular, if scientific criticism is fallible, on what ground can we ever eliminate a theory?

A most intriguing answer is provided by *methodological falsification-ism*. Methodological falsificationism is a brand of conventionalism; therefore in order to understand it, we must first discuss conven-tionalism in general.

There is an important demarcation between '*passivist*' and '*activist*' *theories of knowledge*. 'Passivists' hold that true knowledge is Nature's imprint on a perfectly inert mind: mental *activity* can only result in bias and distortion. The most influential passivist school is classical empiricism. 'Activists' hold that we cannot read the book of Nature without mental activity, without interpreting it in the light of our expectations or theories.[1] Now *conservative* '*activists*' hold that we are born with our basic expectations; with them we turn the world into 'our world' but must then live for ever in the prison of our world. The idea that we live and die in the prison of our 'conceptual frame-work' was developed primarily by Kant: pessimistic Kantians thought that the real world is for ever unknowable because of this prison, while optimistic Kantians thought that God created our conceptual framework to fit the world.[2] But *revolutionary activists* believe that conceptual frameworks can be developed and also replaced by new, *better* ones; it is *we* who create our 'prisons' and we can also, critically, demolish them.[3]

---

[1] This demarcation – and terminology – is due to Popper; cf. especially his [1934], section 19 and his [1945], chapter 23 and n. 3 to chapter 25.

[2] No version of conservative activism explained why Newton's *gravitational* theory should be invulnerable; Kantians restricted themselves to the explanation of the tenacity of Euclidean geometry and Newtonian *mechanics*. About Newtonian *gravi-tation* and *optics* (or other branches of science) they had an ambiguous, and occasionally inductivist position.

[3] I do not include Hegel among 'revolutionary *activists*'. For Hegel and his followers change in conceptual frameworks is a predetermined, inevitable process, where individual creativity or rational criticism plays no essential role. Those who run ahead are equally at fault as those who stay behind in this 'dialectic'. The clever man is not he who creates a better 'prison' or who demolishes critically the old one, but the one who is always in step with history. Thus dialectic accounts for change without criticism.

New steps from conservative to revolutionary activism were made by Whewell and then by Poincaré, Milhaud and Le Roy. Whewell held that theories are developed by trial and error – in the 'preludes to the inductive epochs'. The best ones among them are then 'proved' – during the 'inductive epochs' – by a long primarily *a priori* consideration which he called 'progressive intuition'. The 'inductive epochs' are followed by 'sequels to the inductive epochs': cumulative developments of auxiliary theories.[1] Poincaré, Milhaud and Le Roy were averse to the idea of *proof* by progressive intuition and preferred to explain the continuing historical success of Newtonian mechanics by a *methodological decision* taken by scientists: after a considerable period of initial empirical success scientists may *decide* not to allow the theory to be refuted. Once they have taken this decision, they solve (or dissolve) the apparent anomalies by auxiliary hypotheses or other 'conventionalist stratagems'.[2] This *conservative conventionalism* has, however, the disadvantage of making us unable to get out of our self-imposed prisons, once the first period of trial and error is over and the great decision taken. It cannot solve the problem of the elimination of those theories which have been triumphant for a long period. According to conservative conventionalism, experiments may have sufficient power to refute young theories, but not to refute old, established theories: *as science grows, the power of empirical evidence diminishes.*[3]

Poincaré's critics refused to accept his idea, that, although the scientists build their conceptual frameworks, there comes a time when these frameworks turn into prisons which cannot be demolished. This criticism gave rise to two rival schools of *revolutionary conventionalism*: Duhem's simplicism and Popper's methodological falsificationism.[4]

[1] Cf. Whewell's [1837], [1840] and [1858].

[2] Cf. especially Poincaré [1891] and [1902]; Milhaud [1896]; Le Roy [1899] and [1901]. It was one of the chief philosophical merits of conventionalists to direct the limelight to the fact that any theory can be saved by 'conventionalist stratagems' from refutations. (The term 'conventionalist stratagem' is Popper's: cf. the critical discussion of Poincaré's conventionalism in his [1934], especially sections 19 and 20.)

[3] Poincaré first elaborated his conventionalism only with regard to geometry (cf. his [1891]). Then Milhaud and Le Roy generalized Poincaré's idea to cover all branches of accepted physical theory. Poincaré's [1902] starts with a strong criticism of the Bergsonian Le Roy against whom he defends the empirical (falsifiable or 'inductive') character of all physics *except for* geometry and mechanics. Duhem, in turn, criticized Poincaré: in his view there was a possibility of overthrowing even Newtonian mechanics.

[4] The *loci classici* are Duhem's [1905] and Popper's [1934]. Duhem was not a *consistent* revolutionary conventionalist. Very much like Whewell, he thought that conceptual changes are only *preliminaries* to the final – if perhaps distant – 'natural classification'. 'The more a theory is perfected, the more we apprehend that the logical order in which it arranges experimental laws is the reflection of an ontological order.' In particular, he refused to see Newton's mechanics *actually* 'crumbling' and characterized Einstein's relativity theory as the manifestation of a 'frantic and hectic race in pursuit of a novel idea' which 'has turned physics into a real chaos where logic loses its way and common-sense runs away frightened' (Preface – of 1914 – to the second edition of his [1905]).

Duhem accepts the conventionalists' position that no physical theory ever crumbles merely under the weight of 'refutations', but claims that it still may crumble under the weight of 'continual repairs, and many tangled-up stays' when 'the worm-eaten columns' cannot support 'the tottering building' any longer;[1] then the theory loses its original simplicity and has to be replaced. But falsification is then left to subjective taste or, at best, to scientific fashion, and too much leeway is left for dogmatic adherence to a favourite theory.[2]

Popper set out to find a criterion which is both more objective and more hard-hitting. He could not accept the emasculation of empiricism, inherent even in Duhem's approach, and proposed a methodology which allows experiments to be powerful even in 'mature' science. Popper's methodological falsificationism is both conventionalist and falsificationist, but he 'differs from the [conservative] conventionalists in holding that the statements decided by agreement are *not* [spatio-temporally] universal but [spatio-temporally] singular';[3] and he differs from the dogmatic falsificationist in holding that the truth-value of such statements cannot be proved by facts but, in some cases, may be decided by agreement.[4]

The Duhemian *conservative conventionalist* (or 'methodological justificationist', if you wish) makes unfalsifiable by *fiat* some (spatio-temporally) universal theories, which are distinguished by their explanatory power, simplicity or beauty. Our Popperian *revolutionary conventionalist* (or 'methodological falsificationist') makes unfalsifiable by *fiat* some (spatio-temporally) singular statements which are distinguishable by the fact that there exists at the time a 'relevant technique' such that 'anyone who has learned it' will be able to *decide* that the statement is 'acceptable'.[5] Such a statement may be called an 'observational' or 'basic' statement, but only in inverted commas.[6] Indeed, the very selection of all such statements is a matter of a decision, which is not based on exclusively psychological considerations. This decision is then followed by a second kind of decision concerning the separation of the set of *accepted* basic statements from the rest.

These *two decisions* correspond to the *two assumptions* of dogmatic falsificationism. But there are important differences. Above all, the methodological falsificationist is not a justificationist, he has no illusions about 'experimental proofs' and is fully aware of the fallibility of his decisions and the risks he is taking.

---

[1] Duhem [1905], chapter VI, section 10.
[2] For a further discussion of conventionalism, cf. *below*, pp. 96–101.
[3] Popper [1934], section 30.
[4] *In this section I discuss the 'naive' variant of Popper's methodological falsificationism. Thus, throughout the section 'methodological falsificationism' stands for 'naive methodological falsificationism'; for this 'naivety', cf. below, p. 31.*
[5] Popper [1934], section 27.
[6] *Op. cit.*, section 28. For the non-basicness of these methodologically 'basic' statements, cf. e.g. Popper [1934] *passim* and Popper [1959a], p. 35, n. *2.

The methodological falsificationist realizes that in the 'experimental techniques' of the scientist fallible theories are involved,[1] in the 'light' of which he interprets the facts. In spite of this he 'applies' these theories, he regards them in the given context not as theories under test but as *unproblematic background knowledge* 'which we accept (tentatively) as unproblematic while we are testing the theory'.[2] He may call these theories – and the statements whose truth-value he decides in their light – 'observational': but this is only a manner of speech which he inherited from naturalistic falsificationism.[3] The methodological falsificationist *uses our most successful theories as extensions of our senses* and widens the range of theories which can be applied in testing far beyond the dogmatic falsificationist's range of strictly observational theories. For instance, let us imagine that a big radio-star is discovered with a system of radio-star satellites orbiting it. We should like to test some gravitational theory on this planetary system – a matter of considerable interest. Now let us imagine that Jodrell Bank succeeds in providing a set of space–time coordinates of the planets which is inconsistent with the theory. We shall take these basic statements as falsifiers. Of course, these basic statements are not 'observational' in the usual sense but only "'observational'". They describe planets that neither the human eye nor optical instruments can reach. Their truth-value is arrived at by an 'experimental technique'. This 'experimental technique' is based on the 'application' of a well-corroborated theory of radio-optics. Calling these statements 'observational' is no more than a manner of saying that, in the context of his problem, that is, in testing our gravitational theory, the methodological falsificationist uses radio-optics uncritically, as 'background knowledge'. *The need for decisions to demarcate the theory under test from unproblematic background knowledge is a characteristic feature of this brand of methodological falsificationism.*[4] (This situation does not really differ from Galileo's 'observation' of Jupiter's satellites: moreover, as some of Galileo's contemporaries rightly pointed out, he relied on a virtually non-existent optical theory – which then was less corroborated, and even less articulated, than present-day radio-optics. On the other hand, calling the reports of our human eye 'observational' only indicates that we 'rely' on some vague physiological theory of human vision.[5])

This consideration shows the conventional element in granting – in a given context – (methodologically) 'observational' status to a theory.[6]

---

[1] Cf. Popper [1934], end of section 26 and also his [1968c], pp. 291–2.

[2] Cf. Popper [1963], p. 390.

[3] Indeed, Popper carefully puts 'observational' in quotes; cf. his [1934], section 28.

[4] This demarcation plays a role both in the *first* and in the *fourth* type of decisions of the methodological falsificationist. (For the *fourth* decision, cf. *below*, p. 26.)

[5] For a fascinating discussion, cf. Feyerabend [1969a].

[6] One wonders whether it would not be better to make a break with the terminology of naturalistic falsificationism and rechristen observational theories '*touchstone theories*'.

Similarly, there is a considerable conventional element in the decision concerning the actual truth-value of a basic statement which we take after we have decided which 'observational theory' to apply. One single observation may be the stray result of some trivial error: in order to reduce such risks, methodological falsificationists prescribe some safety control. The simplest such control is to repeat the experiment (it is a matter of convention how many times); thus fortifying the potential falsifier by a 'well-corroborated falsifying hypothesis'.[1]

The methodological falsificationist also points out that, as a matter of fact, these conventions are institutionalized and endorsed by the scientific community; the list of 'accepted' falsifiers is provided by the verdict of the experimental scientists.[2]

This is how the methodological falsificationist establishes his 'empirical basis'. (He uses inverted commas in order 'to give ironical emphasis' to the term.[3]) This 'basis' can hardly be called a 'basis' by justificationist standards: there is nothing proven about it – it denotes 'piles driven into a swamp'.[4] Indeed, if this 'empirical basis' clashes with a theory, the theory may be *called* 'falsified', but it is not falsified in the sense that it is disproved. Methodological 'falsification' is very different from dogmatic falsification. If a theory is falsified, it is proven false; if it is 'falsified', it may still be true. If we follow up this sort of 'falsification' by the actual 'elimination' of a theory, we may well end up by eliminating a true, and accepting a false, theory (a possibility which is thoroughly abhorrent to the old-fashioned justificationist).

Yet the methodological falsificationist advises that exactly this is to be done. The methodological falsificationist realizes that if we want to reconcile fallibilism with (non-justificationist) rationality, we *must* find a way to eliminate *some* theories. If we do not succeed, the growth of science will be nothing but growing chaos.

Therefore the methodological falsificationist maintains that '[if we want] to make the method of selection by elimination work, and to ensure that only the fittest theories survive, their struggle for life must be made severe'.[5] Once a theory has been falsified, in spite of the risk involved, it must be eliminated: '[with theories we work only] as long as they stand up to tests'.[6] The elimination must be methodologically conclusive: 'In general we regard an inter-subjectively testable falsification as final...A corroborative appraisal made at a later date... can replace a positive degree of corroboration by a negative one, but

---

[1] Cf. Popper [1934], section 22. Many philosophers overlooked Popper's important qualification that a basic-statement has no power to refute anything without the support of a well-corroborated falsifying hypothesis.

[2] Cf. Popper [1934], section 30.   [3] Popper [1963a], p. 387.

[4] Popper [1934], section 30; also cf. section 29: 'The Relativity of Basic Statements'.

[5] Popper [1957b], p. 134. Popper, in other places, emphasizes that this method cannot 'ensure' the survival of the fittest. Natural selection may go wrong: the fittest may perish and monsters survive.

[6] Popper [1935].

not *vice versa*'.[1] This is the methodological falsificationist's explanation of how we get out of a rut: 'It is always the experiment which saves us from following a track that leads nowhere'.[1]

The methodological falsificationist separates rejection and disproof, which the dogmatic falsificationist had conflated.[2] He is a fallibilist but his fallibilism does not weaken his critical stance: he turns fallible propositions into a 'basis' for a hard-line policy. On these grounds he proposes a *new demarcation criterion*: only those theories – that is, non-'observational' propositions – which forbid certain 'observable' states of affairs, and therefore may be 'falsified' and rejected, are 'scientific': or, briefly, *a theory is 'scientific' (or 'acceptable') if it has an 'empirical basis'*. This criterion brings out sharply the difference between dogmatic and methodological falsificationism.[3]

This methodological demarcation criterion is much more liberal than the dogmatic one. Methodological falsificationism opens up new avenues of criticism: many more theories may qualify as 'scientific'. We have already seen that there are more 'observational' theories than observational theories,[4] and therefore there are more 'basic' statements than basic statements.[5] Furthermore, probabilistic theories may qualify now as 'scientific': although they are not falsifiable they can be easily made 'falsifiable' by an *additional (third type) decision* which the scientist can make by specifying certain rejection rules which may render statistically interpreted evidence 'inconsistent' with the probabilistic theory.[6]

But even these three decisions are not sufficient to enable us to 'falsify' a theory which cannot explain anything 'observable' without

---

[1] Popper [1934], section 82.

[2] This kind of methodological 'falsification' is, unlike dogmatic falsification (disproof), a pragmatic, methodological idea. But then what exactly are we to mean by it? Popper's answer – which I am going to discard – is that methodological 'falsification' indicates an 'urgent need of replacing a falsified hypothesis by a better one' (Popper [1959a], p. 87, n. *1). This is an excellent illustration of the process I described in my [1963–4] whereby critical discussion shifts the original *problem* without necessarily changing the old *terms*. The byproducts of such processes are *meaning-shifts*. For a further discussion, cf. *below*, p. 37, n. 5, and p. 70, n. 4.

[3] The demarcation criterion of the dogmatic falsificationist was: a theory is 'scientific' if it has an empirical basis (see *above*, p. 16).

[4] See *above*, pp. 14–15.

[5] Incidentally, Popper, in his [1934], does not seem to have seen this point clearly. He writes: 'Admittedly, it is possible to interpret the concept of an *observable event* in a psychologistic sense. But I am using it in such a sense that it might just as well be replaced by "an event involving position and movement of macroscopic physical bodies"' ([1934], section 28.) In the light of our discussion, for instance, we may regard a positron passing through a Wilson chamber at time t_0 as an 'observable' event, in spite of the non-macroscopic character of the positron.

[6] Popper [1934], section 68. Indeed, this methodological falsificationism is the philosophical basis of some of the most interesting developments in modern statistics. The Neyman–Pearson approach rests completely on methodological falsificationism. Also cf. Braithwaite [1953], chapter VI. (Unfortunately, Braithwaite reinterprets Popper's demarcation criterion as separating meaningful from meaningless rather than scientific from non-scientific propositions).

a *ceteris paribus* clause.[1] No finite number of 'observations' is enough to 'falsify' such a theory. However, if this is the case how can one reasonably defend a methodology which claims to 'interpret natural laws or theories as...statements which are partially decidable, i.e. which are, for logical reasons, not verifiable but, in an asymmetrical way, falsifiable...'?[2] How can we interpret theories like Newton's theory of dynamics and gravitation as 'one-sidedly decidable'?[3] How can we make in such cases genuine 'attempts to weed out false theories – to find the weak points of a theory in order to reject it if it is falsified by the test'?[4] How can we draw them into the realm of rational discussion? The methodological falsificationist solves the problem by making a further (*fourth type*) *decision*: when he tests a theory together with a *ceteris paribus* clause and finds that this conjunction has been refuted, he must *decide* whether to take the refutation also as a refutation of the specific theory. For instance, he may accept Mercury's 'anomalous' perihelion as a refutation of the treble conjunction N_3 of Newton's theory, the known initial conditions and the *ceteris paribus* clause. Then he tests the initial conditions 'severely'[5] and may decide to relegate them into the 'unproblematic background knowledge'. This decision implies the refutation of the double conjunction N_2 of Newton's theory and the *ceteris paribus* clause. Now he has to take the crucial decision: whether to relegate also the *ceteris paribus* clause into the pool of 'unproblematic background knowledge'. He will do so if he finds the *ceteris paribus* clause well corroborated.

How can one test a *ceteris paribus* clause severely? By assuming that there *are* other influencing factors, by specifying such factors, and by testing these specific assumptions. If many of them are refuted, the *ceteris paribus* clause will be regarded as well-corroborated.

Yet the decision to 'accept' a *ceteris paribus* clause is a very risky one because of the grave consequences it implies. If it is decided to accept it as part of such background knowledge, the statements describing Mercury's perihelion from the empirical basis of N_2 are turned into the empirical basis of Newton's specific theory N_1 and what was previously a mere 'anomaly' in relation to N_1, becomes now crucial evidence against it, its falsification. (We may call an event described by a statement A an '*anomaly* in relation to a theory T' if A is a potential falsifier of the conjunction of T and a *ceteris paribus* clause but it becomes a potential falsifier of T itself after having decided to relegate the *ceteris paribus* clause into 'unproblematic background knowledge'.[6]) Since, for our savage falsificationist, falsifications are methodologically conclusive,[7] the fateful decision amounts to the

---

[1] Cf. *above*, pp. 18–20. [2] Popper [1933].
[3] Popper [1933]. [4] Popper [1957*b*], p. 133.
[5] For a discussion of this important concept of Popperian methodology, cf. volume 2, chapter 8, pp. 185 ff.
[6] For an improved 'explication', cf. *below*, p. 72, n. 3.
[7] Cf. *above*, p. 24, text to nn. 5 and 6.

methodological elimination of Newton's theory, making further work on it irrational. If the scientist shrinks back from such bold decisions he will 'never benefit from experience', 'believing, perhaps, that it is his business to defend a successful system against criticism as long as it is not *conclusively disproved*'.[1] He will degenerate into an apologist who may always claim that 'the discrepancies which are asserted to exist between the experimental results and the theory are only apparent and that they will disappear with the advance of our understanding'.[2] But for the falsificationist this is 'the very reverse of the critical attitude which is the proper one for the scientist',[3] and is impermissible. To use one of the methodological falsificationist's favourite expressions: the theory 'must be made to stick its neck out'.

The methodological falsificationist is in a serious plight when it comes to deciding where to draw the demarcation, even if only in a well-defined context, between the problematic and unproblematic. The plight is most dramatic when he has to make a decision about *ceteris paribus* clauses, when he has to promote one of the hundreds of 'anomalous phenomena' into a 'crucial experiment', and decide that in such a case the experiment was 'controlled'.[4]

Thus, with the help of this fourth type of decision,[5] our methodological falsificationist has finally succeeded in interpreting even theories like Newton's theory as 'scientific'.[6]

Indeed, there is no reason why he should not go yet another step. Why not decide that a theory – which even these four decisions cannot turn into an empirically falsifiable one – is falsified if it clashes with another theory which is scientific on some of the previously specified grounds and is also well-corroborated?[7] After all, if we reject one theory because one of its potential falsifiers is seen to be true in the light of an observational theory, why not reject another theory because

---

[1] Popper [1934], section 9.   [2] *Ibid.*
[3] *Ibid.*
[4] The problem of '*controlled experiment*' may be said to be nothing else but the problem of arranging experimental conditions in such a way as to minimize the risk involved in such decisions.
[5] This type of decision belongs, in an important sense, to the same category as the first decision: it demarcates, by decision, problematic from unproblematic knowledge. Cf. *above*, p. 23, text to n. 3.
[6] Our exposition shows clearly the complexity of the decisions needed to define the 'empirical content' of a theory – that is, the set of its potential falsifiers. 'Empirical content' depends on our *decision* as to which are our 'observational theories' and which anomalies are to be promoted to counterexamples. If one attempts to compare the empirical content of different scientific theories in order to see which is 'more scientific', then one will get involved in an enormously complex and therefore hopelessly arbitrary system of decisions about their respective classes of 'relatively atomic statements' and their 'fields of application'. (For the meaning of these (very) technical terms, cf. Popper [1934], section 38.) But such comparison is possible only when one theory supersedes another (cf. Popper [1959a], p. 401, n. 7). And even then, there may be difficulties (which would not, however, add up to irremediable 'incommensurability').
[7] This was suggested by J. O. Wisdom: cf. his [1963].

it clashes *directly* with one that may be relegated into unproblematic background knowledge? This would allow us, by a *fifth type decision*, to eliminate even 'syntactically metaphysical' theories, that is, theories, which, like 'all-some' statements or purely existential statements,[1] because of their *logical form* cannot have spatio-temporally singular potential falsifiers.

To sum up: the methodological falsificationist offers an interesting solution to the problem of combining hard-hitting criticism with fallibilism. Not only does he offer a philosophical basis for falsification after fallibilism had pulled the carpet from under the feet of the dogmatic falsificationist, but he also widens the range of such criticism very considerably. By putting falsification in a new setting, he saves the attractive code of honour of the dogmatic falsificationist: that scientific honesty consists in specifying, in advance, an experiment such that, if the result contradicts the theory, the theory has to be given up.[2]

Methodological falsificationism represents a considerable advance beyond both dogmatic falsificationism and conservative conventionalism. It recommends risky decisions. But the risks are daring to the point of recklessness and one wonders whether there is no way of lessening them.

Let us first have a closer look at the risks involved.

Decisions play a crucial role in this methodology – as in any brand of conventionalism. Decisions however may lead us disastrously astray. The methodological falsificationist is the first to admit this. But this, he argues, is the price which we have to pay for the possibility of progress.

One has to appreciate the dare-devil attitude of our methodological falsificationist. He feels himself to be a hero who, faced with two catastrophic alternatives, dared to reflect coolly on their relative merits and choose the lesser evil. One of the alternatives was sceptical fallibilism, with its 'anything goes' attitude, the despairing abandonment of all intellectual standards, and hence of the idea of scientific progress. Nothing can be established, nothing can be rejected, nothing even communicated: the growth of science is a growth of chaos, a veritable Babel. For two thousand years, scientists and scientifically-minded philosophers chose justificationist illusions of some kind to escape this nightmare. Some of them argued that *one has to choose between inductivist justificationism and irrationalism*: 'I do not see any way out of a dogmatic assertion that we know the inductive principle or some equivalent; the only alternative is to throw over almost everything that is regarded as knowledge by science and common sense'.[3] Our methodological

---

[1] For instance: 'All metals have a solvent'; or 'There exists a substance which can turn all metals into gold'. For discussions of such theories, cf. especially Watkins [1957] and Watkins [1960]. But cf. *below*, pp. 42–3 and pp. 95–6.

[2] See *above*, p. 12. [3] Russell [1943], p. 683.

falsificationist proudly rejects such escapism: he dares to measure up to the full impact of fallibilism and yet escape scepticism by a daring and risky conventionalist policy, with no dogmas. He is fully aware of the risks but insists that *one has to choose between some sort of methodological falsificationism and irrationalism.* He offers a game in which one has little hope of winning, but claims that it is still better to play than to give up.[1]

Indeed, those critics of naive falsificationism who offer no alternative method of criticism are inevitably driven to irrationalism. For instance, Neurath's muddled argument, that the falsification and ensuing elimination of a hypothesis may turn out to have been 'an obstacle in the progress of science',[2] carries no weight as long as the only alternative he seems to offer is chaos. Hempel is, no doubt, right in stressing that 'science offers various examples [when] a conflict between a highly-confirmed theory and an occasional recalcitrant experiential sentence may well be resolved by revoking the latter rather than by sacrificing the former;'[3] nevertheless he admits that he can offer no other 'fundamental standard' than that of naive falsificationism.[4] Neurath – and, seemingly, Hempel – reject falsificationism as 'pseudo-rationalism';[5] but where is 'real rationalism'? Popper warned already in 1934 that Neurath's permissive methodology (or rather lack of methodology) would make science unempirical and therefore irrational: 'We need a set of rules to limit the arbitrariness of "deleting" (or else "accepting") a protocol sentence. Neurath fails to give any such rules and thus unwittingly throws empiricism overboard...Every system becomes defensible if one is allowed (as everybody is, in Neurath's view) simply to "delete" a protocol sentence if it is inconvenient'.[6] Popper agrees with Neurath that all propositions are fallible; but he forcefully makes the crucial point that we cannot make progress unless we have a firm rational strategy or method to guide us when they clash.[7]

But is not the firm strategy of the brand of methodological falsifi-

---

[1] I am sure that some will welcome methodological falsificationism as an 'existentialist' philosophy of science.

[2] Neurath [1935], p. 356.

[3] Hempel [1952], p. 621. Agassi, in his [1966], follows Neurath and Hempel, especially pp. 16 ff. It is rather amusing that Agassi, in making this point, thinks that he is taking up arms against 'the whole literature concerning the methods of science'.

Indeed, many scientists were fully aware of the difficulties inherent in the 'confrontation of theory and facts'. (Cf. Einstein [1949], p. 27.) Several philosophers sympathetic to falsificationism emphasized that 'the process of refuting a scientific hypothesis is more complicated than it appears to be at first sight' (Braithwaite [1953], p. 20). But only Popper offered a constructive, rational solution.

[4] Hempel [1952], p. 622. Hempel's crisp 'theses on empirical certainty' do nothing but refurbish Neurath's – and some of Popper's – old arguments (against Carnap, I take it); but, deplorably, he does not mention either his predecessors or his adversaries.

[5] Neurath [1935].   [6] Popper [1934], section 26.

[7] Neurath's [1935] shows that he never grasped Popper's simple argument.

cationism hitherto discussed *too firm*? Are not the decisions it advocates bound to be *too arbitrary*? Some may even claim that all that distinguishes methodological from dogmatic falsificationism is that *it pays lip-service to fallibilism*!

To criticize a theory of criticism is usually very difficult. Naturalistic falsificationism was relatively easy to refute, since it rested on an empirical psychology of perception: one could show that it was simply *false*. But how can methodological falsification be falsified? No disaster can ever disprove a non-justificationist theory of rationality. Moreover, how can we ever recognize an epistemological disaster? We have no means to judge whether the verisimilitude of our successive theories increases or decreases.[1] At this stage we have not yet developed a general theory of criticism even for scientific theories, let alone for theories of rationality:[2] therefore if we want to falsify our methodological falsificationism, we have to do it before having a theory of how to do it.

If we look at the historical details of the most celebrated crucial experiments, we have to come to the conclusion that either they were accepted as crucial for no rational reason, or that their acceptance rested on rationality principles radically different from the ones we just discussed. First of all, our falsificationist must deplore the fact that stubborn theoreticians frequently challenge experimental verdicts and have them reversed. In the falsificationist conception of scientific 'law and order' we have described there is no place for such successful appeals. Further difficulties arise from the falsification of theories to which a *ceteris paribus* clause is appended.[3] Their falsification as it occurs in actual history is *prima facie* irrational by the standards of our falsificationist. By his standards, scientists frequently seem to be irrationally slow: for instance, eighty-five years elapsed between the acceptance of the perihelion of Mercury as an anomaly and its acceptance as a falsification of Newton's theory, in spite of the fact that the *ceteris paribus* clause was reasonably well corroborated. On the other hand, scientists frequently seem to be irrationally rash: for instance, Galileo and his disciples accepted Copernican heliocentric celestial mechanics in spite of the abundant evidence against the rotation of the Earth; or Bohr and his disciples accepted a theory of light emission in spite of the fact that it ran counter to Maxwell's well-corroborated theory.

---

[1] I am using here 'verisimilitude' in Popper's sense: the difference between the truth content and falsity content of a theory. For the risks involved in estimating it, cf. volume 2, chapter 8, especially pp. 183 ff.

[2] I tried to develop such a general theory of criticism in my [1971a], [1971c] and chapter 3.

[3] The falsification of theories depends on the high degree of corroboration of the *ceteris paribus* clause. However, such corroboration is often lacking. This is why the methodological falsificationist may advise us to rely on our 'scientific instinct' (Popper [1934], section 18, n. 2) or 'hunch' (Braithwaite [1953], p. 20).

Indeed, it is not difficult to see at least two crucial characteristics common to both dogmatic and our methodological falsificationism which are clearly dissonant with the actual history of science: that (1) *a test is – or must be made – a two-cornered fight between theory and experiment so that in the final confrontation only these two face each other; and* (2) *the only interesting outcome of such confrontation is (conclusive) falsification: '[the only genuine] discoveries are refutations of scientific hypotheses.'*[1] However, history of science suggests that (1') tests are – at least – three-cornered fights between rival theories and experiment and (2') some of the most interesting experiments result, *prima facie*, in confirmation rather than falsification.

But if – as seems to be the case – the history of science does not bear out our theory of scientific rationality, we have two alternatives. One alternative is to abandon efforts to give a rational explanation of the success of science. Scientific method (or 'logic of discovery'), conceived as the discipline of rational appraisal of scientific theories – and of criteria of *progress* – vanishes. We may, of course, still try to explain *changes* in 'paradigms' in terms of social psychology.[2] This is Polanyi's and Kuhn's way.[3] The other alternative is to try at least to *reduce* the conventional element in falsificationism (we cannot possibly eliminate it) and replace the *naive* versions of methodological falsificationism – characterized by the theses (1) and (2) above – by a *sophisticated* version which would give a new *rationale* of falsification and thereby rescue methodology and the idea of scientific *progress*. This is Popper's way, and the one I intend to follow.

(c) Sophisticated versus naive methodological falsificationism. Progressive and degenerating problemshifts

Sophisticated falsificationism differs from naive falsificationism both in its rules of *acceptance* (or 'demarcation criterion') and its rules of *falsification* or elimination.

For the naive falsificationist any theory which can be interpreted as experimentally falsifiable, is 'acceptable' or 'scientific'.[4] For the sophisticated falsificationist a theory is 'acceptable' or 'scientific' only

---

[1] Agassi [1959]; he calls Popper's idea of science '*scientia negativa*' (Agassi [1968]).

[2] It should be mentioned here that the Kuhnian sceptic is still left with what I would call the '*scientific sceptic's dilemma*': any scientific sceptic will still try to explain changes in beliefs and will regard his own psychological theory as one which is more than simple belief, which, in some sense, is 'scientific'. Hume, while trying to show up science as a mere system of beliefs with the help of his stimulus–response theory of learning, never raised the problem of whether his theory of learning applies also to his own theory of learning. In contemporary terms, we might well ask, does the popularity of Kuhn's philosophy indicate that people recognize its *truth*? In this case it would be refuted. Or does this popularity indicate that people regarded it as an attractive new fashion? In this case it would be 'verified'. But would Kuhn like *this* 'verification'?

[3] Feyerabend who contributed probably more than anybody else to the spread of Popper's ideas, seems now to have joined the enemy camp. Cf. his intriguing [1970b].

[4] Cf. *above*, p. 25.

if it has corroborated excess empirical content over its predecessor (or rival), that is, only if it leads to the discovery of novel facts. This condition can be analysed into two clauses: that the new theory has excess empirical content ('*acceptability₁*') and that some of this excess content is verified ('*acceptability₂*'). The first clause can be checked instantly[1] by *a priori* logical analysis; the second can be checked only empirically and this may take an indefinite time.

For the naive falsificationist a theory is *falsified* by a ('fortified')[2] 'observational' statement which conflicts with it (or which he decides to interpret as conflicting with it). For the sophisticated falsificationist a scientific theory T is *falsified* if and only if another theory T' has been proposed with the following characteristics: (1) T' has excess empirical content over T: that is, it predicts *novel* facts, that is, facts improbable in the light of, or even forbidden, by T;[3] (2) T' explains the previous success of T, that is, all the unrefuted content of T is included (within the limits of observational error) in the content of T'; and (3) some of the excess content of T' is corroborated.[4]

In order to be able to appraise these definitions we need to understand their problem background and their consequences. First, we have to remember the conventionalists' methodological discovery that no experimental result can ever kill a theory: any theory can be saved from counterinstances either by some auxiliary hypothesis or by a suitable reinterpretation of its terms. Naive falsificationists solved this problem by relegating – in crucial contexts – the auxiliary hypotheses to the realm of unproblematic background knowledge, eliminating them from the deductive model of the test-situation and thereby *forcing* the chosen theory into logical isolation, in which it becomes a sitting target for the attack of test-experiments. But since this procedure did not offer a suitable guide for a rational reconstruction of the history of science, we may just as well completely rethink our approach. Why aim at falsification at any price? Why not rather impose certain standards on the theoretical adjustments by which one is allowed to save a theory? Indeed, some such standards have been well-known for centuries, and we find them expressed in age-old wisecracks against *ad hoc* explanations, empty prevarications, face-saving, linguistic tricks.[5] We have already seen that Duhem adumbrated such standards in terms of 'simplicity' and 'good sense'.[6] But *when* does lack of

---

[1] But *cf. below*, pp. 69–70.   [2] Cf. *above*, p. 24, text to n. 1.

[3] I use 'prediction' in a wide sense that includes 'postdiction'.

[4] *For a detailed discussion of these acceptance and rejection rules and for references to Popper's work*, cf. volume 2, chapter 8, pp. 170–81. For some qualifications (concerning continuity and consistency as regulative principles), cf. *below*, pp. 46–7 and 55–60.

[5] Molière, for instance, ridiculed the doctors of his *Malade Imaginaire*, who offered the *virtus dormitiva* of opium as the answer to the question as to why opium produced sleep. One might even argue that Newton's famous dictum *hypotheses non fingo* was really directed against *ad hoc* explanations – like his own explanation of gravitational forces by an aether-model in order to meet Cartesian objections.

[6] Cf. *above*, p. 21.

'simplicity' in the protective belt of theoretical adjustments reach the point at which the theory *must* be abandoned?[1] In what sense was Copernican theory, for instance, 'simpler' than Ptolemaic?[2] The vague notion of Duhemian 'simplicity' leaves, as the naive falsificationist correctly argued, the decision very much to taste and fashion.[3]

Can one improve on Duhem's approach? Popper did. His solution – a sophisticated version of methodological falsificationism – is more objective and more rigorous. Popper agrees with the conventionalists that theories and factual propositions can always be harmonized with the help of auxiliary hypotheses: he agrees that the problem is how to demarcate between scientific and pseudoscientific *adjustments*, between rational and irrational changes of theory. According to Popper, saving a theory with the help of auxiliary hypotheses which satisfy certain well-defined conditions represents scientific progress; but saving a theory with the help of auxiliary hypotheses which do not, represents degeneration. Popper calls such inadmissible auxiliary hypotheses *ad hoc* hypotheses, mere linguistic devices, 'conventionalist stratagems'.[4] But then any scientific theory has to be appraised together with its auxiliary hypotheses, initial conditions, etc., and, especially, together with its predecessors so that we may see by what sort of *change* it was brought about. Then, of course, what we appraise is a *series of theories* rather than isolated *theories*.

Now we can easily understand why we formulated the criteria of acceptance and rejection of sophisticated methodological falsificationism as we did.[5] But it may be worth while to reformulate them slightly, couching them explicitly in terms of *series of theories*.

Let us take a series of theories, $T_1, T_2, T_3, \ldots$ where each subsequent theory results from adding auxiliary clauses to (or from semantical reinterpretations of) the previous theory in order to accommodate some anomaly, each theory having at least as much content as the unrefuted content of its predecessor. Let us say that such a series of theories is *theoretically progressive* (or '*constitutes a theoretically progressive problemshift*') if each new theory has some excess empirical content over its predecessor, that is, if it predicts some novel, hitherto unexpected fact. Let us say that a theoretically progressive series of theories is also

---

[1] Incidentally, Duhem agreed with Bernard that experiments alone – without simplicity considerations – can decide the fate of theories in physiology. But in physics, he argued, they cannot ([1905], chapter VI, section 1).

[2] Koestler correctly points out that only Galileo created the myth that the Copernican theory was simple (Koestler [1959], p. 476); in fact, 'the motion of the earth [had not] done much to simplify the old theories, for though the objectionable equants had disappeared, the system was still bristling with auxiliary circles' (Dreyer [1906], chapter XIII).

[3] Cf. *above*, p. 22.

[4] Popper [1934], sections 19 and 20. I have discussed in some detail – under the heads 'monster-barring', 'exception-barring', 'monster-adjustment' – such stratagems as they appear in informal, quasi-empirical mathematics; cf. my [1963–4].

[5] Cf. *above*, p. 31.

empirically progressive (or '*constitutes an empirically progressive problemshift*') if some of this excess empirical content is also corroborated, that is, if each new theory leads us to the actual discovery of some *new fact*.[1] Finally, let us call a problemshift *progressive* if it is both theoretically and empirically progressive, and *degenerating* if it is not.[2] We '*accept*' problemshifts as 'scientific' only if they are at least theoretically progressive; if they are not, we '*reject*' them as 'pseudoscientific'. Progress is measured by the degree to which a problemshift is progressive, by the degree to which the series of theories leads us to the discovery of novel facts. We regard a theory in the series 'falsified' when it is superseded by a theory with higher corroborated content.[3]

This demarcation between progressive and degenerating problemshifts sheds new light on the appraisal of *scientific – or, rather, progressive – explanations*. If we put forward a theory to resolve a contradiction between a previous theory and a counterexample in such a way that the new theory, instead of offering a content-increasing (scientific) *explanation*, only offers a content-decreasing (linguistic) *reinterpretation*, the contradiction is resolved in a merely semantical, unscientific way. *A given fact is explained scientifically only if a new fact is also explained with it.*[4]

Sophisticated falsificationism thus shifts the problem of how to appraise *theories* to the problem of how to appraise *series of theories*. Not an isolated *theory*, but only a series of theories can be said to be scientific or unscientific: to apply the term 'scientific' to one *single* theory is a category mistake.[5]

---

[1] If I already know P_1: 'Swan A is white', P_ω: 'All swans are white' represents no progress, because it may only lead to the discovery of such further similar facts as P_2: 'Swan B is white'. So-called 'empirical generalizations' constitute no progress. A *new* fact must be improbable or even impossible in the light of previous knowledge. Cf. *above*, p. 31, and *below*, p. 69 ff.

[2] The appropriateness of the term 'problemshift' for a series of theories rather than of problems may be questioned. I chose it partly because I have not found a more appropriate alternative – 'theoryshift' sounds dreadful – partly because theories are always problematical, they never solve all the problems thay have set out to solve. Anyway, in the second half of the paper, the more natural term 'research programme' will replace 'problemshift' in the most relevant contexts.

[3] For the 'falsification' of certain series of theories (of 'research programmes') as opposed to the 'falsification' of one theory within the series, cf. *below*, p. 69 ff.

[4] Indeed, in the original manuscript of volume 2, chapter 8, I wrote: 'A theory without excess corroboration has no excess explanatory power; *therefore, according to Popper, it does not represent growth and therefore it is not "scientific"*; therefore, we should *say, it has no explanatory power*' (p. 178). I cut out the italicized half of the sentence under pressure from my colleagues who thought it sounded too eccentric. I regret it now.

[5] Popper's conflation of 'theories' and 'series of theories' prevented him from getting the basic ideas of sophisticated falsificationism across more successfully. His ambiguous usage led to such confusing formulations as 'Marxism [as the core of a series of theories or of a "research programme"] is irrefutable' and, at the same time, 'Marxism [as a particular conjunction of this core and some specified auxiliary hypotheses, initial conditions and a *ceteris paribus* clause] has been refuted.' (Cf. Popper [1963a].)

The time-honoured empirical criterion for a satisfactory theory was agreement with the observed facts. Our empirical criterion for a series of theories is that it should produce new facts. *The idea of growth and the concept of empirical character are soldered into one.*

This revised form of methodological falsificationism has many new features. First, it denies that 'in the case of a scientific theory, our decision depends upon the results of experiments. If these confirm the theory, we may accept it until we find a better one. If they contradict the theory, we reject it.'[1] It denies that 'what ultimately decides the fate of a theory is the result of a test, *i.e.* an agreement about basic statements'.[2] Contrary to naive falsificationism, *no experiment, experimental report, observation statement or well-corroborated low-level falsifying hypothesis alone can lead to falsification.*[3] *There is no falsification before the emergence of a better theory.*[4] But then the distinctively negative character of naive falsificationism vanishes; criticism becomes more difficult, and also positive, constructive. But, of course, if falsification depends on the emergence of better theories, on the invention of theories which anticipate new facts, then falsification is *not* simply a relation between a theory and the empirical basis, but a multiple relation between competing theories, the original 'empirical basis', and the empirical growth resulting from the competition. Falsification can thus be said to have a '*historical character*'.[5] Moreover, some of the theories which bring about falsification are frequently proposed *after* the 'counterevidence'. This may sound paradoxical for people indoctrinated with naive falsificationism. Indeed, this epistemological theory of the relation between theory and experiment differs sharply from the epistemological theory of naive falsificationism. The very term 'counterevidence' has to be abandoned in the sense that no experimental result must be interpreted directly as 'counterevidence'. If we still want to retain this time-honoured term, we have to redefine it like this: 'counterevidence to T_1' is a corroborating instance to T_2 which

Of course, there is nothing wrong in saying that an isolated, single theory is 'scientific' if it represents an advance on its predecessor, as long as one clearly realizes that in this formulation we appraise the theory as the outcome of – and in the context of – a certain historical development.

[1] Popper [1945], vol. II, p. 233. Popper's more sophisticated attitude surfaces in the remark that 'concrete and practical consequences can be *more* directly tested by experiment' (*ibid.*, my italics).

[2] Popper [1934], section 30.

[3] For the *pragmatic* character of methodological 'falsification', cf. *above*, p. 25, n. 2.

[4] 'In most cases we have, before falsifying a hypothesis, another one up our sleeves' (Popper [1959a], p. 87, n. *1). But, as our argument shows, we *must* have one. Or, as Feyerabend put it: 'The best criticism is provided by those theories which can replace the rivals they have removed' ([1965], p. 227). He notes that in *some* cases 'alternatives will be quite indispensable for the purpose of refutation' (*ibid.*, p. 254). But according to our argument *refutation without an alternative shows nothing but the poverty of our imagination in providing a rescue hypothesis*. Also cf. *below*, p. 37, n. 1.

[5] Cf. volume 2, chapter 8, pp. 178 ff.

is either inconsistent with or independent of T_1 (with the *proviso* that T_2 is a theory which satisfactorily explains the empirical success of T_1). This shows that '*crucial counterevidence*' – or '*crucial experiments*' – can be recognized as such among the scores of anomalies only *with hindsight*, in the light of some superseding theory.[1]

Thus the crucial element in falsification is whether the *new theory* offers any novel, excess information compared with its predecessor and whether some of this excess information is corroborated. Justificationists valued 'confirming' instances of a theory; naive falsificationists stressed 'refuting' instances; for the methodological falsificationists it is the – rather rare – corroborating instances of the *excess* information which are the crucial ones; these receive all the attention. We are no longer interested in the thousands of trivial verifying instances nor in the hundreds of readily available anomalies: the few crucial *excess-verifying instances* are decisive.[2] This consideration rehabilitates – and reinterprets – the old proverb: *Exemplum docet, exempla obscurant.*

'Falsification' in the sense of naive falsificationism (corroborated counter-evidence) is not a *sufficient* condition for eliminating a specific theory: in spite of hundreds of known anomalies we do not regard it as falsified (that is, eliminated) until we have a better one.[3] Nor is 'falsification' in the naive sense *necessary* for falsification in the sophisticated sense: a progressive problemshift does not have to be interspersed with 'refutations'. Science can grow without any 'refutations' leading the way. Naive falsificationists suggest a linear growth of science, in the sense that theories are followed by powerful refutations which eliminate them; these refutations in turn are followed by new theories.[4] It is perfectly *possible* that theories be put forward 'progressively' in such a rapid succession that the 'refutation' of the nth appears only as the corroboration of the $(n+1)$th. The

---

[1] In the distorting mirror of naive falsificationism, new theories which replace old refuted ones, are themselves born unrefuted. Therefore they do not believe that there is a relevant difference between anomalies and crucial counterevidence. For them, anomaly is a dishonest euphemism for counterevidence. But in actual history new theories are born refuted: they inherit many anomalies of the old theory. Moreover, frequently it is *only* the new theory which dramatically predicts that fact which will function as crucial counterevidence against its predecessor, while the 'old' anomalies may well stay on as 'new' anomalies.

All this will be clearer when we introduce the idea of 'research programme': cf. *below*, pp. 50 and 89 ff.

[2] *Sophisticated falsificationism adumbrates a new theory of learning;* cf. *below*, p. 38.

[3] It is clear that the theory T' may have excess corroborated empirical content over another theory T even if both T and T' are refuted. Empirical content has nothing to do with truth or falsity. Corroborated contents can also be compared irrespective of the refuted content. Thus we may see the rationality of the elimination of Newton's theory in favour of Einstein's, even though Einstein's theory may be said to have been born – like Newton's – 'refuted'. We have only to remember that 'qualitative confirmation' is a euphemism for 'quantitative disconfirmation'. (Cf. volume 2, chapter 8, pp. 176–8).

[4] Cf. Popper [1934], section 85, p. 279 of the 1959 English translation.

problem fever of science is raised by proliferation of rival theories rather than counterexamples or anomalies.

This shows that the slogan of *proliferation of theories* is much more important for sophisticated than for naive falsificationism. For the naive falsificationist science grows through repeated experimental overthrow of theories; new rival theories proposed before such 'overthrows' may speed up growth but are not absolutely necessary;[1] constant proliferaton of theories is optional but not mandatory. For the sophisticated falsificationist proliferation of theories cannot wait until the accepted theories are 'refuted' (or until their proponents get into a Kuhnian crisis of confidence).[2] While naive falsificationism stresses 'the urgency of replacing a *falsified* hypothesis by a better one',[3] sophisticated falsificationism stresses the urgency of replacing *any* hypothesis by a better one. Falsification cannot, 'compel the theorist to search for a better theory',[4] simply because falsification cannot precede the better theory.

The problem-shift from naive to sophisticated falsificationism involves a semantic difficulty. For the naive falsificationist a 'refutation' is an experimental result which, by force of his decisions, is made to conflict with the theory under test. But according to sophisticated falsificationism one must not take such decisions before the alleged 'refuting instance' has become the confirming instance of a new, better theory. Therefore whenever we see terms like 'refutation', 'falsification', 'counterexample', we have to check in each case whether these terms are being applied in virtue of decisions by the naive or by the sophisticated falsificationist.[5]

Sophisticated methodological falsificationism offers new standards for intellectual honesty. Justificationist honesty demanded the acceptance of only what was proven and the rejection of everything unproven.

---

[1] It is true that a certain type of *proliferation of rival theories* is allowed to play an accidental heuristic role in falsification. In many cases falsification heuristically 'depends on [the condition] that sufficiently many and sufficiently different theories are offered' (Popper [1940]). For instance, we may have a theory T which is apparently unrefuted. But it may happen that a new theory T', inconsistent with T, is proposed which equally fits the available facts: the differences are smaller than the range of observational error. In such cases the inconsistency prods us into improving our 'experimental techniques', and thus refining the 'empirical basis' so that either T or T' (or, incidentally, both) can be falsified: 'We need [a] new theory in order to find out where the old theory was deficient' (Popper [1963a], p. 246). But the role of this proliferation is *accidental* in the sense that, once the empirical basis is refined, the fight is beween this refined empirical basis and the theory T under test; the rival theory T' acted only as a *catalyst*. (Also cf. *above*, p. 35, n. 4).

[2] Also cf. Feyerabend [1965], pp. 254-5.   [3] Popper [1959a], p. 87, n. *1.

[4] Popper [1934], section 30.

[5] Cf. also *above*, p. 25, n. 2. Possibly it would be better in future to abandon these terms altogether, just as we have abandoned terms like 'inductive (or experimental) proof'. Then we may call (naive) 'refutations' anomalies, and (sophisticatedly) 'falsified' theories 'superseded' ones. Our 'ordinary' language is impregnated not only by 'inductivist' but also by falsificationist dogmatism. A reform is overdue.

Neojustificationist honesty demanded the specification of the probability of any hypothesis in the light of the available empirical evidence. The honesty of naive falsificationism demanded the testing of the falsifiable and the rejection of the unfalsifiable and the falsified. Finally, the honesty of sophisticated falsificationism demanded that one should try to look at things from different points of view, to put forward new theories which anticipate novel facts, and to reject theories which have been superseded by more powerful ones.

Sophisticated methodological falsificationism blends several different traditions. From the empiricists it has inherited the determination to learn primarily from experience. From the Kantians it has taken the activist approach to the theory of knowledge. From the conventionalists it has learned the importance of decisions in methodology.

I should like to emphasize here a further distinctive feature of sophisticated methodological empiricism: the crucial role of excess corroboration. For the inductivist, learning about a new theory is learning how much confirming evidence supports it; about refuted theories one *learns* nothing (learning, after all, is to build up proven or probable *knowledge*). For the dogmatic falsificationist, learning about a theory is learning whether it is refuted or not; about confirmed theories one learns nothing (one cannot prove or probabilify anything), about refuted theories one learns that they are disproved.[1] For the sophisticated falsificationist, learning about a theory is primarily learning which new facts it anticipated: indeed, for the sort of Popperian empiricism I advocate, the only relevant evidence is the evidence anticipated by a theory, and *empiricalness (or scientific character) and theoretical progress are inseparably connected.*[2]

This idea is not entirely new. Leibnitz, for instance, in his famous letter to Conring in 1678, wrote: 'It is the greatest commendation of an hypothesis (next to [proven] truth) if by its help predictions can be made even about phenomena or experiments not tried'.[3] Leibnitz's view was widely accepted by scientists. But since appraisal of a scientific theory, before Popper, meant appraisal of its degree of justification, this position was regarded by some logicians as untenable. Mill, for instance, complains in 1843 in horror that 'it seems to be thought that an hypothesis...is entitled to a more favourable reception, if besides accounting for all the facts previously known, it has led to the anticipation and prediction of others which experience afterwards verified'.[4] Mill had a point: this appraisal was in conflict both with

[1] For a defence of this theory of 'learning from experience', cf. Agassi [1969].

[2] *These remarks show that 'learning from experience' is a normative idea; therefore all purely 'empirical' learning theories miss the heart of the problem.*

[3] Cf. Leibnitz [1678]. The expression in brackets shows that Leibnitz regarded this criterion as second best and thought that the best theories are those which are proved. Thus Leibnitz's position – like Whewell's – is a far cry from fully fledged sophisticated falsificationism.

[4] Mill [1843], vol. II, p. 23.

justificationism and with probabilism: why should an event *prove* more, if it was anticipated by the theory than if it was known already before? As long as *proof* was the only criterion of the scientific character of a theory, Leibnitz's criterion could only be regarded as irrelevant.[1] Also, the *probability* of a theory given evidence cannot possibly be influenced, as Keynes pointed out, by *when* the evidence was produced: the probability of a theory given evidence can depend only on the theory and the evidence,[2] and not upon whether the evidence was produced before or after the theory.

In spite of this convincing justificationist criticism, the criterion survived among some of the best scientists, since it formulated their strong dislike of merely *ad hoc* explanations, which 'though [they] truly express the facts [they set out to explain, are] not borne out by any other phenomena'.[3]

But it was only Popper who recognized that the *prima facie* inconsistency between the few odd, casual remarks against *ad hoc* hypotheses on the one hand and the huge edifice of justificationist philosophy of knowledge must be solved by demolishing justificationism and by introducing new, non-justificationist criteria for appraising scientific theories based on anti-adhocness.

Let us look at a few examples. Einstein's theory is not better than Newton's *because* Newton's theory was 'refuted' but Einstein's was not: there are many known 'anomalies' to Einsteinian theory. Einstein's theory is better than – that is, represents progress compared with – Newton's theory *anno 1916* (that is, Newton's laws of dynamics, law of gravitation, the known set of initial conditions; 'minus' the list of known anomalies such as Mercury's perihelion) *because* it explained everything that Newton's theory had successfully explained, and it explained also *to some extent* some known anomalies and, in addition, forbade events like transmission of light along straight lines near large masses about which Newton's theory had said nothing but which had been permitted by other well-corroborated scientific theories of the day; moreover, *at least some* of the unexpected excess Einsteinian content was in fact *corroborated* (for instance, by the eclipse experiments).

On the other hand, according to these sophisticated standards, Galileo's theory that the natural motion of terrestrial objects was circular, introduced no improvement since it did not forbid anything that had not been forbidden by the relevant theories he intended to

---

[1] This was J. S. Mill's argument (*ibid.*). He directed it against Whewell, who thought that 'consilience of inductions' or successful prediction of improbable events *verifies* (that is, *proves*) a theory. (Whewell [1858], pp. 95–6.) No doubt, *the basic mistake both in Whewell's and in Duhem's philosophy of science is their conflation of predictive power and proven truth. Popper separated the two.*

[2] Keynes [1921], p. 305. But cf. volume 2, chapter 8, p. 183.

[3] This is Whewell's critical comment on an *ad hoc* auxiliary hypothesis in Newton's theory of light (Whewell [1858], vol. II, p. 317).

improve upon (that is, by Aristotelian physics and by Copernican celestial kinematics). This theory was therefore *ad hoc* and therefore – from the heuristic point of view – valueless.[1]

A beautiful example of a theory which satisfied only the first part of Popper's criterion of progress (excess content) but not the second part (corroborated excess content) was given by Popper himself: the Bohr–Kramers–Slater theory of 1924. This theory was refuted in *all* its new predictions.[2]

Let us finally consider how much conventionalism remains in sophisticated falsificationism. Certainly *less* than in naive falsificationism. We need *fewer* methodological decisions. The '*fourth-type decision*' which was essential for the naive version[3] has become completely redundant. To show this we only have to realize that if a scientific theory, consisting of some 'laws of nature', initial conditions, auxiliary theories (but without a *ceteris paribus* clause) conflicts with some factual propositions we do not have to decide which – explicit or 'hidden' – part to replace. We may try to replace *any* part and only when we have hit on an explanation of the anomaly with the help of some content-increasing change (or auxiliary hypothesis), and nature corroborates it, do we move on to eliminate the 'refuted' complex. Thus sophisticated falsification is a slower but possibly safer process than naive falsification.

Let us take an example. Let us assume that the course of a planet differs from the one predicted. Some conclude that this refutes the dynamics and gravitational theory applied: the initial conditions and the *ceteris paribus* clause have been ingeniously corroborated. Others conclude that this refutes the initial conditions used in the calculations: dynamics and gravitational theory have been superbly corroborated in the last two hundred years and all suggestions concerning further factors in play failed. Yet others conclude that this refutes the underlying assumption that there were no other factors in play except for those which were taken into account: these people may possibly be motivated by the metaphysical principle that any explanation is only approximate because of the infinite complexity of the factors involved in determining any single event. Should we praise the first type as '*critical*', scold the second type as '*hack*', and condemn the third as '*apologetic*'? No. We do not need to draw any conclusions about such 'refutation'. We never reject a specific theory simply by *fiat*. If we have an inconsistency like the one mentioned, we do not have to decide which ingredients of the theory we regard as

---

[1] In the terminology of my [1968b], this theory was '*ad hoc*₁' (cf. volume 2, chapter 8, p. 180, n. 1); the example was originally suggested to me by Paul Feyerabend as a paradigm of a *valuable ad hoc theory*. But cf. *below*, p. 56, expecially n. 4.

[2] In the terminology of my [1968b], this theory was not '*ad hoc*₁', but it was '*ad hoc*₂' (cf. volume 2, chapter 8, p. 180, n. 1). For a simple but artificial illustration, see *ibid.* p. 179, n. 1. (For *ad hoc*₃, cf. *below*, p. 88, n. 2.)

[3] Cf. *above*, p. 26.

problematic and which ones as unproblematic: we regard all ingredients as problematic in the light of the conflicting accepted basic statement and try to replace all of them. If we succeed in replacing some ingredient in a 'progressive' way (that is, the replacement has more corroborated empirical content than the original), we call it 'falsified'.

We do not need the *fifth type decision* of the naive falsificationist[1] either. In order to show this let us have a new look at the problem of the appraisal of (syntactically) metaphysical theories – and the problem of their retention and elimination. The 'sophisticated' solution is obvious. We retain a syntactically metaphysical theory as long as the problematic instances can be explained by content-increasing changes in the auxiliary hypotheses appended to it.[2] Let us take, for instance, Cartesian metaphysics C: 'in *all* natural processes *there is* a clockwork mechanism regulated by (*a priori*) animating principles.' This is syntactically irrefutable: it can clash with no – spatiotemporally singular – 'basic statement'. It may, of course, clash with a refutable theory like N: 'gravitation is a force equal to fm_1m_2/r^2 *which acts at a distance*'. But N will only clash with C if 'action at a distance' is interpreted literally and possibly, in addition, as representing an *ultimate* truth, irreducible to any still deeper cause. (Popper would call this an 'essentialist' interpretation.) Alternatively we can regard 'action at a distance' as a mediate cause. Then we interpret 'action at a distance' figuratively, and regard it as a shorthand for some hidden mechanism of action by contact. (We may call this a 'nominalist' interpretation.) In this case we can attempt to explain N by C – Newton himself and several French physicists of the eighteenth century tried to do so. If an auxiliary theory which performs this explanation (or, if you wish, 'reduction') produces novel facts (that is, it is 'independently testable'), Cartesian metaphysics should be regarded as good, scientific, empirical metaphysics, generating a progressive problemshift. A progressive (syntactically) metaphysical theory produces a sustained progressive shift in its protective belt of auxiliary theories. If the reduction of the theory to the 'metaphysical' framework does not produce new empirical content, let alone novel facts, then the reduction represents a degenerating problemshift, it is a mere linguistic exercise. The Cartesian efforts to bolster up their 'metaphysics' in order to explain Newtonian gravitation is an outstanding example of such a merely linguistic reduction.[3]

---

[1] Cf. *above*, p. 28.

[2] *We can formulate this condition with striking clarity only in terms of the methodology of research programmes to be explained in* §3: *we retain a syntactically metaphysical theory as the 'hard core' of a research programme as long as its associated positive heuristic produces a progressive problemshift in the 'protective belt' of auxiliary hypotheses.* Cf. *below*, pp. 51–2.

[3] This phenomenon was described in a beautiful paper by Whewell [1851]; but he could not explain it methodologically. Instead of recognizing the victory of the *progressive* Newtonian programme over the *degenerating* Cartesian programme, he

Thus we do not eliminate a (syntactically) metaphysical theory if it clashes with a well-corroborated scientific theory, as naive falsificationism suggests. We eliminate it if it produces a degenerating shift in the long run and there is a better, rival, metaphysics to replace it. The methodology of a research programme with a 'metaphysical' core does not differ from the methodology of one with a 'refutable' core except perhaps for the logical level of the inconsistencies which are the driving force of the programme.[1]

(It has to be stressed, however, that the very choice of the logical form in which to articulate a theory depends to a large extent on our methodological decision. For instance, instead of formulating Cartesian metaphysics as an 'all-some' statement, we can formulate it as an 'all-statement'; 'all natural processes are clockworks'. A 'basic statement' contradicting this would be: 'a is a natural process and it is not clockwork'. The question is whether according to the 'experimental techniques', or rather, to the interpretative theories of the day, 'x is not a clockwork' can be 'established' or not. Thus the rational choice of the logical form of a theory depends on the state of our knowledge; for instance, a metaphysical 'all-some' statement of today may become, with the change in the level of observational theories, a scientific 'all-statement' tomorrow. I have already argued that only series of theories and not theories should be classified as scientific or non-scientific; now I have indicated that even the logical form of a theory can only be rationally chosen on the basis of a critical appraisal of the state of the research programme in which it is embedded.)

The first, second, and third type decisions of naive falsificationism[2] however cannot be avoided, but as we shall show, the conventional element in the second decision – and also in the third – can be slightly reduced. We cannot avoid the decision which sort of propositions should be the 'observational' ones and which the 'theoretical' ones. We cannot avoid either the decision about the truth-value of some 'observational propositions'. These decisions are vital for the decision whether a problemshift is empirically progressive or degenerating.[3] But the sophisticated falsificationist may at least mitigate the arbitrariness of this second decision by allowing for an *appeal procedure*.

Naive falsificationists do not lay down any such appeal procedure. They accept a basic statement if it is backed up by a well-corroborated falsifying hypothesis,[4] and let it overrule the theory under test – even though they are well aware of the risk.[5] But there is no reason why we should not regard a falsifying hypothesis – and the basic statement it supports – as being just as problematic as a falsified hypothesis. Now

thought this was the victory of proven truth over falsity. For a general discussion of the demarcation between progressive and degenerating reduction cf. Popper [1969a].

[1] Cf. *above*, p. 41, n. 2   [2] Cf. *above*, pp. 22 and 25.
[3] Cf. *above*, p. 33.   [4] Popper [1934], section 22.
[5] Cf. e.g. Popper [1959a], p. 107, n. *2. Also cf. *above*, pp. 28–30.

how exactly can we expose the problematicality of a basic statement? On what grounds can the proponents of the 'falsified' theory appeal and win?

Some people may say that we might go on testing the basic statement (or the falsifying hypothesis) 'by their deductive consequences' until agreement is finally reached. In this testing we deduce – in the same deductive model – further consequences from the basic statement either with the help of the theory under test or some other theory which we regard as unproblematic. Although this procedure 'has no natural end', we always come to a point when there is no further disagreement.[1]

But when the theoretician appeals against the verdict of the experimentalist, the appeal court does not normally cross-question the basic statement directly but rather questions the *interpretative theory* in the light of which its truth-value had been established.

One typical example of a series of successful appeals is the Proutians' fight against unfavourable experimental evidence from 1815 to 1911. For decades Prout's theory T ('all atoms are compounds of hydrogen atoms and thus "atomic weights" of all chemical elements must be expressible as whole numbers') and falsifying 'observational' hypotheses, like Stas's 'refutation' R ('the atomic weight of chlorine is 35·5') confronted each other. As we know, in the end T prevailed over R.[2]

The first stage of any serious criticism of a scientific theory is to reconstruct, improve, its logical deductive articulation. Let us do this in the case of Prout's theory *vis à vis* Stas's refutation. First of all, we have to realize that in the formulation we just quoted, T and R were *not* inconsistent. (Physicists rarely articulate their theories sufficiently to be pinned down and caught by the critic.) In order to show them up as inconsistent we have to put them in the following form. T: 'the atomic weight of all pure (homogeneous) chemical elements are multiples of the atomic weight of hydrogen', and R: 'chlorine is a pure (homogeneous) chemical element and its atomic weight is 35·5'. The last statement is in the form of a falsifying hypothesis which, if well corroborated, would allow us to use basic statements of the form B: 'Chlorine X is a pure (homogeneous) chemical element and its atomic weight is 35·5' – where X is the proper name of a 'piece' of chlorine determined, say, by its space–time coordinates.

But how well-corroborated is R? Its first component depends on R_1: 'Chlorine X is a pure chemical element.' This was the verdict of the experimental chemist after a rigorous application of the 'experimental techniques' of the day.

---

[1] This is argued in Popper [1934], section 29.

[2] Agassi claims that this example shows that we may 'stick to the hypothesis in the face of known facts in the hope that the facts will adjust themselves to theory rather than the other way round' ([1966], p. 18). But *how* can facts 'adjust themselves'? Under which *particular* conditions should the theory win? Agassi gives no answer.

Let us have a closer look at the fine-structure of R_1. In fact R_1 stands for a conjunction of two longer statements T_1 and T_2. The first statement, T_1, could be this: 'If seventeen chemical purifying procedures p_1, $p_2 \ldots p_{17}$ are applied to a gas, what remains will be pure chlorine.' T_2 is then: 'X was subjected to the seventeen procedures p_1, $p_2 \ldots p_{17}$.' The careful 'experimenter' carefully applied all seventeen procedures: T_2 is to be accepted. But the conclusion that therefore what remained *must* be pure chlorine is a 'hard fact' only in virtue of T_1. The experimentalist, while *testing* T, applied T_1. He *interpreted* what he saw in the light of T_1: the result was R_1. *Yet in the monotheoretical deductive model of the test situation this interpretative theory does not appear at all.*

But what if T_1, the interpretative theory, is false? Why not 'apply' T rather than T_1 and claim that atomic weights *must be* whole numbers? Then *this* will be a 'hard fact' in the light of T, and T_1 will be overthrown. Perhaps additional new purifying procedures must be invented and applied.

The problem is then *not* when we should stick to a '*theory*' in the face of '*known facts*' and when the other way round. The problem is *not* what to do when 'theories' clash with 'facts'. Such a 'clash' is only suggested by the '*monotheoretical deductive model*'. Whether a proposition is a '*fact*' or a '*theory*' in the context of a test-situation depends on our methodological decision. 'Empirical basis of a theory' is a monotheoretical notion, it is *relative* to some monotheoretical deductive structure. We may use it as first approximation; but in case of 'appeal' by the theoretician, we must use a *pluralistic model*. In the pluralistic model the clash is not 'between theories and facts' but between two high-level theories: between an *interpretative theory* to provide the facts and an *explanatory theory* to explain them; and the interpretative theory may be on quite as high a level as the explanatory theory. The clash is then not any more between a logically higher-level theory and a lower-level falsifying hypothesis. The problem should not be put in terms of whether a '*refutation*' is real or not. The problem is how to repair an *inconsistency* between the 'explanatory theory' under test and the – explicit or hidden – 'interpretative' theories; or, if you wish, *the problem is which theory to consider as the interpretative one which provides the 'hard' facts and which the explanatory one which 'tentatively' explains them.* In a monotheoretical model we regard the higher-level theory as an *explanatory theory to be judged by the 'facts'* delivered from outside (by the authoritative experimentalist): in the case of a clash we reject the explanation.[1] In a pluralistic model we

---

[1] The decision to use some monotheoretical model is clearly vital for the naive falsificationist to enable him to reject a theory on the *sole* ground of experimental evidence. *It is in line with the necessity for him to divide sharply, at least in a test-situation, the body of science into two: the problematic and the unproblematic.* (Cf. *above* p. 23.) *It is only the theory he decides to regard as problematic which he articulates in his deductive model of criticism.*

may decide, alternatively, to regard the higher-level theory as an *interpretative theory to judge the 'facts'* delivered from outside: in case of a clash we may reject the 'facts' as 'monsters'. In a pluralistic model of testing, several theories – more or less deductively organized – are soldered together.

This argument alone would be enough to show the correctness of the conclusion, which we drew from a different earlier argument, that experiments do not simply overthrow theories, that no theory forbids a state of affairs specifiable in advance.[1] It is not that we propose a theory and Nature may shout NO; rather, we propose a maze of theories, and Nature may shout INCONSISTENT.[2]

The problem is then *shifted* from the old problem of replacing a theory refuted by 'facts' to the new problem of how to resolve inconsistencies between closely associated theories. Which of the mutually inconsistent theories should be eliminated? The sophisticated falsificationist can answer that question easily: one has to try to replace first one, then the other, then possibly both, and opt for that new set-up which provides the biggest increase in corroborated content, which provides the most progressive problemshift.[3]

Thus we have established an appeal procedure in case the theoretician wishes to question the negative verdict of the experimentalist. The theoretician may demand that the experimentalist specify his 'interpretative theory',[4] and he may then replace it – to the experimentalist's annoyance – by a better one in the light of which his originally 'refuted' theory may receive positive appraisal.[5]

[1] Cf. *above*, p. 16.

[2] Let me here answer a possible objection: 'Surely we do not need Nature to tell us that a set of theories is *inconsistent*. Inconsistency – unlike falsehood – can be ascertained without Nature's help'. But Nature's actual 'NO' in a monotheoretical methodology takes the form of a fortified 'potential falsifier', that is a sentence which, in this way of speech, we claim Nature had uttered and which is the *negation of our theory*. Nature's actual 'INCONSISTENCY' in a pluralistic methodology takes the form of a 'factual' statement couched in the light of one of the theories involved, which we claim Nature had uttered and which, if added to our proposed theories, yields an *inconsistent system*.

[3] For instance, in our earlier example (cf. *above*, p. 23 ff) some may try to replace the gravitational theory with a new one and others may try to replace the radio-optics by a new one: we choose the way which offers the more spectacular growth, the more progressive problemshift.

[4] Criticism does not *assume* a fully articulated deductive structure: it creates it. (Incidentally, this is the main message of my [1963–4].)

[5] A classical example of this pattern is Newton's relation to Flamsteed, the first Astronomer Royal. For instance, Newton visited Flamsteed on 1 September 1694, when working full time on his lunar theory; told him to reinterpret some of his data since they contradicted his own theory; and he explained to him exactly how to do it. Flamsteed obeyed Newton and wrote to him on 7 October: 'Since you went home, I examined the observations I employed for determining the greatest equations of the earth's orbit, and considering the moon's places at the times...I find that (*if, as you intimate, the earth inclines on that side the moon then is*) you may abate abt 20″ from it.' Thus Newton constantly criticized and corrected Flamsteed's observational theories. Newton taught Flamsteed, for instance, a better theory of the refractive

But even this appeal procedure cannot do more than *postpone* the conventional decision. For the verdict of the appeal court is not infallible either. When we decide whether it is the replacement of the 'interpretative' or of the 'explanatory' theory that produces novel facts, we again must take a decision about the acceptance or rejection of basic statements. But then we have only *postponed* – and possibly *improved* – the decision, not avoided it.[1] The difficulties concerning the empirical basis which confronted 'naive' falsificationism cannot be avoided by 'sophisticated' falsificationism either. Even if we regard a theory as 'factual', that is, if our slow-moving and limited imagination cannot offer an alternative to it (as Feyerabend used to put it), we have to make, at least occasionally and temporarily, decisions about its truth-value. *Even then, experience still remains, in an important sense, the 'impartial arbiter'*[2] of scientific controversy. We cannot get rid of the problem of the 'empirical basis', if we want to learn from experience;[3] but we can make our learning less dogmatic – but also less fast and less dramatic. By regarding some observational theories as problematic we may make our methodology more flexible; but we cannot articulate and include *all* 'background knowledge' (or 'background ignorance'?) into our critical deductive model. This process is bound to be piecemeal and some conventional line must be drawn at any given time.

There is one objection even to the sophisticated version of methodological falsificationism which cannot be answered without some concession to Duhemian 'simplicism'. The objection is the so-called 'tacking paradox'. According to our definitions, adding to a theory completely disconnected low-level hypotheses may constitute a 'progressive shift'. It is difficult to eliminate such makeshift shifts without demanding that the additional assertions must be connected with the original assertion *more intimately* than by mere conjunction. This, of course, is a sort of simplicity requirement which would assure the continuity in the series of theories which can be said to constitute *one* problemshift.

This leads us to further problems. For one of the crucial features of sophisticated falsificationism is that it replaces the concept of *theory* as the basic concept of the logic of discovery by the concept of *series of*

power of the atmosphere; Flamsteed accepted this and corrected his original 'data'. One can understand the constant humiliation and slowly increasing fury of this great observer, having his data criticized and improved by a man who, on his own confession, made no observations himself: it was this feeling – I suspect – which led finally to a vicious personal controversy.

[1] The same applies to the third type of decision. If we reject a stochastic hypothesis only for one which, in our sense, supersedes it, the exact form of the 'rejection rules' becomes *less* important.

[2] Popper [1945], volume II, chapter 23, p. 218.

[3] Agassi is then wrong in his thesis that 'observation reports may be accepted as false and hence the problem of the empirical basis is thereby disposed of' (Agassi [1966], p. 20).

theories. It is a succession of theories and not one given theory which is appraised as scientific or pseudo-scientific. But the members of such series of theories are usually connected by a remarkable *continuity* which welds them into *research programmes.* This *continuity* – reminiscent of Kuhnian 'normal science' – plays a vital role in the history of science; the main problems of the logic of discovery cannot be satisfactorily discussed except in the framework of a *methodology of research programmes.*

3   A METHODOLOGY OF SCIENTIFIC RESEARCH PROGRAMMES

I have discussed the problem of objective appraisal of scientific growth in terms of progressive and degenerating problemshifts in series of scientific theories. The most important such series in the growth of science are characterized by a certain *continuity* which connects their members. This continuity evolves from a genuine research programme adumbrated at the start. The programme consists of methodological rules: some tell us what paths of research to avoid (*negative heuristic*), and others what paths to pursue (*positive heuristic*).[1]

Even science as a whole can be regarded as a huge research programme with Popper's supreme heuristic rule: 'devise conjectures which have more empirical content than their predecessors.' Such methodological rules may be formulated, as Popper pointed out, as metaphysical principles.[2] For instance, the *universal* anti-conventionalist rule against exception-barring may be stated as the metaphysical principle: 'Nature does not allow exceptions.' This is why Watkins called such rules 'influential metaphysics'.[3]

But what I have primarily in mind is not science as a whole, but rather *particular* research programmes, such as the one known as 'Cartesian metaphysics'. Cartesian metaphysics, that is, the mechanistic theory of the universe – according to which the universe is a huge clockwork (and system of vortices) with push as the only cause of motion – functioned as a powerful heuristic principle. It discouraged work on scientific theories – like (the 'essentialist' version of) Newton's theory of action at a distance – which were inconsistent with it (*negative heuristic*). On the other hand, it encouraged work on auxiliary hypo-

---

[1] One may point out that the negative and positive heuristic gives a rough (implicit) definition of the 'conceptual framework' (and consequently of the language). The recognition that the history of science is the history of research programmes rather than of theories may therefore be seen as a partial vindication of the view that the history of science is the history of conceptual frameworks or of scientific languages.

[2] Popper [1934], sections 11 and 70. I use 'metaphysical' as a technical term of naive falsificationism: a contingent proposition is 'metaphysical' if it has no 'potential falsifiers'.

[3] Watkins [1958]. Watkins cautions that 'the logical gap between statements and prescriptions in the metaphysical-methodological field is illustrated by the fact that a person may reject a [metaphysical] doctrine in its fact-stating form while subscribing to the prescriptive version of it' (*Ibid.*, pp. 356–7).

theses which might have saved it from apparent counterevidence – like Keplerian ellipses (*positive heuristic*).[1]

(a) Negative heuristic: the 'hard core' of the programme

All scientific research programmes may be characterized by their 'hard core'. The negative heuristic of the programme forbids us to direct the *modus tollens* at this 'hard core'. Instead, we must use our ingenuity to articulate or even invent 'auxiliary hypotheses', which form a *protective belt* around this core, and we must redirect the *modus tollens* to *these*. It is this protective belt of auxiliary hypotheses which has to bear the brunt of tests and get adjusted and re-adjusted, or even completely replaced, to defend the thus-hardened core. A research programme is successful if all this leads to a progressive problemshift; unsuccessful if it leads to a degenerating problemshift.

The classical example of a successful research programme is Newton's gravitational theory: possibly the most successful research programme ever. When it was first produced, it was submerged in an ocean of 'anomalies' (or, if you wish, 'counterexamples'[2]), and opposed by the observational theories supporting these anomalies. But Newtonians turned, with brilliant tenacity and ingenuity, one counterinstance after another into corroborating instances, primarily by overthrowing the original observational theories in the light of which this 'contrary evidence' was established. In the process they themselves produced new counter-examples which they again resolved. They 'turned each new difficulty into a new victory of their programme'.[3]

In Newton's programme the negative heuristic bids us to divert the *modus tollens* from Newton's three laws of dynamics and his law of gravitation. This 'core' is 'irrefutable' by the methodological decision of its proponents: anomalies must lead to changes only in the 'protective' belt of auxiliary, 'observational' hypotheses and initial conditions.[4]

I have given a contrived micro-example of a progressive Newtonian problemshift.[5] If we analyse it, it turns out that each successive link in this exercise predicts some new fact; each step represents an increase in empirical content: the example constitutes a *consistently progressive theoretical shift*. Also, each prediction is in the end verified; although on three subsequent occasions they may have seemed momentarily to

---

[1] For this Cartesian research programme, cf. Popper [1960*b*] and Watkins [1958], pp. 350–1.

[2] For the clarification of the concepts of 'counterexample' and 'anomaly' cf. *above*, p. 26, and especially *below*, p. 72, text to n. 3.

[3] Laplace [1824], livre IV, chapter 11.

[4] The actual hard core of a programme does not actually emerge fully armed like Athene from the head of Zeus. It develops slowly, by a long, preliminary process of trial and error. In this paper this process is not discussed.

[5] Cf. *above*, pp. 16–17.

be 'refuted'.[1] While 'theoretical progress' (in the sense here described) may be verified immediately,[2] 'empirical progress' cannot, and in a research programme we may be frustrated by a long series of 'refutations' before ingenious and lucky content-increasing auxiliary hypotheses turn a chain of defeats – *with hindsight* – into a resounding success story, either by revising some false 'facts' or by adding novel auxiliary hypotheses. We may then say that we must require that each step of a research programme be consistently content-increasing: that each step constitute a *consistently progressive theoretical problemshift*. All we need in addition to this is that at least every now and then the increase in content should be seen to be retrospectively corroborated: the programme as a whole should also display an *intermittently progressive empirical shift*. We do not demand that each step produce *immediately* an *observed* new fact. Our term '*intermittently*' gives sufficient *rational* scope for dogmatic adherence to a programme in face of *prima facie* 'refutations'.

The idea of 'negative heuristic' of a scientific research programme rationalizes classical conventionalism to a considerable extent. We may rationally decide not to allow 'refutations' to transmit falsity to the hard core as long as the corroborated empirical content of the protecting belt of auxiliary hypotheses increases. But our approach differs from Poincaré's justificationist conventionalism in the sense that, unlike Poincaré, we maintain that if and when the programme ceases to anticipate novel facts, its hard core might have to be abandoned: that is, *our* hard core, unlike Poincaré's, may crumble under certain conditions. In this sense we side with Duhem who thought that such a possibility must be allowed for;[3] but for Duhem the reason for such crumbling is purely *aesthetic*,[4] while for us it is mainly *logical and empirical*.

(b) Positive heuristic: the construction of the 'protective belt' and the relative autonomy of theoretical science

Research programmes, besides their negative heuristic, are also characterized by their positive heuristic.

Even the most rapidly and consistently progressive research programmes can digest their 'counter-evidence' only piecemeal: anomalies are never completely exhausted. But it should not be thought that yet unexplained anomalies – 'puzzles' as Kuhn might call them – are taken in random order, and the protective belt built up in an eclectic fashion, without any preconceived order. The order is usually decided in the theoretician's cabinet, independently of the *known* anomalies.

---

[1] The 'refutation' was each time successfully diverted to 'hidden lemmas'; that is, to lemmas emerging, as it were, from the *ceteris paribus* clause.
[2] But cf. *below*, pp. 69–71.   [3] Cf. *above*, p. 22.
[4] *Ibid.*

Few theoretical scientists engaged in a research programme pay undue attention to 'refutations'. They have a long-term research policy which anticipates these refutations. This research policy, or order of research, is set out – in more or less detail – in the *positive heuristic* of the research programme. The negative heuristic specifies the 'hard core' of the programme which is 'irrefutable' by the methodological decision of its proponents; the positive heuristic consists of a partially articulated set of suggestions or hints on how to change, develop the 'refutable variants' of the research-programme, how to modify, sophisticate, the 'refutable' protective belt.

The positive heuristic of the programme saves the scientist from becoming confused by the ocean of anomalies. The positive heuristic sets out a programme which lists a chain of ever more complicated *models* simulating reality: the scientist's attention is riveted on building his models following instructions which are laid down in the positive part of his programme. He ignores the *actual* counterexamples, the available '*data*'.[1] Newton first worked out his programme for a planetary system with a fixed point-like sun and one single point-like planet. It was in this model that he derived his inverse square law for Kepler's ellipse. But this model was forbidden by Newton's own third law of dynamics, therefore the model had to be replaced by one in which both sun and planet revolved round their common centre of gravity. This change was not motivated by any observation (the data did not suggest an 'anomaly' here) but by a theoretical difficulty in developing the programme. Then he worked out the programme for more planets as if there were only heliocentric but no interplanetary forces. Then he worked out the case where the sun and planets were not mass-points but mass-*balls*. Again, for this change he did not *need* the observation of an anomaly; infinite density was forbidden by an (inarticulated) touchstone theory, therefore planets *had* to be extended. This change involved considerable mathematical difficulties, held up Newton's work – and delayed the publication of the *Principia* by more than a decade. Having solved this 'puzzle', he started work on *spinning balls* and their wobbles. Then he admitted interplanetary forces and started work on *perturbations*. At this point he started to look more anxiously at the facts. Many of them were beautifully explained (qualitatively) by this model, many were not. It was then that he started to work on *bulging* planets, rather than round planets, etc.

Newton despised people who, like Hooke, stumbled on a first naive model but did not have the tenacity and ability to develop it into a research programme, and who thought that a first version, a mere

---

[1] If a scientist (or mathematician) has a positive heuristic, he refuses to be drawn into observation. He will 'lie down on his couch, shut his eyes and forget about the data'. (Cf. my [1963–4], especially pp. 300 ff, where there is a detailed case study of such a programme.) Occasionally, of course, he will ask Nature a shrewd question: he will then be encouraged by Nature's YES, but not discouraged by its NO.

aside, constituted a 'discovery'. He held up publication until his programme had achieved a remarkable progressive shift.[1]

Most, if not all, Newtonian 'puzzles', leading to a series of new variants superseding each other, were forseeable at the time of Newton's first naive model and no doubt Newton and his colleagues *did* forsee them: Newton must have been fully aware of the blatant falsity of his first variants. Nothing shows the existence of a positive heuristic of a research programme clearer than this fact: this is why one speaks of 'models' in research programmes. A *'model'* is a set of initial conditions (possibly together with some of the observational theories) which one knows is *bound* to be replaced during the further development of the programme, and one even knows, more or less, how. This shows once more how irrelevant 'refutations' of any specific variant are in a research programme: their existence is fully expected, the positive heuristic is there as the strategy both for predicting (producing) and digesting them. Indeed, if the positive heuristic is clearly spelt out, the difficulties of the programme are mathematical rather than empirical.[2]

One may formulate the 'positive heuristic' of a research programme as a 'metaphysical' principle. For instance one may formulate Newton's programme like this: 'the planets are essentially gravitating spinning-tops of roughly spherical shape'. This idea was never *rigidly* maintained: the planets are not *just* gravitational, they have also, for example, electromagnetic characteristics which may influence their motion. Positive heuristic is thus in general more flexible than negative heuristic. Moreover, it occasionally happens that when a research programme gets into a degenerating phase, a little revolution or a *creative shift* in its positive heuristic may push it forward again.[3] It is better therefore to separate the 'hard core' from the more flexible metaphysical principles expressing the positive heuristic.

Our considerations show that the positive heuristic forges ahead with almost complete disregard of 'refutations': it may seem that it is the *'verifications'*[4] rather than the refutations which provide the

---

[1] Reichenbach, following Cajori, gives a different explanation of what delayed Newton in the publication of his *Principia*; 'To his disappointment he found that the observational results disagreed with his calculations. Rather than set any theory, however beautiful, before the facts, Newton put the manuscript of this theory into his drawer. Some twenty years later, after new measurements of the circumference of the earth had been made by a French expedition, Newton saw that the figures on which he had based his test were false and that the improved figures agreed with his theoretical calculation. It was only after this test that he published his law... The story of Newton is one of the most striking illustrations of the method of modern science' (Reichenbach [1951], pp. 101-2). Feyerabend criticizes Reichenbach's account (Feyerabend [1965], p. 229), but does not give an alternative *rationale*.

[2] For this point cf. Truesdell [1960].

[3] Soddy's contribution to Prout's programme or Pauli's to Bohr's (old quantum theory) programme are typical examples of such creative shifts.

[4] A 'verification' is a corroboration of excess content in the expanding programme. But, of course, a 'verification' does not *verify* a programme: it shows only its heuristic power.

contact points with reality. Although one must point out that any 'verification' of the $(n+1)$th version of the programme is a refutation of the nth version, we cannot deny that *some* defeats of the subsequent versions are always foreseen: it is the 'verifications' which keep the programming going, recalcitrant instances notwithstanding.

We may appraise research programmes, even after their 'elimination', for their *heuristic power*: how many new facts did they produce, how great was 'their capacity to explain their refutations in the course of their growth'?[1]

(We may also appraise them for the stimulus they gave to mathematics. The real difficulties for the theoretical scientist arise rather from the *mathematical difficulties* of the programme than from anomalies. The greatness of the Newtonian programme comes partly from the development – by Newtonians – of classical infinitesimal analysis which was a crucial precondition of its success.)

Thus the methodology of scientific research programmes accounts for the *relative autonomy of theoretical science*: a historical fact whose rationality cannot be explained by the earlier falsificationists. Which problems scientists working in powerful research programmes rationally choose, is determined by the positive heuristic of the programme rather than by psychologically worrying (or technologically urgent) anomalies. The anomalies are listed but shoved aside in the hope that they will turn, in due course, into corroborations of the programme. Only those scientists have to rivet their attention on anomalies who are either engaged in trial and error exercises[2] or who work in a degenerating phase of a research programme when the positive heuristic ran out of steam. (All this, of course, must sound repugnant to naive falsificationists who hold that once a theory is 'refuted' by experiment (by *their* rule book), it is irrational (and dishonest) to develop it further: one has to replace the old 'refuted' theory by a new, unrefuted one.)

(c) Two illustrations: Prout and Bohr

The dialectic of positive and negative heuristic in a research programme can best be illuminated by examples. Therefore I am now going to sketch a few aspects of two spectacularly successful research programmes: Prout's programme[3] based on the idea that all atoms are compounded of hydrogen atoms and Bohr's programme based on the idea that light-emission is due to electrons jumping from one orbit to another within the atoms.

(*In writing a historical case study, one should, I think, adopt the following*

---

[1] Cf. my [1963–4], pp. 324–30. Unfortunately in 1963–4 I had not yet made a clear terminological distinction between theories and research programmes, and this impaired my exposition of a research programme in informal, quasi-empirical mathematics.

[2] Cf. *below*, p. 88.   [3] Already mentioned *above*, pp. 43–4.

procedure: (1) *one gives a rational reconstruction;* (2) *one tries to compare this rational reconstruction with actual history and to criticize both one's rational reconstruction for lack of historicity and the actual history for lack of rationality. Thus any historical study must be preceded by a heuristic study: history of science without philosophy of science is blind. In this paper it is not my purpose to go on seriously to the second stage.*)

(*c 1*) *Prout: a research programme progressing in an ocean of anomalies*

Prout, in an anonymous paper of 1815, claimed that the atomic weights of all pure chemical elements were whole numbers. He knew very well that anomalies abounded, but said that these arose because chemical substances as they ordinarily occurred were *impure*: that is, the relevant 'experimental techniques' of the time were unreliable, or, to put it in our terms, the contemporary 'observational' theories in the light of which the truth values of the basic statements of his theory were established, were false.[1] The champions of Prout's theory therefore embarked on a major venture: to overthrow those theories which supplied the counter-evidence to their thesis. For this they had to revolutionize the established analytical chemistry of the time and correspondingly revise the experimental techniques with which pure elements were to be separated.[2] Prout's theory, as a matter of fact, defeated the theories previously applied in purification of chemical substances one after the other. Even so, some chemists became tired of the research programme and gave it up, since the successes were still far from adding up to a final victory. For instance, Stas, frustrated by some stubborn, recalcitrant instances, concluded in 1860 that Prout's theory was 'without foundations'.[3] But others were more encouraged by the progress than discouraged by the lack of complete success. For instance, Marignac immediately retorted that 'although [he is satisfied that] the experiments of Monsieur Stas are perfectly exact, [there is no proof] that the differences observed between his results and those required by Prout's law cannot be explained by the imperfect character

---

[1] Alas, all this is rational reconstruction rather than actual history. Prout denied the existence of any anomalies. For instance, he claimed that the atomic weight of chlorine was exactly 36.

[2] Prout was aware of some of the basic methodological features of his programme. Let us quote the first lines of his [1815]: 'The author of the following essay submits it to the public with the greatest diffidence...He trusts, however, that its importance will be seen, and that some one will undertake to examine it, and thus verify or refute its conclusions. If these should be proved erroneous, still new facts may be brought to light, or old ones better established, by the investigation; but if they should be verified, a new and interesting light will be thrown upon the whole science of chemistry.'

[3] Clerk Maxwell was on Stas's side: he thought it was impossible that there should be two kinds of hydrogen, 'for if some [molecules] were of slightly greater mass than others, we have the means of producing a separation between molecules of different masses, one of which would be somewhat denser than the other. As this cannot be done, we must admit [that all are alike]' (Maxwell [1871]).

of experimental methods'.[1] As Crookes put it in 1886: 'Not a few chemists of admitted eminence consider that we have here [in Prout's theory] an expression of the truth, masked by some residual or collateral phenomena which we have not yet succeeded in eliminating.'[2] That is, there had to be some *further* false hidden assumption in the 'observational' theories on which 'experimental techniques' for chemical purification were based and with the help of which atomic weights were calculated: in Crookes's view even in 1886 'some present atomic weights merely represented a mean value'.[3] Indeed, Crookes went on to put this idea in a scientific (content-increasing) form: he proposed concrete new theories of 'fractionation', a new 'sorting Demon'.[4] But, alas, his new observational theories turned out to be as false as they were bold and, being unable to anticipate any new fact, they were eliminated from the (rationally reconstructed) history of science. As it turned out a generation later, there was a very basic hidden assumption which failed the researchers: that two pure elements must be separable by *chemical* methods. The idea that two different pure elements may behave identically in all *chemical* reactions but can be separated by *physical* methods, required a change, a '*stretching*', of the concept of 'pure element' which constituted a change – a *concept-stretching expansion* – of the research programme itself.[5] This revolutionary highly *creative shift* was taken only by Rutherford's school;[6] and then 'after many vicissitudes and the most convincing apparent disproofs, the hypothesis thrown out so lightly by Prout, an Edinburgh physician, in 1815, has, a century later, become the cornerstone of modern theories of the structure of atoms'.[7] However, this creative step was in fact only a side-result of progress in a different, indeed, distant research programme; Proutians, lacking this *external* stimulus, never dreamt of trying, for instance, to build powerful centrifugal machines to separate elements.

(When an 'observational' or 'interpretative' theory finally gets eliminated, the 'precise' measurements carried out within the discarded framework may look – with hindsight – rather foolish. Soddy made fun of 'experimental precision' for its own sake: 'There is something surely akin to if not transcending tragedy in the fate that has overtaken the life work of that distinguished galaxy of nineteenth-century chemists, rightly revered by their contemporaries as representing the crown and perfection of accurate scientific measurement. Their hard won results, for the moment at least, appears as of as little interest and

---

[1] Marignac [1860].   [2] Crookes [1886].
[3] *Ibid.*   [4] Crookes [1886], p. 491.
[5] For 'concept-stretching', cf. my [1963–4], part IV.
[6] The shift is anticipated in Crookes's fascinating [1888] where he indicates that the solution should be sought in a new demarcation between 'physical' and 'chemical'. But the anticipation remained philosophical; it was left to Rutherford and Soddy to develop it, after 1910, into a scientific theory.
[7] Soddy [1932], p. 50.

significance as the determination of the average weight of a collection of bottles, some of them full and some of them more or less empty.'[1])

Let us stress that in the light of the methodology of research programmes here proposed there never was any rational reason to *eliminate* Prout's programme. Indeed, the programme produced a beautiful, progressive shift, even if, in between, there were considerable hitches.[2] Our sketch shows how a research programme can challenge a considerable bulk of accepted scientific knowledge: it is planted, as it were, in an inimical environment which, step by step, it can override and transform.

Also, the actual history of Prout's programme illustrates only too well how much the progress of science was hindered and slowed down by justificationism and by naive falsificationism. (The opposition to atomic theory in the nineteenth century was fostered by both.) An elaboration of this particular influence of bad methodology on science may be a rewarding research programme for the historian of science.

(c 2) Bohr: a research programme progressing on inconsistent foundations

A brief sketch of Bohr's research programme of light emission (in *early* quantum physics) will illustrate further – and even expand – our thesis.[3]

The story of Bohr's research programme can be characterized by: (1) its initial problem; (2) its negative and positive heuristic; (3) the problems which it attempted to solve in the course of its development; and (4) its degeneration point (or, if you wish, 'saturation point') and, finally, (5) the programme by which it was superseded.

The background problem was the riddle of how Rutherford atoms (that is, minute planetary systems with electrons orbiting round a positive nucleus) can remain stable; for, according to the well-corroborated Maxwell–Lorentz theory of electromagnetism they should collapse. But Rutherford's theory was well corroborated too. Bohr's suggestion was to ignore for the time being the inconsistency and consciously develop a research programme whose 'refutable' versions were inconsistent with the Maxwell–Lorentz theory.[4] He proposed five postulates as the *hard core* of his programme: '(1) that

---

[1] *Ibid.*

[2] These hitches inevitably induce many individual scientists to shelve or altogether jettison the programme and join other research programmes where the positive heuristic happens to offer at the time cheaper successes: the history of science cannot be *fully* understood without mob-psychology. (Cf. *below*, pp. 90–93.)

[3] This section may again strike the historian as more a caricature than a sketch; but I hope it serves its purpose. (Cf. *above*, p. 52.) Some statements are to be taken not with a grain, but with tons, of salt.

[4] This, of course, is a further argument against J. O. Wisdom's thesis that metaphysical theories can be refuted by a conflicting well-corroborated scientific theory (Wisdom [1963].) Also, cf. *above*, p. 27, text to n. 7, and p. 42.

energy radiation [within the atom] is not emitted (or absorbed) in the continuous way assumed in the ordinary electrodynamics, but only during the passing of the systems between different "stationary" states. (2) That the dynamical equilibrium of the systems in the stationary states is governed by the ordinary laws of mechanics, while these laws do not hold for the passing of the systems between the different states. (3) That the radiation emitted during the transition of a system between two stationary states is homogeneous, and that the relation between the frequency v and the total amount of energy emitted E is given by $E = hv$, where h is Planck's constant. (4) That the different stationary states of a simple system consisting of an electron rotating round a positive nucleus are determined by the condition that the ratio between the total energy, emitted during the formation of the configuration, and the frequency of revolution of the electron is an entire multiple of $\frac{1}{2}h$. Assuming that the orbit of the electron is circular, this assumption is equivalent with the assumption that the angular momentum of the electron round the nucleus is equal to an entire multiple of $h/2\pi$. (5) That the "permanent" state of any atomic system, i.e. the state in which the energy emitted is maximum, is determined by the condition that the angular momentum of every electron round the centre of its orbit is equal to $h/2\pi$."[1]

We have to appreciate the crucial methodological difference between the inconsistency introduced by Prout's programme and that introduced by Bohr's. Prout's research programme declared war on the analytical chemistry of his time: its positive heuristic was designed to overthrow it and replace it. But Bohr's research programme contained no analogous design: its positive heuristic, even if it had been completely successful, would have left the inconsistency with the Maxwell–Lorentz theory unresolved.[2] To suggest such an idea required even greater courage than Prout's; the idea crossed Einstein's mind but he found it unacceptable, and rejected it.[3] Indeed, *some of the most important research programmes in the history of science were grafted on to older programmes with which they were blatantly inconsistent.* For instance, Copernican astronomy was 'grafted' on to Aristotelian physics, Bohr's programme on to Maxwell's. Such 'grafts' are irrational for the justificationist and for the naive falsificationist, neither of whom can countenance growth on inconsistent foundations. Therefore they are usually concealed by *ad hoc* stratagems – like Galileo's theory of circular inertia or Bohr's correspondence, and, later, complementarity principle – the only purpose of which is to hide the 'deficiency'.[4] As

---

[1] Bohr [1913a], p. 874.

[2] Bohr held at this time that the Maxwell–Lorentz theory would *eventually* have to be replaced. (Einstein's photon theory had already indicated this need.)

[3] Hevesy [1913]; cf. also *above*, p. 50, text to n. 1.

[4] In our methodology there is no need for such protective *ad hoc* stratagems. But, on the other hand, they are harmless as long as they are clearly seen as problems, not as solutions.

the young grafted programme strengthens, the peaceful co-existence comes to an end, the symbiosis becomes competitive and the champions of the new programme try to replace the old programme altogether.

It may well have been the success of his 'grafted programme' which later misled Bohr into believing that such fundamental inconsistencies in research programmes can and should be put up with *in principle*, that they do not present any serious problem and one merely has to get used to them. Bohr tried in 1922 to lower the standards of scientific criticism; he argued that '*the most* that one can demand of a theory [i.e. programme] is that the classification [it establishes] can be pushed so far that it can contribute to the development of the field of observation by the prediction of *new* phenomena.'[1]

(This statement by Bohr is similar to d'Alembert's when faced with the inconsistency in the foundations of infinitesimal theory: '*Allez en avant et la foi vous viendra.*' According to Margenau, 'it is understandable that, in the excitement over its success, men overlooked a malformation in the theory's architecture; for Bohr's atom sat like a baroque tower upon the Gothic base of classical electrodynamics.'[2] But as a matter of fact, the 'malformation' was not 'overlooked': everybody was aware of it, only they ignored it – more or less – during the progressive phase of the programme.[3] Our methodology of research programmes shows the rationality of this attitude but it also shows the irrationality of the defence of such 'malformations' once the progressive phase is over.

It should be said here that in the thirties and forties Bohr abandoned his demand for 'new phenomena' and was prepared to 'proceed with the immediate task of co-ordinating the multifarious evidence regarding atomic phenomena, which accumulated from day to day in the exploration of this new field of knowledge'.[4] This indicates that Bohr, by this time, had fallen back on 'saving the phenomena', while Einstein sarcastically insisted that 'every theory is true provided that one suitably associates its symbols with observed quantities'.[5])

But *consistency* – in a strong sense of the term[6] – *must remain an*

---

[1] Bohr [1922], my italics.   [2] Margenau [1950], p. 311.
[3] Sommerfeld ignored it more than Bohr: cf. *below*, p. 63, n. 7.
[4] Bohr [1949], p. 206.   [5] Quoted in Schrödinger [1958], p. 170.
[6] Two propositions are inconsistent if their conjunction has no model, that is, there is no interpretation of their descriptive terms in which the conjunction is true. But in informal discourse we use more formative terms than in formal discourse: some descriptive terms are given a fixed interpretation. In this informal sense two propositions may be (weakly) inconsistent given the standard interpretations of some characteristic terms even if formally, in some unintended interpretation, they may be consistent. For instance, the first theories of electron spin were inconsistent with the special theory of relativity if 'spin' was given its ('strong') standard interpretation and thereby treated as a formative term; but the inconsistency disappears if 'spin' is treated as an uninterpreted descriptive term. The reason why we should not give up standard interpretations too easily is that such emasculation of meanings may emasculate the positive heuristic of the programme. (On the other hand, such

important regulative principle (over and above the requirement of progressive problemshift); and inconsistencies (including anomalies) *must* be seen as problems. The reason is simple. If science aims at truth, it must aim at consistency; if it resigns consistency, it resigns truth. To claim that 'we must be modest in our demands',[1] that we must resign ourselves to – weak or strong – inconsistencies, remains a methodological vice. On the other hand, this does not mean that the discovery of an inconsistency – or of an anomaly – must *immediately* stop the development of a programme: it may be rational to put the inconsistency into some temporary, *ad hoc* quarantine, and carry on with the positive heuristic of the programme. This has been done even in mathematics, as the examples of the early infinitesimal calculus and of naive set theory show.[2]

(From this point of view, Bohr's 'correspondence principle' played an interesting double role in his programme. On the one hand it functioned as an important heuristic principle which suggested many new scientific hypotheses which, in turn, led to novel facts, especially in the field of the intensity of spectrum lines.[3] On the other hand it functioned also as a defence mechanism, which 'endeavoured to utilize to the utmost extent the concepts of the classical theories of mechanics and electrodynamics, in spite of the contrast between these theories and the quantum of action',[4] instead of emphasizing the urgency of a unified programme. In this second role it reduced the degree of problematicality of the programme.[5])

Of course, the research programme of quantum theory as a whole was a 'grafted programme' and therefore repugnant to physicists with deeply conservative views like Planck. There are two extreme and equally irrational positions with regard to a grafted programme.

---

meaning shifts may be in some cases progressive: cf. *above*, p. 41.)

For the shifting demarcation between formative and descriptive terms in informal discourse, cf. my [1963–4], 9(*b*), especially p. 335, n. 1.

[1] Bohr [1922], last paragraph.

[2] Naive falsificationists tend to regard this liberalism as a *crime against reason*. Their main argument runs like this: 'If one were to accept contradictions, then one would have to give up any kind of scientific activity: it would mean a complete breakdown of science. This can be shown by proving that *if two contradictory statements are admitted, any statement whatever must be admitted*; for from a couple of contradictory statements any statement whatever can be validly inferred...A theory which involves a contradiction is therefore entirely useless *as a theory*' (Popper [1940]). In fairness to Popper, one has to stress that he is here arguing against Hegelian dialectic, in which inconsistency becomes a *virtue*; and he is absolutely right when he points out its dangers. But Popper never analysed patterns of empirical (or non-empirical) progress on inconsistent foundations; indeed, in section 24 of his [1934] he makes consistency and falsifiability mandatory requirements for any scientific theory. I discuss this problem in more detail in chapter 3.

[3] Cf. e.g. Kramers [1923].   [4] Bohr [1923].

[5] Born, in his [1954], gives a vivid account of the correspondence principle which strongly supports this double appraisal: 'The art of guessing correct formulae, which deviate from the classical ones, yet contain them as a limiting case...was brought to a high degree of perfection.'

The *conservative position* is to halt the new programme until the basic inconsistency with the old programme is somehow repaired: it is irrational to work on inconsistent foundations. The 'conservatives' will concentrate on eliminating the inconsistency by explaining (approximately) the postulates of the new programme in terms of the old programme: they find it irrational to go on with the new programme without a successful *reduction* of the kind mentioned. Planck himself chose this way. He did not succeed, in spite of the decade of hard work he invested in it.[1] Therefore Laue's remark that his lecture on 14 December 1900, was the 'birthday of the quantum theory' is not quite true: that day was the birthday of Planck's reduction programme. The decision to go *ahead* with temporarily inconsistent foundations was taken by Einstein in 1905, but even he wavered in 1913, when Bohr forged forward again.

The *anarchist position* concerning grafted programmes is to extol anarchy in the foundations as a virtue and regard [weak] inconsistency either as some basic property of nature or as an ultimate limitation of human knowledge, as some of Bohr's followers did.

The *rational position* is best characterized by Newton's, who faced a situation which was to a certain extent similar to the one discussed. Cartesian push-mechanics, on which Newton's programme was originally grafted, was (weakly) inconsistent with Newton's theory of gravitation. Newton worked both on his positive heuristic (successfully) *and* on a reductionist programme (unsuccessfully), and disapproved both of Cartesians who, like Huyghens, thought that it was not worth wasting time on an 'unintelligible' programme and of some of his rash disciples who, like Cotes, thought that the inconsistency presented no problem.[2]

The rational position with regard to 'grafted' programmes is then to exploit their heuristic power without resigning oneself to the fundamental chaos on which it is growing. On the whole, this attitude dominated old, pre-1925 quantum theory. In the new, post-1925 quantum theory the 'anarchist' position became dominant and modern quantum physics, in its 'Copenhagen interpretation', became one of the main standard bearers of philosophical obscurantism. In the *new* theory Bohr's notorious 'complementarity principle' enthroned

[1] For the fascinating story of this long series of frustrating failures, cf. Whittaker, [1953], pp. 103-4. Planck himself gives a dramatic description of these years: 'My futile attempts to fit the elementary quantum of action into the classical theory continued for a number of years, and they cost me a great deal of effort. Many of my colleagues saw in this something bordering on a tragedy' (Planck [1947]).

[2] Of course, a reductionist programme is scientific only if it explains more than it has set out to explain; otherwise the reduction is *not* scientific (cf. Popper [1969]). If the reduction does not produce new empirical content, let alone novel facts, then the reduction represents a degenerating problemshift – it is a mere linguistic exercise. The Cartesian efforts to bolster up their metaphysics in order to be able to interpret Newtonian gravitation in its terms, is an outstanding example for such merely linguistic reduction. Cf. *above*, p. 41, n. 3.

[weak] inconsistency as a basic ultimate feature of nature, and merged subjectivist positivism and antilogical dialectic and even ordinary language philosophy into one unholy alliance. After 1925 Bohr and his associates introduced a new and unprecedented lowering of critical standards for scientific theories. This led to a defeat of reason within modern physics and to an anarchist cult of incomprehensible chaos. Einstein protested: 'The Heisenberg–Bohr tranquillizing philosophy – or religion? – is so delicately contrived that, for the time being, it provides a gentle pillow for the true believer'.[1] On the other hand, Einstein's *too* high standards may well have been the reason that prevented him from discovering (or perhaps only from publishing) the Bohr model and wave mechanics.

Einstein and his allies have not won the battle. Physics textbooks are nowadays full of statements like this: 'The two viewpoints, quanta and electromagnetic field strengths, are complementary in the sense of Bohr. This complementarity is one of the great achievements of natural philosophy in which the Copenhagen interpretation of the epistemology of quantum theory has resolved the age-old conflict between the corpuscular and the wave theories of light. From the reflection and rectilinear propagation properties of Hero of Alexandria in the first century A.D., right through to the interference and wave properties of Young and Maxwell in the nineteenth century, this controversy raged. The quantum theory of radiation during the past half century, in a striking Hegelian manner, has *completely* resolved the dichotomy'.[2]

Let us now return to the logic of discovery of *old* quantum theory and, in particular, concentrate on its *positive heuristic*. Bohr's plan was to work out first the theory of the hydrogen atom. His first model was to be based on a fixed proton-nucleus with an electron in a circular orbit; in his second model he wanted to calculate an elliptical orbit in a fixed plane; then he intended to remove the clearly artificial restrictions of the fixed nucleus and fixed plane; after this he thought

---

[1] Einstein [1928]. Among the critics of the Copenhagen 'anarchism' we should mention – besides Einstein – Popper, Landé, Schrödinger, Margenau, Blokhinzev, Bohm, Fényes and Jánossy. For a defence of the Copenhagen interpretation, cf. Heisenberg [1955]; for a hard-hitting recent criticism, cf. Popper [1967]. Feyerabend in his [1968–9], makes use of some inconsistencies and waverings in Bohr's position for a crude apologetic falsification of Bohr's philosophy. Feyerabend misrepresents Popper's, Landé's and Margenau's critical attitude to Bohr, gives insufficient emphasis to Einstein's opposition, and seems to have forgotten completely that in some of his earlier papers he was more Popperian than Popper on this issue.

[2] Power [1964], p. 31 (my italics). '*Completely*' is meant here literally. As we read in *Nature* (**222**, 1969, pp. 1034–5): 'It is absurd to think that any fundamental element of [quantum] theory can be false...The arguments that *scientific* results are always temporary, cannot hold. It is the *philosophers'* conceptions of modern physics that are temporary, because they have not yet realized how profoundly the discoveries of quantum physics affect the whole of epistemology...The assertion that ordinary language is the ultimate source of the unambiguousness of physical description is verified most convincingly by the observational conditions in quantum physics.'

of taking the possible spin of the electron into account,[1] and then he hoped to extend his programme to the structure of complicated atoms and molecules and to the effect of electromagnetic fields on them, etc., etc. All this was planned right at the start: the idea that atoms are analogous to planetary systems adumbrated a long, difficult but optimistic programme and clearly indicated the policy of research.[2] 'It looked at this time – in the year 1913 – as if the authentic key to the spectra had at last been found, as if only time and patience would be needed to resolve their riddles completely.'[3]

Bohr's celebrated first paper of 1913 contained the initial step in the research programme. It contained his first model (I shall call it M_1) which already predicted facts hitherto unpredicted by any previous theory: the wavelengths of hydrogen's line emission spectrum. Though some of these wavelengths were known before 1913 – the Balmer series (1885) and the Paschen series (1908) – Bohr's theory predicted much more than these two known series. And tests soon corroborated its novel content: one additional Bohr series was discovered by Lyman in 1914, another by Brackett in 1922, and yet another by Pfund in 1924.

Since the Balmer and the Paschen series were known before 1913, some historians present the story as an example of a Baconian 'inductive ascent': (1) the chaos of spectrum lines, (2) an 'empirical law' (Balmer), (3) the theoretical explanation (Bohr). This certainly looks like the three 'floors' of Whewell. But the progress of science would hardly have been delayed had we lacked the laudable trials and errors of the ingenious Swiss school-teacher: the speculative mainline of science, carried forward by the bold speculations of Planck, Rutherford, Einstein and Bohr would have produced Balmer's results deductively, as test-statements of their theories, without Balmer's so-called 'pioneering'. In the rational reconstruction of science there is little reward for the pains of the discoverers of 'naive conjectures'.[4]

---

[1] This is rational reconstruction. As a matter of fact, Bohr accepted this idea only in his [1926].

[2] Besides this analogy, there was another basic idea in Bohr's positive heuristic: the 'correspondence principle'. This was indicated by him as early as 1913 (cf. the second of his five postulates quoted above on p. 56), but he developed it only later when he used it as a guiding principle in solving some problems of the later, sophisticated models (like the intensities and states of polarization). The peculiarity of this second part of his positive heuristic was that Bohr did not believe its metaphysical version: he thought it was a temporary rule until the replacement of classical electromagnetics (and possibly mechanics).

[3] Davisson [1937]. A similar euphoria was experienced by MacLaurin in 1748 over Newton's programme: Newton's 'philosophy being founded on experiment and demonstration, cannot fail till reason or the nature of things are changed... [Newton] left to posterity little more to do, but observe the heavens, and compute after his models' (MacLaurin [1748], p. 8).

[4] I use here 'naive conjecture' as a technical term in the sense of my [1963–4]. For a case study and detailed criticism of the myth of the 'inductive basis' of science (natural or mathematical) cf. *ibid.*, section 7, especially pp. 298–307. There I show that

As a matter of fact, Bohr's problem was not to explain Balmer's and Paschen's series, but to explain the paradoxical stability of the Rutherford atom. Moreover, Bohr had not even heard of these formulae before he wrote the first version of his paper.[1]

Not all the novel content of Bohr's first model M_1 was corroborated. For instance, Bohr's M_1 claimed to predict all the lines in the hydrogen emission spectrum. But there was experimental evidence for a hydrogen series where according to Bohr's M_1 there should have been none. The anomalous series was the Pickering–Fowler ultraviolet series.

Pickering discovered this series in 1896 in the spectrum of the star ζ Puppis. Fowler, after having discovered its first line also in the sun in 1898, produced the whole series in a discharge tube containing hydrogen and helium. True, it could be argued that the monster-line had nothing to do with the hydrogen – after all, the sun and ζ Puppis contain many gases and the discharge tube also contained helium. Indeed, the line could *not* be produced in a pure hydrogen tube. But Pickering's and Fowler's 'experimental technique', that led to a falsifying hypothesis of Balmer's law, had a plausible, although never severely tested, theoretical background: (a) their series had the same convergence number as the Balmer series and therefore was taken to be a hydrogen series and (b) Fowler gave a plausible explanation why helium could not possibly be responsible for producing the series.[2]

Bohr was not, however, very impressed by the 'authoritative' experimental physicists. He did not question their 'experimental precision' or the 'reliability of their observations', but questioned their observational theory. Indeed, he proposed an alternative. He first elaborated a new model (M_2) of his research programme: the model of ionized helium, with a double proton orbited by an electron. Now this model predicts an ultra-violet series in the spectrum of ionized

---

Descartes's and Euler's 'naive conjecture' that for all polyhedra $V - E + F = 2$ was irrelevant and superfluous for the later development; as further examples one may mention that Boyle's and his successor's labours to establish $pv = RT$ was irrelevant for the later theoretical development (except for developing some experimental techniques), as Kepler's three laws may have been superfluous for the Newtonian theory of gravitation.

For further discussion of this point cf. *below*, p. 88.

[1] Cf. Jammer [1966], pp. 77 ff.

[2] Fowler [1912]. Incidentally his 'observational' theory was provided by 'Rydberg's theoretical investigations' which 'in the absence of strict experimental proof [he] regarded as justifying [his experimental] conclusion' (p. 65). But his theoretician colleague, Professor Nicholson, referred three months later to Fowler's findings as 'laboratory confirmations of Rydberg's theoretical deduction' (Nicholson [1913]). This little story, I think, bears out my pet thesis that most scientists tend to understand little more *about* science than fish about hydrodynamics.

In the Report of the Council to the Ninety-third Annual General Meeting of the Royal Astronomical Society, Fowler's 'observation in laboratory experiments' of new 'hydrogen lines which have so long eluded the efforts of the physicists' is described as 'an advance of great interest' and as 'a triumph of well-directed experimental work'.

helium which coincides with the Pickering–Fowler series. This constituted a rival theory. Then he suggested a 'crucial experiment': he predicted that Fowler's series can be produced, possibly with even stronger lines, in a tube which is filled with a mixture of helium and chlorine. Moreover, Bohr explained to the experimentalists, without even looking at their apparatus, the catalytic role of the hydrogen in Fowler's experiment and of chlorine in the experiment he suggested.[1] Indeed, he was right.[2] Thus the first apparent defeat of the research programme was turned into a resounding victory.

The victory, however, was immediately questioned. Fowler acknowledged that his series was not a hydrogen, but a helium series. But he pointed out that Bohr's monster-adjustment[3] still failed: the wavelengths in the Fowler series differ significantly from the values predicted by Bohr's M_2. Thus the series, although it does not refute M_1, still refutes M_2, and because of the close connection between M_1 and M_2, it undermines M_1![4]

Bohr brushed off Fowler's argument: *of course* he never meant M_2 to be taken too seriously. His values were based on a crude calculation based on the electron orbiting round a fixed nucleus; but *of course* it orbits round the common centre of gravity; *of course*, as is done when treating two-body problems, one has to substitute reduced mass for mass: $m_{e'} = m_e/[1+(m_e/m_n)]$.[5] This modified model was Bohr's M_3. And Fowler himself had to admit that Bohr was again right.[6]

The apparent refutation of M_2 turned into a victory for M_3; and it was clear that M_2 and M_3 would have been developed within the research programme – perhaps even M_{17} or M_{20} – without *any* stimulus from observation or experiment. It was at this stage that Einstein said of Bohr's theory: 'It is one of the greatest discoveries.'[7]

Bohr's research programme then went on as planned. The next step was to calculate elliptical orbits. This was done by Sommerfeld in 1915, but with the (unexpected) result that the increased number of possible

---

[1] Bohr [1913b].

[2] Evans [1913]. For a similar example of a theoretical physicist teaching a refutation-keen experimentalist what he – the experimentalist – had really observed, cf. *above*, p. 45, n. 5.

[3] Monster-adjustment: turning a counterexample, in the light of some new theory, into an example. Cf. my [1963–4], pp. 127 ff. But Bohr's 'monster-adjustment' was empirically 'progressive': it predicted a new fact (the appearance of the 4686 line in tubes containing no hydrogen).

[4] Fowler [1913a].

[5] Bohr [1913c]. This monster-adjustment was also 'progressive': Bohr predicted that Fowler's observations must be slightly imprecise and the Rydberg 'constant' must have a fine structure.

[6] Fowler [1913b]. But he sceptically noted that Bohr's programme had not yet explained the spectrum lines of *un-ionized*, ordinary helium. However, he soon abandoned his scepticism and joined Bohr's research programme (Fowler [1914]).

[7] Cf. Hevesy [1913]: 'When I told him of the Fowler spectrum, the big eyes of Einstein looked still bigger and he told me: "Then it is one of the greatest discoveries."'

steady orbits did *not* increase the number of possible energy levels, so there seemed to be no possibility of a crucial experiment between the elliptical and circular theory. However, electrons orbit the nucleus with very high velocity so that when they accelerate their mass should change noticeably if Einsteinian mechanics is true. Indeed, calculating such relativistic corrections, Sommerfeld got a new array of energy levels and thus the 'fine-structure' of the spectrum.

The switch to this new relativistic model required much more mathematical skill and talent than the development of the first few models. Sommerfeld's achievement was primarily mathematical.[1]

Curiously, the doublets of the hydrogen spectrum had already been discovered in 1891 by Michelson.[2] Moseley pointed out immediately after Bohr's first publication that 'it fails to account for the second weaker line found in each spectrum'.[3] Bohr was not upset: he was convinced that the positive heuristic of his research programme would, *in due course*, explain and even correct Michelson's observations.[4] And so it did. Sommerfeld's theory was, of course, inconsistent with Bohr's first versions; the fine structure experiments – with the old observations corrected! – provided the crucial evidence in its favour. Many defeats of Bohr's first models were turned by Sommerfeld and his Munich school into victories for Bohr's research programme.

It is interesting that just as Einstein got worried and slowed down in the middle of the spectacular progress of quantum physics by 1913, Bohr got worried and slowed down by 1916; and just as Bohr had, by 1913 taken the initiative from Einstein, Sommerfeld had taken the initiative from Bohr by 1916. The difference between the atmosphere of Bohr's Copenhagen school and Sommerfeld's Munich school was conspicuous: 'In Munich one used more concrete formulations and was therefore more easily understood; one had been successful in the systematization of spectra and in the use of the vector model. In Copenhagen, however, one believed that an adequate language for the new [phenomena] had not yet been found, one was reticent in the face of too definite formulations, one expressed oneself more cautiously and more in general terms, and was therefore much more difficult to understand.'[5]

Our sketch shows how a progressive shift may lend credibility – and a *rationale* – to an inconsistent programme. Born, in his obituary of

[1] For the vital mathematical aspects of research programmes, cf. *above*, p. 52.
[2] Michelson [1891–2], especially pp. 287–9. Michelson does not even mention Balmer.
[3] Moseley [1914].   [4] Sommerfeld [1916], p. 68.
[5] Hund [1961]. This is discussed at some length in Feyerabend [1968–9], pp. 83–7. But Feyerabend's paper is heavily biased. The main aim of his paper is to play down Bohr's methodological anarchism and show that Bohr *opposed* the Copenhagen interpretation of the *new* (post-1925) quantum programme. In order to do so, Feyerabend, on the one hand, overemphasizes Bohr's unhappiness about the inconsistency of the *old* (pre-1925) quantum programme and, on the other hand, makes too much of the fact that Sommerfeld cared less for the problematicality of the inconsistent foundations of the *old* programme than Bohr.

Planck, describes this process forcefully: 'Of course the mere introduction of the quantum of action does not yet mean that a *true* Quantum Theory has been established...The difficulties which the introduction of the quantum of action into the well-established classical theory has encountered from the outset have already been indicated. They have gradually increased rather than diminished; and although research in its forward march has in the meantime passed over some of them, the remaining gaps in the theory are the more distressing to the conscientious theoretical physicist. In fact, what in Bohr's theory served as the basis of the laws of action consists of certain hypotheses which a generation ago would doubtless have been flatly rejected by every physicist. That within the atom certain quantized orbits (i.e. picked out on the quantum principle) should play a special role could well be granted; somewhat less easy to accept is the further assumption that the electrons moving on these curvilinear orbits, and therefore accelerated, radiate no energy. But that the sharply defined frequency of an emitted light quantum should be different from the frequency of the emitting electron would be regarded by a theoretician who had grown up in the classical school as monstrous and almost inconceivable. But numbers [or, rather, *progressive problemshifts*] decide, and in consequence the tables have been turned. While originally it was a question of fitting in with as little strain as possible a new and strange element into an existing system which was generally regarded as settled, *the intruder, after having won an assured position, now has assumed the offensive*; and it now appears certain that it is about to blow up the old system at some point. The only question now is, at what point and to what extent this will happen.'[1]

One of the most important points one learns from studying research programmes is that relatively few experiments are really important. The heuristic guidance the theoretical physicist receives from tests and 'refutations' is usually so trivial that large-scale testing – or even bothering too much with the data already available – may well be a waste of time. In most cases we need no refutations to tell us that the theory is in urgent need of replacement: the positive heuristic of the programme drives us forward anyway. Also, to give a stern 'refutable interpretation' to a fledgling version of a programme is dangerous methodological cruelty. The first versions may even 'apply' only to non-existing 'ideal' cases; it may take decades of theoretical work to arrive at the first novel facts and still more time to arrive at *interestingly testable* versions of the research programmes, at the stage when refutations are no longer foreseeable in the light of the programme itself.

The dialectic of research programmes is then not necessarily an alternating series of speculative conjectures and empirical refutations. The interaction between the development of the programme and the empirical checks may be very varied – which pattern is actually realized

---

[1] Born [1948], p. 180, my italics.

depends only on historical accident. Let us mention three typical variants.

(1) Let us imagine that each of the first three consecutive versions, H_1, H_2, H_3 predict some new facts successfully but others unsuccessfully, that is each version is both corroborated *and* refuted in turn. Finally H_4 is proposed which predicts some novel facts but stands up to the severest tests. The problemshift is progressive, and also we have a beautiful Popperian alternation of conjectures and refutations.[1] People will admire this as a classical example of theoretical and experimental work going hand in hand.

(2) Another pattern could have been a lone Bohr (possibly without Balmer preceding him), working out H_1, H_2, H_3, H_4 but self-critically withholding publication until H_4. Then H_4 is tested: all the evidence will turn up as corroborations of H_4, the first (and only) published hypothesis. The theoretician – at his desk – is here seen to work far ahead of the experimenter: we have a period of relative autonomy of theoretical progress.

(3) Let us now imagine that *all* the empirical evidence mentioned in these three patterns is already there at the time of the invention of H_1, H_2, H_3, H_4. In this case H_1, H_2, H_3, H_4 will not represent an empirically progressive problemshift and therefore, although all the evidence supports his theories, the scientist has to work on further in order to prove the scientific value of his programme.[2] Such a state of affairs may be brought about either by the fact that an older research programme (which has been challenged by the one leading to H_1, H_2, H_3, H_4) had already produced all these facts – or by the fact that too much government money lay around for collecting data about spectrum lines and hacks stumbled upon all the data. However, the latter case is extremely unlikely, for, as Cullen used to say, 'the number of false facts, afloat in the world, infinitely exceeds that of the false theories;'[3] in most such cases the research programme will clash with the available 'facts', the theoretician will look into the 'experimental techniques' of the experimentalist, and having overthrown and replaced his observational theories will correct his facts thereby producing *novel* ones.[4]

---

[1] In the first three patterns we do not involve complications like successful appeals against the verdict of the experimental scientists.

[2] This shows that if exactly the same theories and the same evidence is rationally reconstructed in different time orders, they may constitute either a progressive or a degenerative shift. Also cf. volume 2, chapter 8, p. 178.

[3] Cf. McCulloch [1825], p. 19. For a strong argument on how extremely unlikely such a pattern is, see *below*, p. 70.

[4] Perhaps it should be mentioned that manic data collection – and 'too much' precision – prevents even the formation of naive 'empirical' hypotheses like Balmer's. Had Balmer known of Michelson's fine-spectra, would he have ever found his formula? Or, had Tycho Brahe's data been more precise, would Kepler's elliptical law ever have been put forward? The same applies to the naive first version of the general gas law, etc. The Descartes–Euler conjecture on polyhedra might never have been made but for the scarcity of data; cf. my [1963–4], pp. 298 ff.

After this methodological excursion, let us return to Bohr's programme. Not all developments in the programme were foreseen and planned when the positive heuristic was first sketched. When some curious gaps appeared in Sommerfeld's sophisticated models (some predicted lines never did appear), Pauli proposed a deep auxiliary hypothesis (his 'exclusion principle') which accounted not only for the known gaps but reshaped the shell theory of the periodic system of elements and anticipated facts then unknown.

I do not wish to give here an elaborate account of the development of Bohr's programme. But its detailed study from the methodological viewpoint is a veritable goldmine: its marvellously fast progress – on inconsistent foundations! – was breathtaking, the beauty, originality and empirical success of its auxiliary hypotheses, put forward by scientists of brilliance and even genius, was unprecedented in the history of physics.[1] Occasionally the next version of the programme required only a trivial improvement, like the replacement of mass by reduced mass. Occasionally, however, to arrive at the next version required new sophisticated mathematics, like the mathematics of the many-body problem, or new sophisticated physical auxiliary theories. The additional mathematics or physics was either dragged in from some part of extant knowledge (like relativity theory) or invented (like Pauli's exclusion principle). In the latter case we have a 'creative shift' in the positive heuristic.

But even this great programme came to a point where its heuristic power petered out. *Ad hoc* hypotheses multiplied and could not be replaced by content-increasing explanations. For instance, Bohr's theory of molecular (band) spectra predicted the following formula for diatomic molecules:

$$\nu = \frac{h}{8\pi^2 I}[(m+1)^2 - m^2]$$

But the formula was refuted. Bohrians replaced the term m^2 by $m(m+1)$: this fitted the facts but was sadly *ad hoc*.

Then came the problem of some unexplained doublets in alkali spectra. Landé explained them in 1924 by an *ad hoc* 'relativistic splitting rule', Goudsmit and Uhlenbeck in 1925 by electron spin. If Landé's explanation was *ad hoc*, Goudsmit's and Uhlenbeck's was also inconsistent with special relativity theory: surface points on the largish electron had to travel faster than light, and the electron had even to be bigger than the whole atom.[2] Considerable courage was needed to

---

[1] 'Between the appearance of Bohr's great trilogy in 1913 and the advent of wave mechanics in 1925, a large number of papers appeared developing Bohr's ideas into an impressive theory of atomic phenomena. It was a collective effort and the names of the physicists contributing to it make up an imposing roll-call: Bohr, Born, Klein, Rosseland, Kramers, Pauli, Sommerfeld, Planck, Einstein, Ehrenfest, Epstein, Debye, Schwarzschild, Wilson' (Ter Haar [1967], p. 43).

[2] A footnote in their paper reads: 'It should be observed that [according to our theory] the peripheral velocity of the electron would considerably exceed the velocity of light' (Uhlenbeck and Goudsmit [1925]).

propose it. (Kronig got the idea earlier but refrained from publishing it because he thought it was inadmissible.[1])

But temerity in proposing wild inconsistencies did not reap any more rewards. The programme lagged behind the discovery of 'facts'. Undigested anomalies swamped the field. With ever more sterile inconsistencies and ever more *ad hoc* hypotheses, the degenerating phase of the research programme had set in: it started – to use one of Popper's favourite phrases – 'to lose its empirical character'.[2] Also many problems, like the theory of perturbations, could not even be expected to be solved within it. A rival research programme soon appeared: wave mechanics. Not only did the new programme, even in its first version (de Broglie, 1924), explain Planck's and Bohr's quantum conditions; it also led to an exciting new fact, to the Davisson–Germer experiment. In its later, ever more sophisticated versions it offered solutions to problems which had been completely out of the reach of Bohr's research programme, and explained the *ad hoc* later theories of Bohr's programme by theories satisfying high methodological standards. Wave mechanics soon caught up with, vanquished and replaced Bohr's programme.

De Broglie's paper came at the time when Bohr's programme was degenerating. But this was mere coincidence. One wonders what would have happened if de Broglie had written and published his paper in 1914 instead of 1924.

(d) A new look at crucial experiments: the end of instant rationality

It would be wrong to assume that one must stay with a research programme until it has exhausted all its heuristic power, that one must not introduce a rival programme before everybody agrees that the point of degeneration has probably been reached. (Although one can understand the irritation of a physicist when, in the middle of the progressive phase of a research programme, he is confronted by a proliferation of vague metaphysical theories stimulating no empirical progress.[3]) One must never allow a research programme to become a *Weltanschauung*, or a sort of *scientific rigour*, setting itself up as an arbiter between explanation and non-explanation, as mathematical rigour sets itself up as an arbiter between proof and non-proof. Unfortunately this is the position which Kuhn tends to advocate:

---

[1] Jammer [1966], pp. 146–8 and 151.

[2] For a vivid description of this degenerating phase of Bohr's programme, cf. Margenau [1950], pp. 311–13.

In the progressive phase of a programme the main heuristic stimulus comes from the positive heuristic: anomalies are largely ignored. In the degenerating phase the heuristic power of the programme peters out. In the absence of a rival programme this situation may be reflected in the psychology of the scientists by an unusual hypersensitivity to anomalies and by a feeling of a Kuhnian 'crisis'.

[3] This is what must have irritated Newton most in the 'sceptical proliferation of theories' by Cartesians.

indeed, what he calls 'normal science' is nothing but a research programme that has achieved monopoly. But, as a matter of fact, research programmes have achieved complete monopoly only rarely and then only for relatively short periods, in spite of the efforts of some Cartesians, Newtonians and Bohrians. *The history of science has been and should be a history of competing research programmes (or, if you wish, 'paradigms'), but it has not been and must not become a succession of periods of normal science: the sooner competition starts, the better for progress.* 'Theoretical pluralism' is better than 'theoretical monism': on this point Popper and Feyerabend are right and Kuhn is wrong.[1]

The idea of competing scientific research programmes leads us to the problem: *how are research programmes eliminated?* It has transpired from our previous considerations that a degenerating problemshift is no more a sufficient reason to eliminate a research programme than some old-fashioned 'refutation' or a Kuhnian 'crisis'. *Can there be any objective* (as opposed to socio-psychological) *reason to reject a programme, that is, to eliminate its hard core and its programme for constructing protective belts?* Our answer, in outline, is that such an objective reason is provided by a rival research programme which explains the previous success of its rival and supersedes it by a further display of *heuristic power.*[2]

However, the criterion of 'heuristic power' strongly depends on how we construe '*factual novelty*'. Until now we have assumed that it is immediately ascertainable whether a new theory predicts a novel fact or not.[3] But *the novelty of a factual proposition can frequently be seen only after a long period has elapsed.* In order to show this, I shall start with an example.

Bohr's theory logically implied Balmer's formula for hydrogen lines as a consequence.[4] Was this a novel fact? One might have been tempted to deny this, since after all, Balmer's formula was well-known. But this is a half-truth. Balmer merely 'observed' B_1: that *hydrogen lines obey the Balmer formula.* Bohr predicted B_2: that *the differences in the energy levels in different orbits of the hydrogen electron obey the Balmer formula.* Now one may say that B_1 already contains all the purely 'observational' content of B_2. But to say this presupposes that there

---

[1] Nevertheless there is something to be said for at least *some* people sticking to a research programme until it reaches its 'saturation point'; a new programme is then challenged to account for the full success of the old. It is no argument against this that the rival may, when it was first proposed, already have explained all the success of the first programme; the growth of a research programme cannot be predicted – it may stimulate important unforeseeable auxiliary theories of its own. Also, if a version A_n of a research programme P_1 is mathematically equivalent to a version A_m of a rival P_2, one should develop both: their heuristic strength can still be very different.

[2] I use '*heuristic power*' here as a technical term to characterize the power of a research programme to anticipate theoretically novel facts in its growth. I could of course use '*explanatory power*': cf. *above*, p. 34, n. 4.

[3] Cf. *above*, p. 31, text to n. 4, and p. 49, text to n. 2.

[4] Cf. *above*, p. 61.

can be a pure 'observational level', untainted by theory, and impervious to theoretical change. In fact, B_1 was accepted only because the optical, chemical and other theories *applied* by Balmer were well corroborated and accepted as interpretative theories; and these theories could always be questioned. It might be argued that we can 'purge' even B_1 of its theoretical presuppositions, and arrive at what Balmer really 'observed', which might be expressed in the more modest assertion, B_0: that *the lines emitted in certain tubes in certain well-specified circumstances (or in the course of a 'controlled experiment'[1]) obey the Balmer formula*. Now some of Popper's arguments show that we can *never* arrive at any hard 'observational' rock-bottom in this way; 'observational' theories can easily be shown to be involved in B_0.[2] On the other hand, given that Bohr's programme after a long progressive development, had shown its heuristic power, its hard core would itself have become well corroborated[3] and therefore qualified as an 'observational' or interpretative theory. But then B_2 will be seen not as a mere theoretical reinterpretation of B_1, but as a *new fact* in its own right.

These considerations lend new emphasis to the hindsight element in our appraisals and lead to a further liberalizaton of our standards. A new research programme which has just entered the competition may start by explaining 'old facts' in a novel way but may take a very long time before it is seen to produce 'genuinely novel' facts. For instance, the kinetic theory of heat *seemed* to lag behind the results of the phenomenological theory for decades before it finally overtook it with the Einstein–Smoluchowski theory of Brownian motion in 1905. After this, what had previously seemed a speculative reinterpretation of old facts (about heat, etc.) turned out to be a discovery of novel facts (about atoms).

All this suggests that we must not discard a budding research programme simply because it has so far failed to overtake a powerful rival. We should not abandon it if, supposing its rival were not there, it would constitute a progressive problemshift.[4] And we should certainly regard a newly interpreted

---

[1] Cf. *above*, p. 27, n. 4.

[2] One of Popper's arguments is particularly important: 'There is a widespread belief that the statement "I see that this table here is white", possesses some profound advantage over the statement "This table here is white", from the point of view of epistemology. But from the point of view of evaluating its possible objective tests, the first statement, in speaking about me, does not appear more secure than the second statement, which speaks about the table here' ([1934], section 27). Neurath makes a characteristically blockheaded comment on this passage: 'For us such protocol statements have the advantage of *having more stability.* One may retain the statement: "People in the 16th century saw fiery swords in the sky" while crossing out "There were fiery swords in the sky"' (Neurath [1935], p. 362).

[3] *This remark, incidentally, defines a 'degree of corroboration' for the 'irrefutable' hard cores of research programmes. Newton's theory (in isolation) had no empirical content, yet it was, in this sense, highly corroborated.*

[4] Incidentally, in the methodology of research programmes, the pragmatic meaning of 'rejection' [of a programme] becomes crystal clear: it means *the decision to cease working on it.*

fact as a new fact, ignoring the insolent priority claims of amateur fact collectors. As long as a budding research programme can be rationally reconstructed as a progressive problemshift, it should be sheltered for a while from a powerful established rival.[1]

These considerations, on the whole, stress the importance of methodological tolerance, and leave the question of how research programmes are eliminated still unanswered. The reader may even suspect that laying this much stress on fallibility liberalizes or, rather, softens up, our standards to the extent that we will be landed with radical scepticism. Even the celebrated *'crucial experiments'* will then have no force to overthrow a research programme; anything goes.[2]

But this suspicion is unfounded. *Within* a research programme *'minor crucial experiments'* between subsequent versions are quite common. Experiments easily 'decide' between the nth and $(n+1)$th scientific version, since the $(n+1)$th is not only inconsistent with the nth, but also supersedes it. If the $(n+1)$th version has more corroborated content in the light of the *same* programme and in the light of the *same* well corroborated observational theories elimination is a relatively routine affair (only relatively, for even here this decision may be subject to appeal). Appeal procedures too are occasionally easy: in many cases the challenged observational theory, far from being well corroborated, is in fact an unarticulated, naive, 'hidden' assumption; it is only the challenge which reveals the existence of this hidden assumption, and brings about its articulation, testing and downfall. Time and again, however, the observational theories are themselves embedded in some research programme and then the appeal procedure leads to a clash between two research programmes: in such cases we may need a *'major crucial experiment'*.

When two research programmes compete, their first 'ideal' models usually deal with different aspects of the domain (for example, the first model of Newton's semi-corpuscular optics described light-refraction, the first model of Huyghens's wave optics light-interference). As the rival research programmes expand, they gradually encroach on each other's territory and the nth version of the first will be blatantly, dramatically inconsistent with the mth version of the second.[3] An experiment is repeatedly performed, and as a result, the first is defeated in *this battle*, while the second wins. But *the war* is not over: any research programme is allowed a few such defeats. All it needs

---

[1] Some might regard – cautiously – this sheltered period of development as *'pre-scientific'* (or 'theoretical'); and be prepared only when it starts producing 'genuinely novel' facts to recognize its truly *scientific* (or 'empirical') character – but then their recognition will have to be retroactive.

[2] Incidentally, *this conflict between fallibility and criticism can be rightly said to be the main problem – and driving force – of the Popperian research programme in the theory of knowledge.*

[3] An especially interesting case of such competition is *competitive symbiosis*, when a new programme is grafted on to an old one which is inconsistent with it; cf. *above*, p. 57.

for a comeback is to produce an $(n+1)$th (or $(n+k)$th) content-increasing version and a verification of some of its novel content.

If such a comeback, after sustained effort, is not forthcoming, the war is lost and the original experiment is seen, *with hindsight*, to have been 'crucial'. But especially if the defeated programme is a young, fast-developing programme, and if we decide to give sufficient credit to its 'pre-scientific' successes, allegedly crucial experiments dissolve one after the other in the wake of its forward surge. Even if the defeated programme is an old, established and 'tired' programme, near its 'natural saturation point',[1] it may continue to resist for a long time and hold out with ingenious content-increasing innovations even if these are unrewarded with empirical success. It is very difficult to defeat a research programme supported by talented, imaginative scientists. Alternatively, stubborn defenders of the defeated programme may offer *ad hoc* explanations of the experiments or a shrewd *ad hoc* 'reduction' of the victorious programme to the defeated one. But such efforts we should reject as unscientific.[2]

Our considerations explain why crucial experiments are seen to be crucial only decades later. Kepler's ellipses were generally admitted as crucial evidence for Newton and against Descartes only about one hundred years after Newton's claim. The anomalous behaviour of Mercury's perihelion was known for decades as one of the many yet unsolved difficulties in Newton's programme; but only the fact that Einstein's theory explained it better transformed a dull anomaly into a brilliant 'refutation' of Newton's research programme.[3] Young claimed that his double-slit experiment of 1802 was a crucial experiment between the corpuscular and the wave programmes of optics; but his claim was only acknowledged much later, after Fresnel developed the wave programme much further 'progressively' and it became clear that the Newtonians could not match its heuristic power. The anomaly, which had been known for decades, received the honorific title of refutation, the experiment the honorific title of 'crucial experiment' only after a

---

[1] There is no such thing as a *natural* 'saturation point'; in my [1963–4], especially on pp. 327–8, I was more of a Hegelian, and I thought there was; now I use the expression with an ironical emphasis. There is no predictable or ascertainable limitation on human imagination in inventing new, content-increasing theories or on the 'cunning of reason' (*List der Vernunft*) in rewarding them with some empirical success even if they are false or even if the new theory has less verisimilitude – in Popper's sense – than its predecessor. (Probably all scientific theories ever uttered by men will be false: they still may be rewarded by empirical successes and even have increasing verisimilitude.)

[2] For an example, cf. *above*, p. 41, n. 3.

[3] *Thus an anomaly in a research programme is a phenomenon which we regard as something to be explained in terms of the programme. More generally, we may speak, following Kuhn, about 'puzzles': a 'puzzle' in a programme is a problem which we regard as a challenge to that particular programme. A 'puzzle' can be resolved in three ways: by solving it within the original programme (the anomaly turns into an example); by neutralizing it, i.e. solving it within an independent, different programme (the anomaly disappears); or, finally, by solving it within a rival programme (the anomaly turns into a counterexample).*

long period of uneven development of the two rival programmes. Brownian motion was in the middle of the battlefield for nearly a century before it was *seen* to defeat the phenomenological research programme and turn the war in favour of the atomists. Michelson's 'refutation' of the Balmer series was ignored for a generation until Bohr's triumphant research programme backed it up.

It may be worthwhile to discuss in detail some examples of experiments whose 'crucial' character became evident only retrospectively. First I shall take the celebrated Michelson–Morley experiment of 1887 which allegedly falsified the ether theory and 'led to the theory of relativity', then the Lummer–Pringsheim experiments which allegedly falsified the classical theory of radiation and 'led to the quantum theory'.[1] Finally I shall discuss an experiment which many physicists thought would turn out to decide against the conservation laws but which, in fact, ended up as their most triumphant corroboration.

(d 1) The Michelson–Morley experiment

Michelson first devised an experiment in order to test Fresnel's and Stokes's contradictory theories about the influence of the motion of the earth on the ether,[2] during his visit to Helmholtz's Berlin institute in 1881. According to Fresnel's theory, the earth moves through an ether at rest, but the ether within the earth is *partially* carried along with the earth; Fresnel's theory therefore entailed that the velocity of the ether outside the earth relative to the earth was positive (i.e. Fresnel's theory implied the existence of an 'ether wind'). According to Stokes's theory, the ether was dragged along by the earth and immediately on the surface of the earth the velocity of the ether was equal to that of the earth: therefore its relative velocity was zero (i.e. there was no ether wind on the surface). Stokes originally thought that the two theories were observationally equivalent: for instance, with suitable auxiliary assumptions both theories explained the aberration of light. But Michelson claimed that his 1881 experiment was a crucial experiment between the two and that it *proved* Stokes's theory.[3] He claimed that the velocity of the earth relative to the ether is far less than Fresnel's theory would have it. Indeed, he concluded that from his experiment 'the *necessary conclusion* follows that the hypothesis [of a stationary ether] is erroneous. This conclusion *directly contradicts* the explanation of the phenomenon of aberration which... presupposes that the earth moves through the ether, the latter remaining at rest'.[4] As often happens, Michelson the experimenter was then taught a lesson by a theoretician. Lorentz, the leading theoretical physicist of the period,

---

[1] Cf. Popper [1934], section 30.
[2] Cf. Fresnel [1818], Stokes [1845] and [1846]. For an excellent brief exposition cf. Lorentz [1895].
[3] This transpires, obliquely, from the concluding section of his [1881].
[4] Michelson [1881], p. 128, my italics.

in what Michelson later described as 'a very searching analysis...of the entire experiment',[1] showed that Michelson 'misinterpreted' the facts and that what he observed did *not* in fact contradict the hypothesis of the stationary ether. Lorentz showed that Michelson's calculations were wrong; Fresnel's theory predicted only half of the effect Michelson had calculated. Lorentz concluded that Michelson's experiment did *not* refute Fresnel's theory, and that it certainly did not prove Stokes's theory either. Lorentz went on to show that Stokes's theory was inconsistent: that it assumed the ether at the earth's surface to be at rest with regard to the latter *and* required that the relative velocity have a potential; but these two conditions are incompatible. But even if Michelson *had* refuted *one* theory of the stationary ether, the programme is untouched: one can easily devise several other versions of the ether programme, which predict very small values for the ether winds and he, Lorentz, immediately produced one. This theory was testable and Lorentz proudly submitted it to the verdict of experiment.[2] Michelson, jointly with Morley, took up the challenge. The relative velocity of the earth to the ether again seemed to be zero, in conflict with Lorentz's theory. By this time, Michelson had become more cautious in interpreting his data and even thought of the possibility that the solar system as a whole might have moved in the opposite direction to the earth; therefore he decided to repeat the experiment 'at intervals of three months and thus avoid all uncertainty'.[3] Michelson, in his second paper, does not talk any more about 'necessary conclusions' and 'direct contradictions'. He only thinks that from his experiment 'it appears, from all that precedes, *reasonably certain* that if there be any relative motion between the earth and the luminiferous ether, it must be *small*; quite small enough entirely to refute *Fresnel's* explanation of aberration'.[4] Thus in this paper Michelson still claims to have refuted Fresnel's theory (and also Lorentz's new theory); but there is not a word about his old 1881 claim that he refuted 'the theory of stationary ether' in general. (Indeed, he believed that in order to do so, he would have to test the ether wind also at high altitudes, 'at the top of an isolated mountain peak, for instance'.[5])

While some ether-theorists – like Kelvin – did not trust Michelson's 'experimental skill',[6] Lorentz pointed out that, in spite of Michelson's

[1] Michelson and Morley [1887], p. 335.
[2] Lorentz [1886]. For the inconsistency of Stokes's theory also cf. his [1892*b*].
[3] Michelson and Morley [1887], p. 341. But Pearce Williams points out that he never did. (Pearce Williams [1968], p. 34.)
[4] *Ibid.*, p. 341, my italics.
[5] Michelson and Morley [1887]. This remark shows that Michelson realized that his 1887 experiment was completely consistent with an ether wind higher up. Max Born, in his [1920], that is, thirty-three years later, asserted that from the 1887 experiment 'we *must* conclude that the ether wind does not exist' (my italics).
[6] Kelvin said in the 1900 International Congress of Physics that 'the only cloud in the clear sky of the [ether] theory was the null result of the Michelson–Morley experiment'

naive claim, even his *new* experiment 'furnishes no evidence for the question for which it was undertaken'.[1] One can perfectly well regard Fresnel's theory as an *interpretative* theory, which interprets facts, rather than is refutable by them, and then, Lorentz showed, 'the significance of the Michelson–Morley experiment lies rather in the fact that it can teach us something about *the changes in the dimensions*':[2] the dimensions of bodies is affected by their movement through the ether. Lorentz elaborated this 'creative shift' within Fresnel's programme with great ingenuity and thereby claimed to have 'removed the contradiction between Fresnel's theory and Michelson's result'.[3] But he admitted that 'since the nature of the molecular forces is entirely unknown to us, it is impossible to test the hypothesis':[4] *at least for the time being* it could predict no novel facts.[5]

In the meanwhile, in 1897, Michelson carried out his long planned experiment to measure the velocity of ether wind on mountain tops. He found none. Since he had thought earlier that he had proved Stokes's theory which predicted an ether wind higher up, he was dumbfounded. If Stokes's theory was still correct, the gradient of the velocity of the ether had to be very small. Michelson had to conclude that 'the earth's influence upon the ether extended to distances of the order of the earth's diameter'.[6] He thought that this was an 'improbable' result, and decided that in 1887 he had drawn the wrong conclusion from his experiment: it was Stokes's theory which had to be rejected and Fresnel's which had to be accepted; and he decided that he would accept *any* reasonable auxiliary hypothesis to have it saved,

(cf. Miller [1925]) and immediately persuaded Morley and Miller, who were there, to repeat the experiment.

[1] Lorentz [1892a].

[2] *Ibid*, my italics.

[3] Lorentz [1895].

[4] Lorentz [1892b].

[5] Fitzgerald at the same time, independently of Lorentz, produced a testable version of this 'creative shift' which was quickly refuted by Trouton's, Rayleigh's and Brace's experiments: it was theoretically but not empirically progressive. Cf. Whittaker [1947], p. 53 and Whittaker [1953], pp. 28–30.

There is a widespread view that Fitzgerald's theory was ad hoc. What contemporary physicists meant was that the theory was *ad hoc*$_2$ (cf. *above*, p. 40, n. 1): that there was '*no independent* [*positive*] *evidence*' for it. (Cf. e.g. Larmor [1904], p. 624.) Later, under Popper's influence the term '*ad hoc*' was primarily used in the sense of *ad hoc*$_1$, that there was *no independent test* possible for it. But, as the refuting experiments show, it is a mistake to claim, as Popper does, that Fitzgerald's theory was *ad hoc*$_1$ (cf. Popper [1934], section 20). This shows again how important it is to separate *ad hoc*$_1$ and *ad hoc*$_2$.

When Grünbaum, in his [1959a], pointed out Popper's mistake, Popper admitted it but replied that Fitzgerald's theory was certainly *more ad hoc* than Einstein's (Popper [1959b]), and that this provides yet another 'excellent example of "degrees of *ad-hocness*" and of one of the main theses of [his] book – that *degrees* of ad-hocness are related (inversely) to degrees of testability and significance'. But the difference is *not* simply a matter of degrees of a unique *ad-hocness* which can be measured by testability. Also cf. *below*, p. 88.

[6] Michelson [1897], p. 478.

including Lorentz's 1892 theory.[1] He *now* seemed to prefer the Fitzgerald–Lorentz contraction and by 1904 his colleagues at Case were trying to find out whether this contraction varies with different materials.[2]

While most physicists tried to interpret Michelson's experiments within the framework of the ether programme, Einstein, unaware of Michelson, Fitzgerald and Lorentz, but stimulated primarily by Mach's criticism of Newtonian mechanics, arrived at a new, progressive research programme.[3] This new programme not only 'predicted' and explained the outcome of the Michelson–Morley experiment but also predicted a huge array of previously undreamt-of facts, which obtained dramatic corroborations. It was *only then*, twenty-five years later, that the Michelson–Morley experiment came to be seen as 'the greatest negative experiment in the history of science'.[4] But this could not be seen instantly. Even if the experiment was negative, it was not clear, negative exactly to *what*? Moreover, Michelson in 1881 thought that it was also *positive*: he held that he had *refuted* Fresnel's but had *verified* Stokes's theory. Michelson himself and then Fitzgerald and Lorentz explained the result also *positively* within the ether programme.[5] As it is with all experimental results, its negativity for the old programme was established *only later*, by the slow accumulation of *ad hoc* attempts to account for it within the degenerating old programme and by the gradual establishment of a new *progressive* victorious programme in which it has become a positive instance. But the possibility of the rehabilitation of some part of the 'degenerating' old programme could never be rationally excluded.

Only an extremely difficult and – indefinitely – long process can establish a research programme as superseding its rival; and it is unwise to use the term 'crucial experiment' too rashly. Even when a research programme is seen to be swept away by its predecessor, it is not swept away by some 'crucial' experiment; and even if some such crucial experiment is later called in doubt, the new research programme cannot be stopped without a powerful progressive upsurge of the old programme.[6] The negativity – and importance – of the

---

[1] Lorentz, indeed, immediately commented: 'While [Michelson] considers so far-reaching an influence of the earth improbable, I should, on the contrary, *expect* it' (Lorentz [1897], my italics).

[2] Morley and Miller [1904].

[3] There has been a considerable controversy about the historico-heuristic background of Einstein's theory, in the light of which this statement may turn out to be false.

[4] Bernal [1965], p. 530. For Kelvin, in 1905, it was only a 'cloud in the clear sky': cf. *above*, p. 74, n. 6.

[5] Indeed, Chwolson's excellent physics textbook said in 1902 that the probability of the ether hypothesis borders on certainty. (Cf. Einstein [1909], p. 817.)

[6] Polanyi tells us with *gusto* how, in 1925, in his presidential address to the American Physical Society, Miller announced that Michelson's and Morley's reports notwithstanding, he had 'overwhelming evidence' for an ether-drift; yet the audience remained committed to Einstein's theory. Polanyi draws the conclusion that no '"objectivist" framework' can account for the scientist's acceptance or rejection of

Michelson–Morley experiment lies primarily in the progressive shift in the *new* research programme to which it came to lend powerful support, and its 'greatness' is only a reflection of the greatness of the two *programmes* involved.

It would be interesting to give a detailed analysis of the rival shifts involved in the waning fortunes of the ether theory. But under the influence of naive falsificationism the most interesting degenerating phase in the ether theory after Michelson's 'crucial experiment' is simply ignored by most Einsteinians. They believe that the Michelson–Morley experiment single-handedly defeated the ether theory, the tenacity of which was only due to obscurantist conservativism. On the other hand, this post-Michelson period of the ether theory is not scrutinized *critically* by the anti-Einsteinians, who believe that the ether theory suffered no setback whatsoever: what is good in Einstein's theory was essentially in Lorentz's ether theory and Einstein's victory is only due to positivist fashion. But, in fact, Michelson's long series of experiments from 1881 to 1935, conducted in order to test subsequent versions of the ether programme provides a fascinating example of a degenerating problemshift.[1] (But research programmes may get out of degenerating troughs. It is well known that Lorentz's ether theory can easily be strengthened in such a way that it becomes, in an interesting sense, equivalent with Einstein's no-ether theory.[2] The ether may, in the context of a major 'creative shift', still return.[3])

The fact that we need hindsight to evaluate experiments explains why, between 1881 and 1886, Michelson's experiment was not even mentioned in the literature. Indeed, when a French physicist, Potier,

theories (Polanyi [1958], pp. 12–14). But my reconstruction makes the tenacity of the Einsteinian research programme in the face of alleged contrary evidence a completely *rational* phenomenon and thereby undermines Polanyi's 'post-critical'-mystical message.

[1] *One typical sign of the degeneration of a programme which is not discussed in this paper is the proliferation of contradictory 'facts'. Using a false theory as an interpretative theory, one may get – without committing any 'experimental mistake' – contradictory factual propositions, inconsistent experimental results.* Michelson, who stuck to the ether to the bitter end, was primarily frustrated by the inconsistency of the 'facts' he arrived at by his ultra-precise measurements. His 1887 experiment 'showed' that there was no ether wind on the earth's surface. But aberration 'showed' that there was. Moreover, his own 1925 experiment (either never mentioned or, as in Jaffé's [1960], misrepresented) also 'proved' that there was one (cf. Michelson and Gale [1925] and, for a sharp criticism, Runge [1925]).

[2] Cf. e.g. Ehrenfest [1913], pp. 17–18, quoted and discussed by Dorling in his [1968]. But one should not forget that *two specific theories, while being mathematically (and observationally) equivalent, may still be embedded into different rival research programmes, and the power of the positive heuristic of these programmes may well be different.* This point has been overlooked by proposers of such equivalence proofs (a good example is the equivalence proof between Schrödinger's and Heisenberg's approach to quantum physics). Also cf. *above*, p. 69, n. 1.

[3] Cf. e.g. Dirac [1951]: 'If one reexamines the question in the light of present-day knowledge, one finds that the aether is no longer ruled out by relativity, and good reasons can now be advanced for postulating an aether.' Also cf. the concluding paragraph of Rabi [1961] and Prokhovnik [1967].

pointed out to Michelson his 1881 mistake, Michelson decided not to publish a correction note. He explains the reason for this decision in a letter to Rayleigh in March 1887: 'I have repeatedly tried to interest my scientific friends in this experiment without avail, and the reason for my never publishing the correction (I am ashamed to confess it) was that I was discouraged at the slight attention the work received, and did not think it worthwhile.'[1] This letter, incidentally, was a reply to a letter from Rayleigh which drew Michelson's attention to Lorentz's paper. This letter triggered off the 1887 experiment. But even after 1887, and even after 1905, the Michelson–Morley experiment was not yet generally regarded as disproving the existence of the ether, and with good reason. This may explain why Michelson was awarded his Nobel Prize (in 1907), not for 'refuting the ether theory', but 'for his optical precision *instruments* and the spectro-scopic and method-ological investigations carried out with their aid'[2]; and why the Michel-son–Morley experiment was not even mentioned in the presentation speeches. Michelson, in his *Nobel Lecture*, did not mention it; and he kept quiet about the fact that although he might have originally devised his instruments to measure precisely the velocity of light, he was compelled to improve them for testing some specific ether theories and that the 'precision' of his 1887 experiment was largely motivated by Lorentz's theoretical criticism: a fact which standard contemporary literature never mentions.[3]

Finally, one tends to forget that even if the Michelson–Morley experiment had shown an 'ether wind', Einstein's programme might have been victorious nonetheless. When Miller,[4] an ardent champion of the classical ether programme, published his sensational claim that the Michelson–Morley experiment was sloppily conducted and in fact there *was* an ether wind, the news correspondent of *Science* crowed that 'Professor Miller's results knock out the relativity theory radi-cally'. In Einstein's view, however, even if Miller had reported the true state of affairs '[only] the *present form* of relativity theory' would have to be abandoned.[5] In fact, Synge pointed out that Miller's results, even if taken at their face value, do not conflict with Einstein's theory: only Miller's explanation of them does. One can easily replace the extant auxiliary theory of rigid bodies by a new, Gardner–Synge theory, and then Miller's results are fully digested within Einstein's programme.[6]

---

[1] Shankland [1964], p. 29.   [2] My italics.

[3] Einstein himself tended to believe that Michelson devised his interferometer in order to test Fresnel's theory. (Cf. Einstein [1931].) Incidentally, Michelson's early experiments on spectrum lines – like his [1881–2] – were also relevant to the ether theories of his day. Michelson over-emphasized his success in 'precise measurements' only when he was frustrated by his lack of success in evaluating their relevance for theories. Einstein, who disliked precision for its own sake, asked him why he devoted so much energy to it. Michelson's answer was 'because he found it fun.' (Cf. Einstein [1931].)

[4] In 1925.   [5] Einstein [1927], my italics.

[6] Synge [1952–4].

(d 2) The Lummer–Pringsheim experiments

Let us discuss another alleged crucial experiment. Planck claimed that Lummer's and Pringsheim's experiments, which ' *refuted* ' Wien's and Rayleigh's and Jeans's laws of radiation at the turn of the century, ' *led to* ' – or 'even brought about' – the quantum theory.[1] But again the role of these experiments is much more complicated and is very much in line with our approach. It is not simply that Lummer's and Pringsheim's experiments put an end to the classical approach but were neatly explained by quantum physics. On the one hand, some early versions of quantum theory by Einstein *entail* Wien's law and therefore were no less refuted by Lummer's and Pringsheim's experiments than the classical theory.[2] On the other hand, several classical explanations of the Planck formula were offered. For instance, at the 1913 meeting of the British Association for the Advancement of Science, there was a special meeting on radiation, attended by among others Jeans, Rayleigh, J. J. Thomson, Larmor, Rutherford, Bragg, Poynting, Lorentz, Pringsheim and Bohr. Pringsheim and Rayleigh were studiedly neutral about quantum theoretical speculations, but Professor Love 'represented the older views, and maintained the possibility of explaining facts about radiation without adopting the theory of quanta. He criticized the application of the equi-partition of energy theory, on which part of the quantum theory rests. The evidence for the quantum theory of most weight is the agreement with experiment of Planck's formula for the emissivity of a black body. From the mathematical point of view, there may be many more formulae which would agree equally well with the experiments. A formula due to A. Korn was dealt with, which gave results over a wide range, showing just about as good agreement with experiment as the Planck formula. In further contention that *the resources of ordinary theory are not exhausted,* he pointed out that it may be possible to extend the calculation for the emissivity of a thin plate due to Lorentz to other cases. For this calculation no simple analytical expression represents the results over the whole range of wavelengths, and it may well be that in the general case no simple formula exists which is applicable to all wavelengths. Planck's formula may, in fact, be nothing more than an empirical formula.'[3] One example of classical explanations was due to Callendar: 'The disagreement with experiment of Wien's well-known formula for the partition of energy in full radiation, is readily explained if we assume that it represents only the intrinsic energy. The corresponding value of the pressure is very easily deduced by reference to Carnot's

---

[1] Planck [1929]. Popper, in his [1934], section 30, and Gamow, in his [1966] (p. 37), take over this locution. Of course, observation statements do not 'lead' to some uniquely determined theory.

[2] Cf. Ter Haar [1967], p. 18. A budding research programme usually starts by explaining already refuted 'empirical laws' – and this, in the light of my approach, may be *rationally* regarded as a success.   [3] *Nature* [1913–14], p. 306, my italics.

principle, as Lord Rayleigh has indicated. The formula which I have proposed (*Phil. Mag.*, October 1913) is simply the sum of the pressure and energy-density thus obtained, and gives very satisfactory agreement with experiment, both for radiation and specific heat. I prefer it to Planck's formula (among other reasons) on the ground that the latter cannot be reconciled with the classical thermodynamics, and involves the conception of a *quantum*, or indivisible unit of action, which is unthinkable. The corresponding physical magnitude on my theory, which I have elsewhere called a molecule of caloric, is not necessarily indivisible, but bears a very simple relation to the intrinsic energy of an atom, which is all that is required to explain the facts that radiation may in special cases be emitted in atomic units which are multiples of a particular magnitude.'[1]

These quotations may have been tediously long but at least they show again convincingly the absence of instant crucial experiments. Lummer's and Pringsheim's refutations did not eliminate the classical approach to the radiation problem. The situation can be better described by pointing out that Planck's original '*ad hoc*' formula[2] – which fitted (and corrected) Lummer's and Pringsheim's data – could be explained *progressively* within the new quantum theoretical programme,[3] while neither his '*ad hoc*' formula, nor its 'semi-empirical' rivals could be explained within the classical programme except at the price of a degenerating problemshift. The 'progressive' development, incidentally, hinged on a 'creative shift': the replacement (by Einstein) of the Boltzman–Maxwell by the Bose–Einstein statistics.[4]

[1] Callendar [1914].

[2] I am referring to Planck's formula as given in his [1900a] in which he admitted that after having tried for a long time to prove that 'Wien's law must be *necessarily* true', the 'law' was refuted. So he switched from proving lofty eternal laws to 'constructing completely arbitrary expressions'. But of course any physical theory turns out to be 'completely arbitrary' by justificationist standards. In fact, Planck's arbitrary formula contradicted – and victoriously corrected – contemporary empirical evidence. (Planck told this part of the story in his scientific autobiography.) Of course, in an important sense, Planck's *original* radiation formula was 'arbitrary', 'formal', '*ad hoc*': it was a rather isolated formula which was not part of a research programme. (Cf. *below*, p. 88, n. 2). As he himself put it: 'Even if the absolutely precise validity of the radiation formula is taken for granted, so long as it had merely the standing of a law disclosed by a lucky intuition, it could not be expected to possess more than a formal significance. For this reason, on the very day when I formulated this law, I began to devote myself to the task of investing it with a true physical meaning' ([1948], p. 41). But the primary importance of 'investing the formula with a physical meaning' – not necessarily '*true* physical meaning' – is that such interpretation frequently leads to a suggestive research programme and *growth*.

[3] First by Planck himself, in his [1900b] which 'founded' the research programme of quantum theory.

[4] This had already been done by Planck, but only inadvertently, as it were by mistake. Cf. Ter Haar [1967], p. 18. Indeed, one role of Pringsheim's and Lummer's results was to stimulate the critical analysis of the informal deductions in the quantum theory of radiation, deductions which were loaded with vital 'hidden lemmas' articulated only in the later development. A most important step in this 'articulating process' was Ehrenfest's [1911].

The progressiveness of the new development was abundantly clear: in Planck's version it predicted correctly the value of the Boltzman–Planck constant and in Einstein's version it predicted a stunning series of further novel facts.[1] But before the invention of the new – but sadly *ad hoc* – auxiliary hypotheses in the old programme, before the unfolding of the new programme, and before the discovery of the new facts indicating a progressive problemshift in the latter, the objective relevance of the Lummer–Pringsheim experiments was very limited.

(d 3) Beta-decay versus conservation laws

Finally, I shall tell a story of an experiment which very nearly, but not quite, became 'the greatest negative experiment in the history of science'. The story again illustrates the supreme difficulties in deciding exactly *what* one learns from experience, what it 'proves' and what it 'disproves'. The piece of experience under scrutiny will be Chadwick's 'observation' of beta decay in 1914. The story shows how an experiment may first be regarded as presenting a routine puzzle within a research programme, then nearly promoted to the rank of 'crucial experiment', and then again downgraded to presenting a (*new*) routine puzzle, all this depending on the *whole* changing theoretical and empirical landscape. Most conventional accounts are confused by these changes and prefer to falsify history.[2]

When Chadwick discovered the continuous spectrum of radioactive beta-emission in 1914, nobody thought that this curious phenomenon had anything to do with conservation laws. Two ingenious rival explanations were offered in 1922, both within the framework of the atomic physics of the day, one by L. Meitner, the other by C. D. Ellis. According to Miss Meitner, the electrons were partly primary electrons from the nucleus, partly secondary electrons from the electron shell. According to Mr Ellis, they were all primary electrons. Both theories contained sophisticated auxiliary hypotheses, but both predicted novel facts. The predicted facts contradicted each other and the experimental testimony supported Ellis against Meitner.[3] Miss Meitner appealed; the experimental 'appeal court' refused to support her, but ruled that one crucial auxiliary hypothesis in Ellis's theory had to be rejected.[4] The result of the contest was a draw.

Still nobody would have thought that Chadwick's experiment defied the law of conservation of energy, had not Bohr and Kramers arrived exactly at the time of the Ellis–Meitner controversy at the idea that a consistent theory could be developed only if they renounced the principle of conservation of energy in single processes. One of the

[1] Cf. e.g. Joffé's 1910 list (Joffé [1911], p. 547).
[2] A notable partial exception is Pauli's account (Pauli [1958]). In what follows I am trying both to correct Pauli's story and to show that its rationality can be easily seen in the light of our approach.
[3] Ellis and Wooster [1927].   [4] Meitner and Orthmann [1930].

main features of the fascinating Bohr–Kramers–Slater theory in 1924 was that the classical laws of conservation of energy and momentum were replaced by statistical ones.[1] This theory (or, rather, 'programme') was immediately 'refuted' and none of its consequences corroborated; indeed, it was never sufficiently developed to explain beta-decay. But in spite of the immediate abandonment of this programme (not simply because of its 'refutations' by the Compton–Simon and Bothe–Geiger experiments but because of the emergence of a powerful rival: the Heisenberg–Schrödinger programme[2]), Bohr remained convinced that the non-statistical conservation laws would finally have to be abandoned and that the beta-decay anomaly would never be explained until these laws were replaced; at which time beta-decay would be seen as a crucial experiment against the conservation laws. Gamow tells us how Bohr tried to use the idea of non-conservation of energy in beta-decay for an ingenious explanation of the seemingly eternal production of energy in stars.[3] Only Pauli, in his Mephistophelian urge to defy the Lord, remained conservative[4] and devised, in 1930, his neutrino theory in order to explain beta-decay and in order to save the principle of conservation of energy. He communicated his idea in a jocular letter to a conference in Tübingen – he himself preferred to stay in Zürich to attend a ball.[5] He first mentioned it in a public lecture in 1931 in Pasadena, but he did not allow the lecture to be published because he felt 'unsure' about it. Bohr, at that time (in 1932), still thought that – at least in nuclear physics – one may have 'to renounce the very idea of energy balance'.[6] Pauli finally decided to publish his talk on the neutrino which he delivered to the 1933 Solvay conference, in spite of the fact that 'the reception at the Congress, except for two young physicists, was sceptical'.[7] But Pauli's theory had some methodological merits. It saved not only the principle of conservation of energy but also the principle of conservation of spin and statistics:

[1] Slater co-operated only reluctantly in sacrificing the conservation principle. He wrote to van der Waerden in 1964: 'As you suspected, the idea of statistical conservation of energy and momentum was put into the theory by Bohr and Kramers, quite against my better judgment.' Van der Waerden does his amusing best to exonerate Slater from the terrible crime of being responsible for a false theory (van der Waerden [1967], p. 13).

[2] Popper is wrong to suggest that these 'refutations' were sufficient to bring about the downfall of this theory. (Popper [1963a], p. 242.)

[3] Gamow [1966], pp. 72–4. Bohr never published this theory (it was untestable as it stood) but 'it looked' – writes Gamow – 'as if he would not be greatly surprised if it were true'. Gamow does not date this unpublished theory but it seems that Bohr entertained it in 1928–9 when Gamow was working in Copenhagen.

[4] Cf. the amusing play 'Faust' produced in Bohr's institute in 1932; published by Gamow as an appendix to his [1966].

[5] Cf. Pauli [1961], p. 160.

[6] Bohr [1932]. Ehrenfest too sided firmly with Bohr against the neutrino. Chadwick's discovery of the neutron in 1932 only slightly shook their opposition: they still dreaded the idea of a particle which has neither charge nor, possibly, even (rest) mass, but only 'disembodied' spin.

[7] Wu [1966].

it explained not only the beta-decay spectrum but, at the same time, the 'nitrogen anomaly'.[1] By Whewellian standards this 'consilience of inductions' should have been sufficient to establish the respectability of Pauli's theory. But on our criteria, the successful prediction of some *novel* fact was needed. This too was provided by Pauli's theory. For Pauli's theory had an interesting observable consequence: if it was right, the β-spectra had to have a clear upper bound. This question was *at the time* undecided, but Ellis and Mott became interested[2] and soon, Ellis's student, Henderson, showed that the experiments supported Pauli's programme.[3] Bohr was not impressed. He knew that if a major programme based on *statistical* conservation of energy ever got going, the growing belt of auxiliary hypotheses would take proper care of the most negative-looking evidence.

Indeed, in these years most leading physicists thought that in nuclear physics the laws of conservation of energy and momentum break down.[4] The reason was stated clearly by Lise Meitner who admitted defeat only in 1933: 'All the attempts to uphold the validity of the law of conservation of energy also for *single* processes demanded a second process [in the beta-decay]. But no such process was found':[5] that is, the conservation programme for the nucleus showed an empirically degenerating problemshift. There were several ingenious attempts to account for the continuous beta-emission spectrum without assuming a 'thief particle'.[6] These attempts were discussed with great interest,[7] but they were abandoned because they failed to establish a progressive shift.

At this point, Fermi entered on the scene. In 1933–4 he reinterpreted the beta-emission problem in the framework of the research programme of the new quantum theory. Thus he initiated a small new research programme of the neutrino (which later grew into the programme of weak interactions). He calculated some first crude models.[8] Although his theory did not yet predict any new fact, he made it clear that this was only a matter of some further work.

Two years passed and Fermi's promise was still not fulfilled. But the new programme of quantum physics developed fast, at least as far as the non-nuclear phenomena were concerned. Bohr became convinced that some of the basic original ideas of the Bohr–Kramers–Slater programme were now firmly embedded in the new

---

[1] For a fascinating discussion of the open problems presented by the beta-decay and by the nitrogen anomaly, cf. Bohr's Faraday Lecture in 1930, read before, but published after, Pauli's solution (Bohr [1932], especially pp. 380–3).
[2] Ellis and Mott [1933].   [3] Henderson [1934].
[4] Mott [1933], p. 823. Heisenberg, in his celebrated [1932], in which he introduced the proton-neutron model of the nucleus, pointed out that 'because of the breakdown of the conservation of energy in the beta-decay one cannot give a unique definition of the binding energy of the electron within the neutron' (p. 164).
[5] Meitner [1933], p. 132.   [6] E.g. Thomson [1929] and Kudar [1929–30].
[7] For a most interesting discussion cf. Rutherford, Chadwick and Ellis [1930], pp. 335–6.
[8] Fermi [1933] and [1934].

quantum programme and that the new programme solved the intrinsic theoretical problems of the old quantum programme without touching the conservation laws. Therefore Bohr followed Fermi's work with sympathy, and in 1930, in an unusual sequence of events, gave it, by our standards prematurely, public support.

In 1936 Shankland devised a new test of rival theories of photon scattering. His results seemed to support the discarded Bohr–Kramers–Slater theory and undermine the reliability of experiments which, more than a decade earlier, refuted it.[1] Shankland's paper created a sensation. Those physicists who abhorred the new trend were quick to hail Shankland's experiment. Dirac, for instance, immediately welcomed back the 'refuted' Bohr–Kramers–Slater programme, wrote a very sharp article against the 'so-called quantum electrodynamics' and demanded 'a profound alteration in current theoretical ideas, involving a departure from the conservation laws [in order] to get a satisfactory relativistic quantum mechanics'.[2] In the article Dirac suggested again that beta-decay may well turn out to be a piece of crucial evidence against the conservation laws and made fun of the 'new unobservable particle, the neutrino, specially postulated by some investigators in an attempt formally to preserve conservation of energy by assuming the unobservable particle to carry off the balance'.[3] Immediately afterwards Peierls joined the discussion. Peierls suggested that Shankland's experiment may turn out to refute even the statistical conservation of energy. He added: 'That, too, seems satisfactory, once detailed conservation has been abandoned.'[4]

In Bohr's Copenhagen institute, Shankland's experiments were immediately repeated and discarded. Jacobsen, a colleague of Bohr reported this in a letter to *Nature*. Jacobsen's results were accompanied by a letter from Bohr himself, who firmly came out against the rebels, and in defence of Heisenberg's new quantum programme. In particular, he came out in defence of the neutrino against Dirac: 'It may be remarked that the grounds for serious doubts as regards the strict validity of the conservation laws in the problem of the emission of β-rays from atomic nuclei are now largely removed by the suggestive agreement between the rapidly increasing experimental evidence regarding β-ray phenomena and the consequences of the neutrino hypotheses of Pauli so remarkably developed in Fermi's theory.'[5]

Fermi's theory, in its first versions, had no striking empirical success. Indeed, even the available data, especially in the case of *RaE*, on which beta emission research then centred, sharply contradicted Fermi's 1933–4 theory. He wanted to deal with these in the second part of his paper which, however, was never published. Even if one construes Fermi's 1933–4 theory as a first version of a flexible programme, by

---

[1] Shankland [1936].
[2] Dirac [1936].
[3] Dirac [1936].
[4] Peierls [1936].
[5] Bohr [1936].

1936 one could not possibly detect any serious sign of a progressive shift.[1] But Bohr wanted to put his *authority* behind Fermi's daring application of Heisenberg's new big programme to the nucleus; and since Shankland's experiment and Dirac's and Peierls's attack brought the beta-decay into the focus of the criticism of the new big programme, he over-praised Fermi's neutrino programme which promised to fill in a sensitive gap. No doubt, the later development spared Bohr from a dramatiç humiliation: the programmes based on conservation principles progressed, while no progress was made in the rival camp.[2]

The moral of this story is again that the status of an experiment as 'crucial' depends on the status of the theoretical competition in which it is embedded. As the fortunes of the competing camps wax or wane, the interpretation and appraisal of the experiment may change.

Our scientific folklore however is impregnated with theories of instant rationality. The story which I described is falsified in most accounts and reconstructed in terms of some wrong theory of rationality. Even the very best popular expositions teem with such falsifications. Let me mention two examples.

In one paper we learn this about beta-decay: 'When this situation was faced for the first time, the alternatives seemed grim. Physicists *either* had to accept a breakdown of the law of energy conservation, *or* they had to suppose the existence of a new and unseen particle. Such a particle, emitted along with the proton and the electron in the disintegration of the neutron, could save the central pillar of physics by carrying off the missing energy. This was in the early 1930s, when the introduction of a new particle was not the casual matter it is today. Nevertheless, *after only the briefest vacillation*, physicists chose the second alternative.'[3] Of course, even the *discussed* alternatives were many more than two and the 'vacillation' was certainly not 'the briefest'.

In a well-known textbook of philosophy of science we learn that (1)

---

[1] Several physicists between 1933 and 1936 offered alternatives or proposed *ad hoc* changes of Fermi's theory; cf. e.g. Beck and Sitte [1933], Bethe and Peierls [1934], Konopinski and Uhlenbeck [1934]. Wu and Moszkowski write in 1966 that 'the Fermi theory [i.e. programme] of β-decay is *now* known to predict with remarkable accuracy both the relation between the rate of β-decay and the energy of disintegration, and also the shape of β-spectra'. But they stress that 'at the very beginning the Fermi theory unfortunately met an unfair test. Until the time when artificial radioactive nuclei could be copiously produced, *RaE* was the only candidate that beautifully fulfilled many experimental requirements as a β source for the investigation of its spectrum shape. How could we have known then that the β spectrum of *RaE* would turn out to be only a very special case, one whose spectrum has, in fact, been understood only very recently. Its peculiar energy dependence defied what was expected of the simple Fermi theory of β decay and greatly slackened the pace of the theory's [i.e. programme's] initial progress' (Wu and Moszkowski [1966], p. 6).

[2] It is very doubtful whether Fermi's neutrino programme was progressive or degenerating even between 1936 and 1950; and after 1950 the verdict is still not crystal clear. But this I shall try to discuss on some other occasion. (Incidentally, Schrödinger stood up for the statistical interpretation of the conservation principles in spite of his crucial role in the development of new quantum physics; cf. his [1958].)

[3] Treiman [1959], my italics.

'the law (or principle) of the conservation of energy was seriously challenged by experiments on beta-ray decay whose outcome could not be denied'; that (2) 'nevertheless, the law was not abandoned, and the existence of a new kind of entity (called a "neutrino") was assumed in order to bring the law into concordance with experimental data'; and that (3) 'the rationale for this assumption is that the rejection of the conservation law would deprive a large part of our physical knowledge of its systematic coherence'.[1] But all the three points are wrong. (1) is wrong because no law can be 'seriously challenged' by experiments only; (2) is wrong because new *scientific* hypotheses are assumed not simply in order to patch up gaps between data and theory but in order to predict novel facts; and (3) is wrong because at the time it seemed that *only* the rejection of the conservation law would secure the 'systematic coherence' of our physical knowledge.

(d4) Conclusion. The requirement of continuous growth

There are no such things as crucial experiments, at least not if these are meant to be experiments which can *instantly* overthrow a research programme. In fact, when one research programme suffers defeat and is superseded by another one, we may – *with long hindsight* – call an experiment crucial if it turns out to have provided a spectacular corroborating instance for the victorious programme and a failure for the defeated one (in the sense that it was never 'explained progressively' – or, briefly, 'explained'[2] – within the defeated programme). But scientists, of course, do not always judge heuristic situations correctly. A rash scientist may *claim* that his experiment defeated a programme, and parts of the scientific community may even, rashly, accept his claim. But if a scientist in the 'defeated' camp puts forward a few years later a scientific explanation of the allegedly 'crucial experiment' within (or consistent with) the allegedly defeated programme, *the honorific title may be withdrawn and the 'crucial experiment' may turn from a defeat into a new victory for the programme.*

Examples abound. There were many experiments in the eighteenth century which were, as a matter of historico-sociological fact, widely accepted as 'crucial' evidence against Galileo's law of free fall, and Newton's theory of gravitation. In the nineteenth century there were several 'crucial experiments' based on measurements of light velocity which 'disproved' the corpuscular theory and which turned out later to be erroneous in the light of relativity theory. These 'crucial experiments' were later deleted from the justificationist textbooks as manifestations of shameful short-sightedness or even of envy. (Recently they reappeared in some new textbooks, this time to illustrate the inescapable irrationality of scientific fashions.) However, in those cases in which ostensibly 'crucial experiments' were indeed *later* borne out

---

[1] Nagel [1961], pp. 65–6.   [2] Cf. *above,* p. 34, n. 4.

by the defeat of the programme, historians charged those who resisted them with stupidity, jealousy, or unjustified adulation of the father of the research programme in question. (Fashionable 'sociologists of knowledge' – or 'psychologists of knowledge' – tend to explain positions in purely social or psychological terms when, as a matter of fact, they are determined by rationality principles. A typical example is the explanation of Einstein's opposition to Bohr's complementarity principle on the ground that 'in 1926 Einstein was forty-seven years old. Forty-seven may be the prime of life, but not for physicists'.[1])

In the light of my considerations, the idea of instant rationality can be seen to be utopian. But this utopian idea is a hallmark of most brands of epistemology. Justificationists wanted scientific theories to be proved even before they were published; probabilists hoped a machine could flash up instantly the value (degree of confirmation) of a theory, given the evidence; naive falsificationists hoped that elimination at least was the instant result of the verdict of *experiment*.[2] I hope I have shown that *all these theories of instant rationality – and instant learning – fail.* The case studies of this section show that rationality works much slower than most people tend to think, and, even then, fallibly. Minerva's owl flies at dusk. I also hope I have shown that the *continuity* in science, the *tenacity* of some theories, the rationality of a certain amount of dogmatism, can only be explained if we construe science as a battleground of research programmes rather than of isolated theories. One can understand very little of the growth of science when our paradigm of a chunk of scientific knowledge is an isolated theory like 'All swans are white', standing aloof, without being embedded in a major research programme. *My account implies a new criterion of demarcation between 'mature science', consisting of research programmes, and 'immature science' consisting of a mere patched up pattern of trial and error.*[3] For instance, we may have a conjecture, have it

---

[1] Bernstein [1961], p. 129. In order to appraise progressive and degenerating elements in rival problemshifts one must understand the *ideas* involved. But the sociology of knowledge frequently serves as a successful cover for illiteracy: most sociologists of knowledge do not understand – or even care for – the ideas; they watch the socio-psychological patterns of behaviour. Popper used to tell a story about a 'social psychologist', Dr X, studying scientists' group behaviour. He went into a physics seminar to study the psychology of science. He observed the 'emergence of a leader', the 'rallying round effect' in some and the 'defence-reaction' in others, the correlation between age, sex and aggressive behaviour, etc. (Dr X claimed to have used some sophisticated small-sample techniques of modern statistics.) At the end of the enthusiastic account Popper asked Dr X: 'What was the *problem* the group was discussing?' Dr X was surprised: 'Why do you ask? I did not listen to the *words*! Anyway, what has *that* to do with the psychology of knowledge?'

[2] Of course, naive falsificationists may take some time to reach the 'verdict of experiment': the experiment has to be repeated and critically considered. But once the discussion ends up in an agreement among the experts, and thus a 'basic statement' becomes 'accepted', and it has been decided which specific theory was hit by it, the naive falsificationist will have little patience with those who still 'prevaricate'.

[3] The elaboration of this demarcation in the two following paragraphs was improved in the press, following invaluable discussions with Paul Meehl in Minneapolis in 1969.

refuted and then rescued by an auxiliary hypothesis which is not *ad hoc* in the senses which we had earlier discussed. It may predict novel facts some of which may even be corroborated.[1] Yet one may achieve such 'progress' with a patched up, arbitrary series of disconnected theories. Good scientists will not find such makeshift progress satisfactory; they may even reject it as not genuinely scientific. They will call such auxiliary hypotheses merely 'formal', 'arbitrary', 'empirical', 'semi-empirical', or even '*ad hoc*'.[2]

Mature science consists of research programmes in which not only novel facts but, in an important sense, also novel auxiliary theories, are anticipated; mature science – unlike pedestrian trial-and-error – has 'heuristic power'. Let us remember that in the positive heuristic of a powerful programme there is, right at the start, a general outline of how to build the protective belts: this heuristic power generates *the autonomy of theoretical science.*[3]

This *requirement of continuous growth* is my rational reconstruction of the widely acknowledged requirement of 'unity' or 'beauty' of science. It highlights the weakness of *two* – apparently very different – types of theorizing. First, it shows up the weakness of programmes which, like Marxism or Freudism, are, no doubt, 'unified', which give a major sketch of the sort of auxiliary theories they are going to use in absorbing anomalies, but which unfailingly devise their actual auxiliary theories in the wake of facts without, at the same time, anticipating others. (What *novel* fact has Marxism *predicted* since, say, 1917?) Secondly, it hits patched-up, unimaginative series of pedestrian 'empirical' adjustments which are so frequent, for instance, in modern social psychology. Such adjustments may, with the help of so-called 'statistical techniques', make some 'novel' predictions and may even conjure up some irrelevant grains of truth in them. But this theorizing has no unifying idea, no heuristic power, no continuity. They do not add up to a genuine research programme and are, on the whole, worthless.[4]

---

[1] Earlier, in my [1968*b*] (volume 2, chapter 8), I distinguished, following Popper, two criteria of *adhocness*. I called *ad hoc₁*, those theories which had no excess content over their predecessors (or competitors) that is, which did not predict any *novel* facts; I called *ad hoc₂*, those theories which predicted novel facts but completely failed: none of their excess content got corroborated (also, cf. *above*, p. 40, nn. 1 and 2).

[2] Planck's radiation formula – given in his [1900*a*] – is a good example: cf. *above*, p. 80, n. 2. We may call such hypotheses which are not *ad hoc₁*, not *ad hoc₂*, but still unsatisfactory in the sense specified in the text, *ad hoc₃*. These three – unfailingly pejorative – usages of *ad hoc* may provide a satisfactory entry in the *Oxford English Dictionary*.

It is intriguing to note that 'empirical' and 'formal' are both used as synonyms for our *ad hoc₃*.

Meehl, in his brilliant [1967], reports that in contemporary psychology – especially in social psychology – many alleged 'research programmes' in fact consist of chains of such *ad hoc₃* stratagems.

[3] Cf. *above*, p. 52.

[4] After reading Meehl [1967] and Lykken [1968] one wonders whether the function of statistical techniques in the social sciences is not primarily to provide a machinery for

My account of scientific rationality, although based on Popper's, leads away from some of his general ideas. I endorse to some extent both Le Roy's conventionalism with regard to theories and Popper's conventionalism with regard to basic propositions. In this view scientists (and as I have shown, mathematicians too[1]) are not irrational when they tend to ignore counterexamples or as they prefer to call them, 'recalcitrant' or 'residual' instances, and follow the sequence of problems as prescribed by the positive heuristic of their programme, and elaborate – and apply – their theories regardless.[2] Contrary to Popper's falsificationist morality, scientists frequently and *rationally* claim 'that the experimental results are not reliable, or that the discrepancies which are asserted to exist between the experimental results and the theory are only apparent and that they will disappear with the advance of our understanding'.[3] When doing so, they may *not* be 'adopting the very reverse of that critical attitude which...is the proper one for the scientist'.[4] Indeed, Popper is right in stressing that 'the dogmatic attitude of sticking to a theory as long as possible is of considerable significance. Without it we could never find out what is in a theory – we should give the theory up before we had a real opportunity of finding out its strength; and in consequence no theory would ever be able to play its role of bringing order into the world, of preparing us for future events, of drawing our attention to events we should otherwise never observe'.[5] Thus the 'dogmatism' of 'normal

producing phoney corroborations and thereby a semblance of 'scientific progress' where, in fact, there is nothing but an increase in pseudo-intellectual garbage. Meehl writes that 'in the physical sciences, the usual result of an improvement in experimental design, instrumentation, or numerical mass of data, is to increase the difficulty of the "observational hurdle" which the physical theory of interest must successfully surmount; whereas, in psychology and some of the allied behaviour sciences, the usual effect of such improvement in experimental precision is to provide an easier hurdle for the theory to surmount'. Or, as Lykken put it: 'Statistical significance [in psychology] is perhaps the least important attribute of a good experiment; it is never a sufficient condition for claiming that a theory has been usefully corroborated, that a meaningful empirical fact has been established, or that an experimental report ought to be published.' It seems to me that most theorizing condemned by Meehl and Lykken may be *ad hoc₃*. Thus the methodology of research programmes might help us in devising laws for stemming this intellectual pollution which may destroy our cultural environment even earlier than industrial and traffic pollution destroys our physical environment.   [1] Cf. my [1963–4].

[2] Thus the *methodological* asymmetry between universal and singular statements vanishes. We may adopt either by convention: in the 'hard core' we decide to 'accept' universal, in the 'empirical basis' singular, statements. The *logical* asymmetry between universal and singular statements is fatal only for the dogmatic inductivist who wants to learn only from hard experience and logic. The conventionalist can, of course, 'accept' this *logical* asymmetry: he does not have to be (although he *may* be) also an inductivist. He 'accepts' some universal statements, but not because he claims to deduce (or induce) them from singular ones.

[3] Popper [1934], section 9.   [4] *Ibid.*

[5] Popper [1940], first footnote. We find a similar remark in his [1963a], p. 49. But these remarks are in *prima facie* contradiction with some of his remarks in [1934] (quoted *above*, p. 27), and therefore may only be interpreted as signs of a growing awareness by Popper of an undigested anomaly in his own research programme.

science' does not prevent growth as long as we combine it with the Popperian recognition that there is good, progressive normal science and that there is bad, degenerating normal science, and as long as we retain the *determination* to eliminate, under certain objectively defined conditions, some research programmes.

The dogmatic attitude in science – which would explain its stable periods – was described by Kuhn as a prime feature of 'normal science'.[1] But Kuhn's conceptual framework for dealing with continuity in science is socio-psychological: mine is normative. I look at continuity in science through 'Popperian spectacles'. Where Kuhn sees 'paradigms', I *also* see rational 'research programmes'.

4 THE POPPERIAN VERSUS THE KUHNIAN RESEARCH PROGRAMME

Let us now sum up the Kuhn–Popper controversy.

We have shown that Kuhn is right in objecting to naive falsificationism, and also in stressing the *continuity* of scientific growth, the *tenacity* of some scientific theories. But Kuhn is wrong in thinking that by discarding naive falsificationism he has discarded thereby all brands of falsificationism. Kuhn objects to the entire Popperian research programme, and he excludes *any* possibility of a rational reconstruction of the growth of science. In a succinct comparison of Hume, Carnap and Popper, Watkins points out that the growth of science is inductive and irrational according to Hume, inductive and rational according to Carnap, non-inductive and rational according to Popper.[2] But Watkins's comparison can be extended by adding that it is non-inductive and irrational according to Kuhn. *In Kuhn's view there can be no logic, but only psychology of discovery.*[3] For instance, in Kuhn's conception, anomalies, inconsistencies *always* abound in science, but in 'normal' periods the dominant paradigm secures a pattern of growth which is eventually overthrown by a 'crisis'. There is no particular rational cause for the appearance of a Kuhnian 'crisis'. 'Crisis' is a psychological concept; it is a contagious panic. Then a new 'paradigm' emerges, incommensurable with its predecessor. There are no rational standards for their comparison. Each paradigm contains its own standards. The crisis sweeps away not only the old theories and rules but also the standards which made us respect them. The new

---

[1] Indeed, my demarcation criterion between mature and immature science can be interpreted as a Popperian absorption of Kuhn's idea of 'normality' as a hallmark of [mature] science; and it also reinforces my earlier argument against regarding highly falsifiable statements as eminently scientific. (Cf. *above*, p. 19.)

Incidentally, this demarcation between mature and immature science appears already in my [1963–4], where I called the former 'deductive guessing' and the latter 'naive trial and error'. (See e.g. [1963–4], section 7(*c*): 'Deductive guessing versus naive guessing.')

[2] Watkins [1968], p. 281.

[3] Kuhn [1970]. But this position is already implicit in his [1962].

paradigm brings a totally new rationality. There are no super-paradigmatic standards The change is a bandwagon effect. Thus *in Kuhn's view scientific revolution is irrational, a matter for mob psychology.*

The reduction of philosophy of science to psychology of science did not start with Kuhn. An earlier wave of 'psychologism' followed the breakdown of justificationism. For many, justificationism represented the only possible form of rationality: the end of justificationism meant the end of rationality. The collapse of the thesis that scientific theories are provable, that the progress of science is cumulative, made justificationists panic. If 'to discover is to prove', but nothing is provable, then there can be no discoveries, only discovery-claims. Thus disappointed justificationists – ex-justificationists – thought that the elaboration of rational standards was a hopeless enterprise and that all one can do is to study – and imitate – the Scientific Mind, as it is exemplified in famous scientists. After the collapse of Newtonian physics, Popper elaborated new, non-justificationist critical standards. Now some of those who had already learned of the collapse of justi-ficationist rationality now learned, mostly by hearsay, of Popper's colourful slogans which suggested naive falsificationism. Finding them untenable, they identified the collapse of naive falsificationism with the end of rationality itself. The elaboration of rational standards was again regarded as a hopeless enterprise; the best one can do is to study, they thought once again, the Scientific Mind.[1] Critical philosophy was to be replaced by what Polanyi called a 'post-critical' philosophy. But the Kuhnian research programme contains a new feature: we have to study not the mind of the individual scientist but the mind of the Scientific Community. Individual psychology is now replaced by social psychology; imitation of the great scientists by submission to the collective wisdom of the community.

But Kuhn overlooked Popper's sophisticated falsificationism and the research programme he initiated. Popper replaced the central problem of classical rationality, *the old problem of foundations*, with *the new problem of fallible-critical growth*, and started to elaborate objective standards of this growth. In this paper I have tried to develop his programme a step further. I think this small development is sufficient to escape Kuhn's strictures.[2]

[1] Incidentally, just as some earlier ex-justificationists led the wave of sceptical irra-tionalism, so now some ex-falsificationists lead the *new* wave of sceptical irrationalism and anarchism. This is best exemplified in Feyerabend [1970b].

[2] Indeed, as I had already mentioned, *my concept of a 'research programme' may be construed as an objective, 'third world' reconstruction of Kuhn's socio-psychological concept of 'paradigm'*: thus the Kuhnian 'Gestalt-switch' can be performed without removing one's Popperian spectacles.

(I have not dealt with Kuhn's and Feyerabend's claim that theories cannot be eliminated on any *objective* grounds because of the 'incommensurability' of rival theories. Incommensurable theories are neither inconsistent with each other, nor comparable for content. But we can *make* them, by a dictionary, inconsistent and their content comparable. If we want to eliminate a programme, we need some

The reconstruction of scientific progress as proliferation of rival research programmes and progressive and degenerative problem-shifts gives a picture of the scientific enterprise which is in many ways different from the picture provided by its reconstruction as a succession of bold theories and their dramatic overthrows. Its main aspects were developed from Popper's ideas and, in particular, from his ban on 'conventionalist', that is, content-decreasing, stratagems. The main difference from Popper's original version is, I think, that in my conception criticism does not – and must not – kill as fast as Popper imagined. *Purely negative, destructive criticism, like 'refutation' or demonstration of an inconsistency does not eliminate a programme. Criticism of a programme is a long and often frustrating process and one must treat budding programmes leniently.*[1] One may, of course, show up the degeneration of a research programme, but it is only *constructive criticism* which, with the help of rival research programmes, can achieve real success; and dramatic spectacular results become visible only with hindsight and rational reconstruction.

Kuhn certainly showed that the psychology of science can reveal important and, indeed, sad truths. But the psychology of science is not autonomous; for *the – rationally reconstructed – growth of science takes place essentially in the world of ideas, in Plato's and Popper's 'third world'*, in the world of articulated knowledge which is independent of knowing subjects.[2] *Popper's research programme* aims at a description of this objective scientific *growth*.[3] Kuhn's research programme seems to aim at a description of *change* in the ('normal') scientific mind (whether individual or communal).[4] But the mirror-image of the

methodological determination. This determination is the heart of methodological falsificationism; for instance, no result of statistical sampling is ever inconsistent with a statistical theory unless we *make them* inconsistent with the help of Popperian rejection rules, cf. *above*, p. 25.)

[1] The reluctance of economists and other social scientists to accept Popper's methodology may have been partly due to the destructive effect of naive falsificationism on budding research programmes.

[2] The *first* world is the material world, the *second* is the world of consciousness, the *third* is the world of propositions, truth, standards: the world of objective knowledge. The modern *loci classici* on this subject are Popper [1968a] and Popper [1968b]; also, cf. Toulmin's impressive programme set out in his [1967]. It should be mentioned here that many passages of Popper [1934] and even of [1963a] sound like descriptions of a psychological contrast between the Critical Mind and the Inductivist Mind. But Popper's psychologistic terms can be, to a large extent, reinterpreted in third-world terms: see Musgrave [1974].

[3] In fact, Popper's programme extends beyond science. The concepts of 'progressive' and 'degenerating' problemshifts, the idea of proliferation of theories can be generalized to any sort of rational discussion and thus serve as tools for a general theory of criticism; cf. *below*, chapters 2 and 3. (My [1963–4] can be seen as the story of a non-empirical progressive research programme; volume 2, chapter 8, contains the story of the non-empirical degenerating programme of inductive logic.)

[4] *Actual* state of minds, beliefs, etc., belong to the second world; states of the *normal* mind belong to a limbo between the second and third. The study of actual scientific minds belongs to *psychology*; the study of the 'normal' (or 'healthy' etc.) mind belongs to a *psychologistic philosophy of science*. There are *two kinds of psychologistic philosophies*

third world in the mind of the individual – even in the mind of the 'normal' – scientists is usually a caricature of the original; and to describe this caricature without relating it to the third-world original might well result in a caricature of a caricature. One cannot understand the history of science without taking into account the interaction of the three worlds.

APPENDIX POPPER, FALSIFICATIONISM AND THE
'DUHEM–QUINE THESIS'

Popper began as a dogmatic falsificationist in the 1920s; but he soon realized the untenability of this position and published nothing before he invented *methodological falsificationism*. This was an entirely new idea in the philosophy of science and it clearly originates with Popper, who put it forward as a solution to the difficulties of dogmatic falsificationism. Indeed, the conflict between the theses that science is both critical and fallible is one of the central problems in Popperian philosophy. While Popper offered a coherent formulation and criticism of dogmatic falsificationism, he never made a sharp distinction between naive and sophisticated falsificationism. In an earlier paper,[1] I distinguished three Poppers: *Popper₀*, *Popper₁* and *Popper₂*. Popper₀ is the dogmatic falsificationist who never published a word: he was invented – and 'criticized' – first by Ayer and then by many others.[2] This paper will, I hope, finally kill this ghost. Popper₁ is the naive falsificationist, Popper₂ the sophisticated falsificationist. The *real* Popper developed from dogmatic to a naive version of methodological falsificationism in the twenties; he arrived at the '*acceptance rules*' of

of science. According to one kind there can be no philosophy of science: only a psychology of individual scientists. According to the other kind there is a psychology of the 'scientific', 'ideal' or 'normal' mind: this turns philosophy of science into a psychology of this ideal mind and, in addition, offers a psychotherapy for turning one's mind into an ideal one. I discuss this second kind of psychologism in detail elsewhere. Kuhn does not seem to have noticed this distinction.

[1] Cf. my [1968c].

[2] Ayer seems to have been the first to attribute dogmatic falsificationism to Popper. (Ayer also invented the myth that according to Popper 'definite confutability' was a criterion not only of the empirical but also of the meaningful character of a proposition: cf. his [1936], chapter 1, p. 38 of the second edition.) Even today, many philosophers (cf. Juhos [1966] or Nagel [1967]) criticize the strawman Popper₀. Medawar, in his [1967], called *dogmatic* falsificationism 'one of the strongest ideas' in Popper's methodology. Nagel, reviewing Medawar's book, criticized Medawar for 'endorsing' what he too believes to be 'Popper's claims' (Nagel [1967], p. 70). Nagel's criticism convinced Medawar that 'the act of falsification is not immune to human error' (Medawar [1969], p. 54). But Medawar and Nagel misread Popper: his *Logik der Forschung* is the strongest ever criticism of dogmatic falsificationism.

One may take a charitable view of Medawar's mistake: for brilliant scientists whose speculative talent was thwarted under the tyranny of an inductivist logic of discovery, falsificationism, even in its dogmatic form, was bound to have a tremendous liberating effect. (Besides Medawar, another Nobel Prize winner, Eccles, learned from Popper to replace his original caution by bold falsifiable speculation: cf. Eccles [1964], pp. 274–5.)

sophisticated falsificationism in the fifties. The transition was marked by his adding to the original requirement of testability the 'second' requirement of 'independent testability',[1] and then the 'third' requirement that some of these independent tests should result in corroborations.[2] But the real Popper never abandoned his earlier (naive) *falsification rules.* He has demanded, until this day, that '*criteria of refutation* have to be laid down beforehand: it must be agreed, which observable situations, if actually observed, mean that the theory is refuted'.[3] He still construes 'falsification' as the result of a duel between theory and observation, without another, better theory *necessarily* being involved. The real Popper has never explained in detail the appeal procedure by which some 'accepted basic statements', may be eliminated. Thus the real Popper consists of Popper₁ together with some elements of Popper₂.

The idea of a demarcation between progressive and degenerating problemshifts, as discussed in this paper, is based on Popper's work: indeed this demarcation is almost identical with his celebrated demarcation criterion between science and metaphysics.[4]

Popper originally had only the *theoretical* aspect of problemshifts in mind, which is hinted at in section 20 of his [1934] and developed in his [1957a].[5] He added a discussion of the *empirical* aspect of problemshifts only later, in his [1963a].[6] However, Popper's ban on 'conventionalist stratagems' is in some respects too strong, in others too weak. It is too *strong,* for, according to Popper, a new version of a progressive programme *never* adopts a content-decreasing stratagem to absorb an anomaly, it *never* says things like 'all bodies are Newtonian except for seventeen anomalous ones'. But since unexplained anomalies always abound, I allow such formulations; an explanation is a step forward (that is, 'scientific') if it explains at least *some* previous anomalies which were not explained 'scientifically' by its predecessor. As long as anomalies are regarded as genuine (though not necessarily urgent)

[1] Popper [1957a].   [2] Popper [1963a], pp. 242 ff.
[3] Popper [1963a], p. 38, n. 3.
[4] If the reader is in doubt about the authenticity of my reformulation of Popper's demarcation criterion, he should re-read the relevant parts of Popper [1934] with Musgrave [1968] as a guide. Musgrave wrote his [1968] against Bartley who, in his [1968], mistakenly attributed to Popper the demarcation criterion of naive falsificationism, as formulated *above,* p. 25.
[5] In his [1934], Popper was primarily concerned with a ban on *surreptitious ad hoc* adjustments. Popper (Popper₁) demands that the design of a potentially negative crucial experiment must be presented together with the theory, and then the verdict of the experimental jury humbly accepted. It follows that conventionalist stratagems, which *after* the verdict give a retrospective twist to the original theory in order to escape the verdict, are *eo ipso* ruled out. But if we admit the refutation and *then* reformulate the theory with the help of an *ad hoc* stratagem, we may admit it as a '*new*' theory; and if it is testable, then Popper₁ accepts it for new criticism: 'Whenever we find that a system has been rescued by a conventionalist stratagem, we shall test it afresh, and reject it, as circumstances may require' (Popper [1934], section 20).
[6] For details, cf. volume 2, chapter 8, especially, pp. 179–80.

problems, it does not matter much whether we dramatize them as 'refutations' or de-dramatize them as 'exceptions': the difference *then* is only a linguistic one. (This degree of tolerance of *ad hoc* stratagems allows us to progress even on inconsistent foundations. Problemshifts may then be progressive in spite of inconsistencies.[1]) However, Popper's ban on content-decreasing stratagems is also too *weak*: it cannot deal for instance, with the 'tacking paradox',[2] and does not ban *ad hoc₃* stratagems.[3] These can be eliminated only by the requirement that *the auxiliary hypotheses should be formed in accordance with the positive heuristic of a genuine research programme.* This new requirement brings us to the problem of *continuity in science.*

The problem of *continuity* in science was raised by Popper and his followers long ago. When I proposed my theory of growth based on the idea of competing research programmes, I again followed, and tried to improve, Popperian tradition. Popper himself, in his [1934], had already stressed the heuristic importance of 'influential metaphysics',[4] and was regarded by some members of the Vienna Circle as a champion of dangerous metaphysics.[5] When his interest in the role of metaphysics revived in the 1950s, he wrote a most interesting 'Metaphysical Epilogue' about 'metaphysical research programmes' to his *Postscript: After Twenty Years* – in galleys since 1957.[6] But Popper

---

[1] Cf. *above*, pp. 57 ff. This tolerance is rarely, if ever, found in textbooks of scientific method.

[2] Cf. *above*, p. 46.   [3] Cf. *above*, p. 88, n. 2.

[4] Cf. e.g. his [1934], end of section 4; also cf. his [1968c], p. 93. One should remember that such importance was denied to metaphysics by Comte and Duhem. The people who did most to reverse the anti-metaphysical tide in the philosophy and the historiography of science were Burtt, Popper and Koyré.

[5] Carnap and Hempel tried, in their reviews of the book, to defend Popper against this charge (cf. Carnap [1935] and Hempel [1937]). Hempel wrote: '[Popper] stresses strongly certain features of his approach which are common with the approach of somewhat metaphysically oriented thinkers. It is to be hoped that this valuable work will not be misinterpreted as if it meant to allow for a new, perhaps even logically defensible, metaphysics.'

[6] A passage of this *Postscript* is worth quoting here: 'Atomism is an...excellent example of a non-testable metaphysical theory whose influence upon science exceeded that of many testable theories...The latest and greatest so far was the programme of Faraday, Maxwell, Einstein, de Broglie, and Schrödinger, of conceiving the world ...in terms of continuous fields...Each of these metaphysical theories functioned, long before it became testable, as a programme for science. It indicated the direction in which satisfactory explanatory theories of science may be found, and it made possible something like an appraisal of the depth of a theory. In biology, the theory of evolution, the theory of the cell, and the theory of bacterial infection, have all played similar parts, at least for a time. In psychology, sensualism, atomism (that is, the theory that all experiences are composed of last elements, such as, for example, sense data) and psycho-analysis should be mentioned as metaphysical research programmes...Even purely existential assertions have sometimes proved suggestive and even fruitful in the history of science even if they never became part of it. Indeed, few metaphysical theories exerted a greater influence upon the development of science than the purely metaphysical one: "There exists a substance which can turn base metals into gold (that is, a philosopher's stone)", although it is non-falsifiable, was never verified, and is now believed by nobody.'

associated tenacity not with *methodological irrefutability* but rather with *syntactical irrefutability*. By 'metaphysics' he meant syntactically specifiable statements like 'all–some' statements and purely existential statements. No basic statements could conflict with them because of their logical form. For instance, 'for all metals there is a solvent' would, in this sense, be 'metaphysical', while Newton's theory of gravitation, taken in isolation, would not be.[1] Popper, in the 1950s, also raised the problem of how to criticize metaphysical theories and suggested solutions.[2] Agassi and Watkins published several interesting papers on the role of this sort of 'metaphysics' in science, which all connected 'metaphysics' with the continuity of scientific progress.[3] My treatment differs from theirs first because I go much further than they in blurring the demarcation between [Popper's] 'science' and [Popper's] 'metaphysics': I do not even use the term 'metaphysical' any more. I only talk about *scientific* research programmes whose hard core is irrefutable not necessarily because of syntactical but possibly because of methodological reasons which have nothing to do with logical form. Secondly, separating sharply the *descriptive problem* of the psychologico-historical role of metaphysics from the *normative problem* of how to distinguish progressive from degenerating research programmes, I elaborate the latter problem further than they had done.

Finally, I should like to discuss the '*Duhem–Quine thesis*', and its relation to falsificationism.[4]

According to the 'Duhem–Quine thesis', given sufficient imagination, any theory (whether consisting of one proposition or of a finite conjunction of many) can be permanently saved from 'refutation' by some suitable adjustment in the background knowledge in which it is embedded. As Quine put it: 'Any statement can be held true come what may, if we make drastic enough adjustments elsewhere in the system...Conversely, by the same token, no statement is immune to revision.'[5] Moreover, the 'system' is nothing less than 'the whole of science'. 'A recalcitrant experience can be accommodated by any of various alternative reëvaluations in various alternative quarters of the total system [including the possibility of reëvaluating the recalcitrant experience itself].'[6]

This thesis has two very different interpretations. In its *weak interpretation* it only asserts the impossibility of a direct experimental hit

---

[1] Cf. especially Popper [1934], section 66. In the 1959 edition he added a clarifying footnote (n. *2) in order to stress that in *metaphysical* 'all–some' statements the existential quantifier must be interpreted as 'unbounded'; but of course, he had made this absolutely clear already in section 15 of the original text.

[2] Cf. especially his [1958], pp. 198–9.

[3] Cf. Watkins [1957] and [1958] and Agassi [1962] and [1964].

[4] This concluding part of the *Appendix* was added in the press.

[5] Quine [1953], chapter II.

[6] *Ibid.* The clause in the square brackets is mine.

on a narrowly specified theoretical target and the logical possibility of shaping science in indefinitely many different ways. The weak interpretation hits only dogmatic, not methodological, falsificationism: it only denies the possibility of a *disproof* of any *separate* component of a theoretical system.

In its *strong interpretation* the Duhem–Quine thesis excludes any *rational* selection rule among the alternatives; this version is inconsistent with all forms of methodological falsificationism. The two interpretations have not been clearly separated, although the difference is methodologically vital. Duhem seems to have held only the weak interpretation: for him the selection is a matter of 'sagacity': we must always make the right choices in order to get nearer to 'natural classification'.[1] On the other hand, Quine, in the tradition of the American pragmatism of James and Lewis, seems to hold a position very near to the strong interpretation.[2]

Let us now have a closer look at the weak Duhem–Quine thesis. Let us take a 'recalcitrant experience' expressed in an 'observation statement' O' which is inconsistent with a conjunction of theoretical (and 'observational') statements $h_1, h_2 \ldots h_n, I_1, I_2 \ldots I_n$, where h_i are theories and I_i the corresponding initial conditions. In the 'deductive model', $h_1 \ldots h_n, I_1 \ldots I_n$ logically imply O; but O' is observed which implies *not-O*. Let us also assume that the premisses are independent and are all necessary for deducing O.

In this case we may restore consistency by altering *any* of the sentences in our deductive model. For instance, let h_1 be: 'whenever a thread is loaded with a weight exceeding that which characterizes the tensile strength of the thread, then it will break'; let h_2 be: 'the weight characteristic for this thread is 1 *lb.*'; let h_3 be: 'the weight put on this thread was 2 *lbs*'. Let, finally, O be: 'an iron weight of 2 *lbs* was put on the thread located in the space-time position P and it did not break'. One may solve the problem in many ways. To give a few examples: (1) We reject h_1; we replace the expression 'is loaded with a weight' by 'is pulled by a force'; we introduce a new initial condition: there was a hidden magnet (or hitherto unknown force) located in the laboratory ceiling. (2) We reject h_2; we propose that the tensile strength *does* depend on how moist threads are; the tensile strength of the actual

---

[1] An experiment, for Duhem, can never *alone* condemn an isolated theory (such as the hard core of a research programme): for such 'condemnation' we *also* need 'common sense', 'sagacity', and, indeed, good metaphysical instinct which leads us towards (or *to*) 'a certain supremely eminent order'. (See the end of the *Appendix* of the second edition of his [1906].)

[2] Quine speaks of statements having 'varying distances from a sensory periphery', and thus more or less exposed to change. But both the sensory periphery and the metric are hard to define. According to Quine 'the considerations which guide [man] in warping his scientific heritage to fit his continuing sensory peripheries are, where rational, pragmatic' (Quine [1953]). But 'pragmatism' for Quine, as for James or LeRoy, is only psychological comfort; and I find it irrational to call this 'rational'.

thread, since it got moist, was 2 *lbs*. (3) We reject h_3; the weight was only 1 *lb*; the scales went wrong. (4) We reject O; the thread *did* break; it was only *observed* not to break, but the professor who proposed h_1 & h_2 & h_3 was a well-known bourgeois liberal and his revolutionary laboratory assistants consistently *saw* his hypotheses refuted when in fact they were confirmed. (5) We reject h_3; the thread was not a 'thread', but a 'superthread', and 'superthreads' never break.[1] We could go on indefinitely. Indeed, there are infinitely many possibilities of how to replace – given sufficient imagination – any of the premisses (*in the deductive model*) by invoking a change in some *distant* part of our total knowledge (*outside the deductive model*) and thereby restore consistency.

Can we formulate this trivial observation by saying that '*each test is a challenge to the whole of our knowledge*'? I do not see any reason why not. The resistance of some falsificationists to this 'holistic dogma of the "global" character of all tests'[2] is due only to a semantic conflation of two different notions of 'test' (or 'challenge') which a recalcitrant experimental result presents to our knowledge.

The Popperian interpretation of a 'test' (or 'challenge') is that the result (O) contradicts ('challenges') a finite, well-specified conjunction of premisses (T): O & T cannot be true. But no proponent of the Duhem–Quine argument would deny this point.

The Quinean interpretation of 'test' (or 'challenge') is that the *replacement* of O & T may invoke some change also outside O and T. The successor to O & T may be inconsistent with some H in some distant part of knowledge. But no Popperian would deny this point.

The conflation of the two notions of testing led to some misunderstandings and logical blunders. Some people felt intuitively that the *modus tollens* from refutation may 'hit' very distant premisses in our total knowledge and therefore were trapped in the idea that the '*ceteris paribus* clause' is a premiss which is joined *conjunctively* with the obvious premisses. But this 'hit' is achieved not by *modus tollens* but as a result of our subsequent replacement of our original deductive model.[3]

Thus 'Quine's weak thesis' trivially holds. But 'Quine's strong thesis' will be strenuously opposed, both by the naive and the sophisticated falsificationist.

The naive falsificationist insists that if we have an inconsistent set

---

[1] For such 'concept-narrowing defences' and 'concept-stretching refutations', cf. my [1963–4].

[2] Popper [1963a], chapter 10, section xvi.

[3] The *locus classicus* of this confusion is Canfield's and Lehrer's wrongheaded criticism of Popper in their [1961]; Stegmüller followed them into the logical morass ([1966], p. 7). Coffa contributed to the clarification of the issue ([1968]).

Unfortunately, my own phraseology in this paper in places suggests that the '*ceteris paribus* clause' must be an independent premiss in the theory under test. My attention was drawn to this easily repairable defect by Colin Howson.

of scientific statements, we first must select from among them (1) a theory under test (to serve as a *nut*); then we must select (2) an accepted basic statement (to serve as a *hammer*) and the rest will be uncontested background knowledge (to provide an *anvil*). And in order to put teeth into this position, we must offer a method of 'hardening' the 'hammer' and the 'anvil' in order to enable us to crack the 'nut', and thus perform a 'negative crucial experiment'. But naive 'guessing' of this division is too arbitrary, it does not give us any serious hardening. (Grünbaum, on the other hand, applies Bayes's theorem in order to show that, at least in some sense, the 'hammer' and the 'anvil' have high posterior probabilities and therefore are 'hard' enough to be used as a nutcracker.[1])

The sophisticated falsificationist allows *any* part of the body of science to be replaced *but* only on the condition that it is replaced in a 'progressive' way, so that the replacement successfully anticipates novel facts. In his rational reconstruction of falsification, 'negative crucial experiments' play no role. He sees nothing wrong with a group of brilliant scientists conspiring to pack everything they can into their favourite research programme ('conceptual framework', if you wish) with a sacred hard core. As long as their genius – and luck – enables them to expand their programme '*progressively*', while sticking to its hard core, they are allowed to do it. And if a genius comes determined to *replace* ('progressively') a most uncontested and corroborated theory which he happens to dislike on philosophical, aesthetic or personal grounds, good luck to him. If two teams, pursuing rival research programmes, compete, the one with more creative talent is likely to succeed – unless God punishes them with an extreme lack of empirical success. The direction of science is determined primarily by human creative imagination and not by the universe of facts which surrounds us. Creative imagination is likely to find corroborating novel evidence even for the most 'absurd' programme, if the search has sufficient drive.[2] This look-out for *new confirming evidence* is perfectly permis-

---

[1] Grünbaum previously took a position which was one of dogmatic falsificationism and claimed, by reference to his thought-provoking and challenging case-studies in physical geometry, that we *can* ascertain the falsity of *some* scientific hypotheses (e.g. Grünbaum [1959*b*] and [1960]). His [1959*b*] was followed by Feyerabend's [1961], in which Feyerabend argued that 'refutations are final only as long as ingenious and nontrivial alternative explanations of the evidence are missing'. In his [1966], Grünbaum modified his position, and then, in response to criticisms by Mary Hesse (Hesse [1968]) and others, he qualified it further: 'At least in some cases, we can ascertain the falsity of a component hypothesis to all scientific intents and purposes, although we cannot falsify it beyond any and all possibility of subsequent rehabilitation' (Grünbaum [1969], p. 1092).

[2] A typical such example is Newton's principle of gravitational attraction according to which bodies attract each other instantly from immense distances. Huyghens described this idea as 'absurd', Leibnitz as 'occult', and the best scientists of the age 'wondered how [Newton] could have given himself all the trouble of making such a number of investigations and difficult calculations that had no other foundation than this very principle' (cf. Koyré [1965], pp. 117–18). I had argued earlier that it is not

sible. Scientists dream up phantasies and then pursue a highly selective hunt for new facts which fit these phantasies. This process may be described as 'science creating its own universe' (as long as one remembers that 'creating' here is used in a provocative, idiosyncratic sense). A brilliant school of scholars (backed by a rich society to finance a few well-planned tests) might succeed in pushing any fantastic programme ahead, or, alternatively, if so inclined, in overthrowing any arbitrarily chosen pillar of 'established knowledge'.

The *dogmatic* falsificationist will throw up his hands in horror at this approach. He will see the spectre of Bellarmino's instrumentalism arising from the rubble under which Newtonian success of 'proven science' had buried it. He will accuse the sophisticated falsificationist of building arbitrary Procrustean pigeon hole systems and forcing the facts into them. He may even brand it as a revival of the unholy irrationalist alliance of James's crude pragmatism and of Bergson's voluntarism, triumphantly vanquished by Russell and Stebbing.[1] But our sophisticated falsificationism combines 'instrumentalism' (or 'conventionalism') with a strong empiricist requirement, which neither medieval 'saviours of phenomena' like Bellarmino, nor pragmatists like Quine and Bergsonians like Le Roy, had appreciated: the Leibnitz–Whewell–Popper requirement that *the – well planned – building of pigeon holes must proceed much faster than the recording of facts which are to be housed in them.* As long as this requirement is met, it does not matter whether we stress the 'instrumental' aspect of imaginative research programmes for finding novel facts and for making trustworthy predictions, or whether we stress the putative growing Popperian 'verisimilitude' (that is, the estimated difference between the truth-content and falsity-content) of their successive versions.[2] Sophisticated falsificationism thus combines the best elements of voluntarism, pragmatism and of the realist theories of empirical growth.

The sophisticated falsificationist sides neither with Galileo nor with Cardinal Bellarmino. He does not side with Galileo, for he claims that our basic theories may all be equally absurd and unverisimilar for the divine mind; and he does not side with Bellarmino, unless the Cardinal were to agree that scientific theories may yet lead, in the long run, to

so that theoretical progress is the merit of the theoretician but empirical success is *merely* a matter of luck. If the theoretician is *more* imaginative, it is likelier that his theoretical programme will achieve at least *some* empirical success. Cf. volume 2, chapter 8, pp. 178–81.

[1] Cf. Russell [1914], Russell [1946] and Stebbing [1914]. Russell, a justificationist, despised conventionalism: 'As will has gone up in the scale, knowledge has gone down. This is the most notable change that has come over the temper of philosophy in our age. It was prepared by Rousseau and Kant' ([1946], p. 787). Popper, of course, got some of his inspiration from Kant and Bergson. (Cf. his [1934], sections 2 and 4.)

[2] For '*verisimilitude*' cf. Popper [1963a], chapter 10 and *below* the next footnote; for '*trustworthiness*' cf. this volume chapter 3, and volume 2 chapter 8.

ever more true and ever fewer false consequences and, *in this strictly technical sense*, may have increasing 'verisimilitude'.[1]

[1] '*Verisimilitude*' has two distinct meanings which must not be conflated. First, it may be used to mean intuitive truthlikeness of the theory; in this sense, in my view, all scientific theories created by the human mind are equally unverisimilar and 'occult'. Secondly, it may be used to mean a quasi-measure-theoretical difference between the true and false consequences of a theory which we can never know but certainly may guess. It was Popper who used 'verisimilitude' as a technical term to denote this sort of difference ([1963], chapter 10). But his claim that this explication corresponds closely to *the* original meaning is mistaken and misleading. In the original pre-popperian usage 'verisimilitude' could mean either *intuitive* truthlikeness or a naive proto-version of Popper's *empirical* truthlikeness. Popper gives interesting quotations for the latter ([1963a], pp. 399 ff) but none for the former. But Bellarmino might have agreed that Copernican theory had high 'verisimilitude' in Popper's technical sense but not that it had verisimilitude in the first, intuitive sense. Most 'instrumentalists' are 'realists' in the sense that they agree that the [Popperian] 'verisimilitude' of scientific theories is likely to be growing; but they are not 'realists' in the sense that they would agree that, for instance, the Einsteinian field approach is *intuitively* closer to the Blueprint of the Universe than the Newtonian action at a distance. *The 'aim of science' may then be increasing Popperian 'verisimilitude', but does not have to be also increasing classical verisimilitude.* The latter, as Popper himself said, is, unlike the former, a 'dangerously vague and metaphysical' idea ([1963a], p. 231).

Popper's 'empirical verisimilitude' in a sense rehabilitates the idea of *cumulative growth* in science. But the driving force of cumulative growth in 'empirical verisimilitude' is revolutionary conflict in 'intuitive verisimilitude'.

When Popper was writing his 'Truth, rationality and the growth of knowledge', I had an uneasy feeling about his identification of the two concepts of verisimilitude. Indeed, it was I who asked him: 'Can we really speak about *better* correspondence? Are there such things as *degrees* of truth? Is it not dangerously misleading to talk as if Tarskian truth were located somewhere in a kind of metrical or at least topological space so that we can sensibly say of two theories – say an earlier theory t_1 and a later theory t_2, that t_2 has superseded t_1, or progressed beyond t_1, by approaching more closely to the truth than t_1?' (Popper [1963a], p. 232). Popper rejected my vague misgivings. He felt – rightly – that he was proposing a very important new idea. But he was mistaken in believing that his new, technical conception of 'verisimilitude' completely absorbed the problems centred on the old *intuitive* 'verisimilitude'. Kuhn says: 'To say, for example, of a field theory that it "approaches more closely to the truth" than an older matter-and-force theory should mean, *unless words are being oddly used*, that the ultimate constituents of nature are more like fields than like matter and force' (Kuhn [1970b], p. 265, my italics). Indeed, Kuhn is right, except that words are *normally* 'oddly used'. I hope that this note may contribute to the clarification of the problem involved. (* For some fundamental difficulties with Popper's 'technical' conception of verisimilitude see, e.g. Miller [1975]. – (*Eds*).)

2

History of science and its rational reconstructions*

INTRODUCTION

'Philosophy of science without history of science is empty; history of science without philosophy of science is blind.' Taking its cue from this paraphrase of Kant's famous dictum, this paper intends to explain *how* the historiography of science should learn from the philosophy of science and *vice versa*. It will be argued that (*a*) philosophy of science provides normative methodologies in terms of which the historian reconstructs 'internal history' and thereby provides a rational explanation of the growth of objective knowledge; (*b*) two competing methodologies can be evaluated with the help of (normatively interpreted) history; (*c*) any rational reconstruction of history needs to be supplemented by an empirical (socio-psychological) 'external history'.

The vital demarcation between normative–internal and empirical–external is different for each methodology. Jointly, internal and external historiographical theories determine to a very large extent the choice of problems for the historian. But some of external history's most crucial problems can be formulated only in terms of one's methodology; thus internal history, so defined, is primary, and external history only secondary. Indeed, in view of the autonomy of internal (but not of external) history, external history is irrelevant for the understanding of science.[1]

---

* This paper was first published as Lakatos [1971a]. His own acknowledgment there reads 'Earlier versions were read and criticized by Colin Howson, Alan Musgrave, John Watkins, Elie Zahar and especially John Worrall.' The paper appeared in 1971 together with some critical remarks (by Feigl, Hall, Koertge and Kuhn) and a 'Reply to Critics' by Lakatos. These are not republished here. (*Eds.*)

[1] 'Internal history' is usually defined as intellectual history; 'external history' as social history (cf. e.g. Kuhn [1968]). My unorthodox, new demarcation between 'internal' and 'external' history constitutes a considerable problemshift and may sound dogmatic. But my definitions form the hard core of a historiographical research programme; their evaluation is part and parcel of the evaluation of the fertility of the whole programme.

I   RIVAL METHODOLOGIES
OF SCIENCE; RATIONAL RECONSTRUCTIONS AS
GUIDES TO HISTORY

There are several methodologies afloat in contemporary philosophy of science; but they are all very different from what used to be understood by 'methodology' in the seventeenth or even eighteenth century. Then it was hoped that methodology would provide scientists with a mechanical book of rules for solving problems. This hope has now been given up: modern methodologies or 'logics of discovery' consist merely of a set of (possibly not even tightly knit, let alone mechanical) rules for the *appraisal* of ready, articulated theories.[1] Often these rules, or systems of appraisal, also serve as 'theories of scientific rationality', 'demarcation criteria' or 'definitions of science'.[2] Outside the legislative domain of these normative rules there is, of course, an empirical psychology and sociology of discovery.

I shall now sketch four different 'logics of discovery'. Each will be characterized by rules governing the (scientific) *acceptance* and *rejection* of theories or research programmes.[3] These rules have a double function. Firstly, they function as *a code of scientific honesty* whose violation is intolerable; secondly, as hard cores of (*normative*) *historio-graphical research programmes*. It is their second function on which I should like to concentrate.

(a) Inductivism

One of the most influential methodologies of science has been in-ductivism. According to inductivism only those propositions can be accepted into the body of science which either describe hard facts or are infallible inductive generalizations from them.[4] When the inductivist *accepts* a scientific proposition, he accepts it as provenly true; he *rejects* it if it is not. His scientific rigour is strict: a proposition must be either proven from facts, or – deductively or inductively – derived from other propositions already proven.

Each methodology has its specific epistemological and logical prob-lems. For example, inductivism has to establish with certainty the truth of 'factual' ('basic') propositions and the validity of inductive

---

[1] This is an all-important shift in the problem of normative philosophy of science. The term 'normative' no longer means rules for arriving at solutions, but merely directions for the appraisal of solutions already there. Thus methodology is separated from *heuristics*, rather as value judgments are from 'ought' statements. (I owe this analogy to John Watkins.)

[2] This profusion of synonyms has proved to be rather confusing.

[3] The epistemological significance of scientific 'acceptance' and 'rejection' is, as we shall see, far from being the same in the four methodologies to be discussed.

[4] '*Neo*-inductivism' demands only (provably) highly probable generalizations. In what follows I shall only discuss classical inductivism; but the watered down neo-inductivist variant can be similarly dealt with.

inferences. Some philosophers get so preoccupied with their epistemological and logical problems that they never get to the point of
becoming interested in actual history; if actual history does not fit their
standards they may even have the temerity to propose that we start
the whole business of science anew. Some others take some crude
solution of these logical and epistemological problems for granted and
devote themselves to a rational reconstruction of history without being
aware of the logico-epistemological weakness (or, even, untenability)
of their methodology.[1]

Inductivist criticism is primarily sceptical: it consists in showing that
a proposition is unproven, that is, pseudoscientific, rather than in
showing that it is false.[2] When the inductivist historian writes the
prehistory of a scientific discipline, he may draw heavily upon such
criticisms. And he often explains the early dark age – when people
were engrossed by 'unproven ideas' – with the help of some 'external'
explanation, like the socio-psychological theory of the retarding influence of the Catholic Church.

The inductivist historian recognizes only two sorts of *genuine scientific discoveries: hard factual propositions* and inductive *generalizations.*
These and only these constitute the backbone of his *internal history.*
When writing history, he looks out for them – finding them is quite
a problem. Only when he finds them, can he start the construction of
his beautiful pyramids. Revolutions consist in unmasking (irrational)
errors which then are exiled from the history of science into the
history of pseudoscience, into the history of mere beliefs: genuine
scientific progress starts with the latest scientific revolution in any given
field.

Each internal historiography has its characteristic victorious paradigms.[3] The main paradigms of inductivist historiography were
Kepler's generalizations from Tycho Brahe's careful observations;
Newton's discovery of his law of gravitation by, in turn, inductively
generalizing Kepler's 'phenomena' of planetary motion; and Ampère's discovery of his law of electrodynamics by inductively generalizing his observations of electric currents. Modern chemistry too is
taken by some inductivists as having really started with Lavoisier's
experiments and his 'true explanations' of them.

But the inductivist historian cannot offer a *rational* 'internal' explanation for *why* certain facts rather than others were selected in the
first instance. For him this is a *non-rational, empirical, external* problem.
Inductivism as an 'internal' theory of rationality is compatible with
many different supplementary empirical or external theories of
problem-choice. It is, for instance, compatible with the vulgar-Marxist

[1] Cf. *below*, p. 120.
[2] For a detailed discussion of inductivist (and, in general, justificationist) criticism
cf. my [1970*b*].
[3] I am now using the term 'paradigm' in its pre-Kuhnian sense.

view that problem-choice is determined by social needs;[1] indeed, some vulgar-Marxists identify major phases in the history of science with the major phases of economic development.[2] But choice of facts need not be determined by social factors; it may be determined by extra-scientific intellectual influences. And inductivism is equally compatible with the 'external' theory that the choice of problems is primarily determined by inborn, or by arbitrarily chosen (or traditional) theo-retical (or 'metaphysical') frameworks.

There is a radical brand of inductivism which condemns all external influences, whether intellectual, psychological or sociological, as creating impermissible bias: radical inductivists allow only a [random] selection by the empty mind. Radical inductivism is, in turn, a special kind of *radical internalism*. According to the latter once one establishes the existence of some external influence on the acceptance of a scientific theory (or factual proposition) one must withdraw one's acceptance: proof of external influence means invalidation;[3] but since external influences always exist, radical internalism is utopian, and, as a theory of rationality, self-destructive.[4]

When the radical inductivist historian faces the problem of why some great scientists thought highly of metaphysics and, indeed, why they thought that their discoveries were great for reasons which, in the light of inductivism, look very odd, he will refer these problems of 'false consciousness' to psychopathology, that is, to external history.

(b) Conventionalism

Conventionalism allows for the building of any system of pigeon holes which organizes facts into some coherent whole. The conventionalist decides to keep the centre of such a pigeonhole system intact as long as possible: when difficulties arise through an invasion of anomalies, he only changes and complicates the peripheral arrangements. But the conventionalist does not regard any pigeonhole system as provenly true, but only as 'true by convention' (or possibly even as neither true nor false). In *revolutionary* brands of conventionalism one does not have to adhere forever to a given pigeonhole system: one may abandon it if it becomes unbearably clumsy and if a simpler one is offered to replace it.[5] This version of conventionalism is epistemologically, and

---

[1] This compatibility was pointed out by Agassi on pp. 23–7 of his [1963]. But did he not point out the analogous compatibility within his own falsificationist historio-graphy; cf. *below*, pp. 109–10.

[2] Cf. e.g. Bernal [1965], p. 377.

[3] Some logical positivists belonged to this set: one recalls Hempel's horror at Popper's casual praise of certain external metaphysical influences upon science (Hempel [1937]).

[4] When German obscurantists scoff at 'positivism', they frequently mean radical internalism, and in particular, radical inductivism.

[5] For what I here call *revolutionary conventionalism*, see chapter 1, pp. 21–2 and 100–2.

especially logically, much simpler than inductivism: it is in no need of valid inductive inferences. Genuine *progress* of science is cumulative and takes place on the ground level of 'proven' facts;[1] the *changes* on the theoretical level are merely instrumental. Theoretical 'progress' is only in convenience ('simplicity'), and not in truth-content.[2] One may, of course, introduce revolutionary conventionalism also at the level of 'factual' propositions, in which case one would accept 'factual' propositions by decision rather than by experimental 'proofs'. But then, if the conventionalist is to retain the idea that the growth of 'factual' science has anything to do with objective, factual truth, he must devise some metaphysical principle which he then has to superimpose on his rules for the game of science.[3] If he does not, he cannot escape scepticism or, at least, some radical form of instrumentalism.

(It is important to clarify the *relation between conventionalism and instrumentalism*. Conventionalism rests on the recognition that false assumptions may have true consequences; therefore false theories may have great predictive power. Conventionalists had to face the problem of comparing rival false theories. Most of them conflated truth with its signs and found themselves holding some version of the pragmatic theory of truth. It was Popper's theory of truth-content, verisimilitude and corroboration which finally laid down the basis of a philosophically flawless version of conventionalism. On the other hand some conventionalists did not have sufficient logical education to realize that some propositions may be true whilst being unproven; and others false whilst having true consequences, and also some which are both false and approximately true. These people opted for 'instrumentalism': they came to regard theories as neither true nor false but merely as 'instruments' for prediction. Conventionalism, as here defined, is a philosophically sound position; instrumentalism is a degenerate version of it, based on a mere philosophical muddle caused by lack of elementary logical competence.)

[1] I mainly discuss here only one version of revolutionary conventionalism, the one which Agassi, in his [1966], called 'unsophisticated': the one which assumes that factual propositions – unlike pigeonhole systems – can be 'proven'. (Duhem, for instance, draws no clear distinction between facts and factual propositions.)

[2] It is important to note that most conventionalists are reluctant to give up inductive generalizations. They distinguish between the '*floor of facts*', the '*floor of laws*' (i.e. inductive generalizations from 'facts') and the '*floor of theories*' (or of pigeonhole systems) which classify, conveniently, both facts and inductive laws. (Whewell, the conservative conventionalist, and Duhem, the revolutionary conventionalist, differ less than most people imagine.)

[3] One may call such metaphysical principles 'inductive principles'. For an 'inductive principle' which – roughly speaking – makes Popper's 'degree of corroboration' (a conventionalist appraisal) the measure of Popper's verisimilitude (truth-content minus falsity-content) see volume 2, chapter 8, pp. 181–93 and this volume, chapter 3, §2. (Another widely held 'inductive principle' may be formulated like this: 'What the group of trained – or up-to-date, or suitably purged – scientists decide to *accept* as "true", is true.')

Revolutionary conventionalism was born as the Bergsonians' philosophy of science: free will and creativity were the slogans. The code of scientific honour of the conventionalist is less rigorous than that of the inductivist: it puts no ban on unproven speculation, and allows a pigeonhole system to be built around *any* fancy idea. Moreover, conventionalism does not brand discarded systems as unscientific: the conventionalist sees much more of the actual history of science as rational ('internal') than does the inductivist.

For the conventionalist historian, major discoveries are primarily inventions of new and simpler pigeonhole systems. Therefore he constantly compares for simplicity: the complications of pigeonhole systems and their revolutionary replacement by simpler ones constitute the backbone of his internal history.

The paradigmatic case of a scientific revolution for the conventionalist has been the Copernican revolution.[1] Efforts have been made to show that Lavoisier's and Einstein's revolutions too were replacements of clumsy theories by simple ones.

Conventionalist historiography cannot offer a *rational* explanation of why certain facts were selected in the first instance or of why certain particular pigeonhole systems were tried rather than others at a stage when their relative merits were yet unclear. Thus conventionalism, like inductivism, is compatible with various supplementary empirical–'externalist' programmes.

Finally, the conventionalist historian, like his inductivist colleague, frequently encounters the problem of 'false consciousness'. According to conventionalism for example, it is a 'matter of fact' that great scientists arrive at their theories by flights of their imaginations. Why then do they often claim that they derived their theories from facts? The conventionalist's rational reconstruction often differs from the great scientists' own reconstruction – the conventionalist historian relegates these problems of false consciousness to the externalist.[2]

---

[1] Most historical accounts of the Copernican revolution are written from the conventionalist point of view. Few claimed that Copernicus' theory was an 'inductive generalization' from some 'factual discovery'; or that it was proposed as a bold theory to replace the Ptolemaic theory which had been 'refuted' by some celebrated 'crucial' experiment.

For a further discussion of the historiography of the Copernican revolution, cf. chapter 4, *below*.

[2] For example, for non-inductivist historians Newton's '*Hypotheses non fingo*' represents a major problem. Duhem, who unlike most historians did not over-indulge in Newton-worship, dismissed Newton's inductivist methodology as logical nonsense; but Koyré whose many strong points did not include logic, devoted long chapters to the 'hidden depths' of Newton's muddle.

(c) Methodological falsificationism

Contemporary falsificationism arose as a logico-epistemological criticism of inductivism and of Duhemian conventionalism. Inductivism was criticized on the grounds that its two basic assumptions, namely, that factual propositions can be 'derived' from facts and that there can be valid inductive (content-increasing) inferences, are themselves unproven and even demonstrably false. Duhem was criticized on the grounds that comparison of intuitive simplicity can only be a matter for subjective taste and that it is so ambiguous that no hard-hitting criticism can be based on it. Popper, in his *Logik der Forschung*, proposed a new 'falsificationist' methodology.[1] This methodology is another brand of revolutionary conventionalism: the main difference is that it allows factual, spatio-temporally singular 'basic statements', rather than spatio-temporally universal theories, to be accepted by convention. In the code of honour of the falsificationist a theory is scientific only if it can be *made* to conflict with a basic statement; and a theory must be eliminated if it conflicts with an accepted basic statement. Popper also indicated a further condition that a theory must satisfy in order to qualify as scientific: it must predict facts which are *novel*, that is, unexpected in the light of previous knowledge. Thus, it is against Popper's code of scientific honour to propose unfalsifiable theories or '*ad hoc*' hypotheses (which imply no *novel* empirical predictions) – just as it is against the (classical) inductivist code of scientific honour to propose unproven ones.

The great attraction of Popperian methodology lies in its clarity and force. Popper's deductive model of scientific criticism contains empirically falsifiable spatio-temporally universal propositions, initial conditions and their consequences. The weapon of criticism is the *modus tollens*: neither inductive logic nor intuitive simplicity complicate the picture.[2]

(Falsificationism, though logically impeccable, has epistemological difficulties of its own. In its 'dogmatic' proto-version it assumes the provability of propositions from facts and thus the disprovability of theories – a false assumption.[3] In its Popperian 'conventionalist' version it needs some (extra-methodological) 'inductive principle' to lend epistemological weight to its decisions to accept 'basic' statements, and in general to connect its rules of the scientific game with verisimilitude.[4])

The Popperian historian looks for great, 'bold', falsifiable theories

---

[1] In this paper I use this term to stand exclusively for one version of falsificationism, namely for '*naïve methodological falsificationism*', as defined in chapter 1, pp. 10–31.

[2] Since in his methodology the *concept* of intuitive simplicity has no place, Popper was able to use the term 'simplicity' for 'degree of falsifiability'. But there is more to simplicity than this: cf. chapter 1, 46 ff.

[3] For a discussion cf. chapter 1, especially pp. 16–17.

[4] For further discussion cf. *below*, pp. 121–2.

and for great negative crucial experiments. These form the skeleton of his rational reconstruction. The Popperians' favourite paradigms of great falsifiable theories are Newton's and Maxwell's theories, the radiation formulas of Rayleigh, Jeans and Wien, and the Einsteinian revolution; their favourite paradigms for crucial experiments are the Michelson–Morley experiment, Eddington's eclipse experiment, and the experiments of Lummer and Pringsheim. It was Agassi who tried to turn this naive falsificationism into a systematic historiographical research programme.[1] In particular he predicted (or 'postdicted', if you wish) that behind each great experimental discovery lies a theory which the discovery contradicted; the importance of a factual discovery is to be measured by the importance of the theory refuted by it. Agassi seems to accept at face value the value judgments of the scientific community concerning the importance of factual discoveries like Galvani's, Oersted's, Priestley's, Roentgen's and Hertz's; but he denies the 'myth' that they were chance discoveries (as the first four were said to be) or confirming instances (as Hertz first thought his discovery was).[2] Thus Agassi arrives at a bold prediction: all these five experiments were successful refutations – in some cases even *planned* refutations – of theories which he proposes to unearth, and, indeed, in most cases, claims to have unearthed.[3]

Popperian internal history, in turn, is readily supplemented by external theories of history. Thus Popper himself explained that (on the positive side) (1) the main *external* stimulus of scientific theories comes from unscientific 'metaphysics', and even from myths (this was later beautifully illustrated, mainly by Koyré); and that (on the negative side) (2) facts do *not* constitute such external stimulus – factual discoveries belong completely to internal history, emerging as refutations of some scientific theory, so that facts are only noticed if they conflict with some previous expectation. Both theses are cornerstones of Popper's *psychology* of discovery.[4] Feyerabend developed another interesting *psychological* thesis of Popper's, namely, that proliferation of rival theories may – *externally* – speed up *internal* Popperian falsification.[5]

---

[1] Agassi [1963].

[2] An experimental discovery is *a chance discovery in the objective sense* if it is neither a confirming nor a refuting instance of some theory in the objective body of knowledge of the time; it is *a chance discovery in the subjective sense* if it is made (or recognized) by the discoverer neither as a confirming nor as a refuting instance of some theory he personally had entertained at the time.

[3] Agassi [1963], pp. 64–74. * See also volume 2, chapter 9. (*Eds.*)

[4] Within the Popperian circle, it was Agassi and Watkins who particularly emphasized the importance of unfalsifiable or barely testable '*metaphysical*' theories in providing an *external* stimulus to later properly *scientific* developments. (Cf. Agassi [1964b] and Watkins [1958].) This idea, of course, is already there in Popper's [1934] and [1960b]. Cf. chapter 1, p. 95; but the new formulation of the difference between their approach and mine which I am going to give in this paper will, I hope, be much clearer.

[5] Popper occasionally – and Feyerabend systematically – stressed the catalytic (*external*) role of alternative theories in devising so-called 'crucial experiments'. But

But the external supplementary theories of falsificationism need not be restricted to purely intellectual influences. It has to be emphasized (*pace* Agassi) that falsificationism is no less compatible with a vulgar-Marxist view of what makes science progress than is inductivism. The only difference is that while for the latter Marxism might be invoked to explain the discovery of *facts*, for the former it might be invoked to explain the invention of *scientific theories*; while the choice of facts (that is, for the falsificationist, the choice of 'potential falsifiers') is primarily determined internally by the theories.

'False awareness' – 'false' from the point of view of *his* rationality theory – creates a problem for the falsificationist historian. For instance, why do some scientists believe that crucial experiments are positive and verifying rather than negative and falsifying? It was the falsificationist Popper who, in order to solve these problems, elaborated better than anybody else before him the cleavage between objective knowledge (in his 'third world') and its distorted reflections in individual minds.[1] Thus he opened up the way for my demarcation between internal and external history.

(d) Methodology of scientific research programmes

According to my methodology the great scientific achievements are research programmes which can be evaluated in terms of progressive and degenerating problemshifts; and scientific revolutions consist of one research programme superseding (overtaking in progress) another.[2] This methodology offers a new rational reconstruction of science. It is best presented by contrasting it with falsificationism and conventionalism, from both of which it borrows essential elements.

From conventionalism, this methodology borrows the licence rationally to accept by convention not only spatio-temporally singular 'factual statements' but also spatio-temporally universal theories: indeed, this becomes the most important clue to the continuity of scientific growth.[3] The basic unit of appraisal must be not an isolated theory or conjunction of theories but rather a '*research programme*', with a conventionally accepted (and thus by provisional decision 'irrefutable') '*hard core*' and with a '*positive heuristic*' which defines problems, outlines the construction of a belt of auxiliary hypotheses, foresees

---

alternatives are not merely catalysts, which can be later removed in the rational reconstruction, they are *necessary* parts of the falsifying process. Cf. Popper [1940] and Feyerabend [1965]; but cf. also chapter 1, especially p. 37, n. 1.

[1] Cf. Popper [1968a] and [1968b].

[2] The terms 'progressive' and 'degenerating problemshifts', 'research programmes' 'superseding' will be crudely defined in what follows – for more elaborate definitions see my [1968c], and especially this volume, chapter 1.

[3] Popper does not permit this: 'There is a vast difference between my views and conventionalism. I hold that what characterises the empirical method is just this: our conventions determine the acceptance of the *singular*, not of the *universal* statements' (Popper [1934], section 30).

anomalies and turns them victoriously into examples, all according to a preconceived plan. The scientist lists anomalies, but as long as his research programme sustains its momentum, he may freely put them aside. *It is primarily the positive heuristic of his programme, not the anomalies, which dictate the choice of his problems.*[1] Only when the driving force of the positive heuristic weakens, may more attention be given to anomalies. The methodology of research programmes can explain in this way *the high degree of autonomy of theoretical science*; the naive falsificationist's disconnected chains of conjectures and refutations cannot. What for Popper, Watkins and Agassi is *external*, influential metaphysics, here turns into the *internal* 'hard core' of a programme.[2]

The methodology of research programmes presents a very different picture of the game of science from the picture of the methodological falsificationist. The best opening gambit is not a falsifiable (and therefore consistent) hypothesis, but a research programme. Mere 'falsification' (in Popper's sense) must not imply rejection.[3] Mere 'falsifications' (that is, anomalies) are to be recorded but need not be acted upon. Popper's great negative crucial experiments disappear; 'crucial experiment' is an honorific title, which may, of course, be conferred on certain anomalies, but only *long after the event*, only when one programme has been defeated by another one. According to Popper, a crucial experiment is described by an accepted basic statement which is inconsistent with a theory – according to the methodology of scientific research programmes, no accepted basic statement *alone* entitles the scientist to reject a theory. Such a clash may present a problem (major or minor), but in no circumstance a 'victory'. Nature may shout *no*, but human ingenuity – contrary to Weyl and Popper[4] – may always be able to shout louder. With sufficient resourcefulness and some luck, any theory can be defended 'progressively' for a long time, even if it is false. The Popperian pattern of 'conjectures and refutations', that is the pattern of trial-by-hypothesis followed by error-shown-by-experiment, is to be abandoned: no experiment is crucial at the time – let alone before – it is performed (except, possibly, psychologically).

[1] The falsificationist hotly denies this: 'Learning from experience is learning from a refuting instance. The refuting instance then becomes a problematic instance' (Agassi [1964b], p. 201). In his [1969] Agassi attributed to Popper the statement that 'we learn from experience by refutations' (p. 169), and adds that according to Popper one can learn *only* from refutation but not from corroboration (p. 167). Feyerabend, even in his [1969b], says that '*negative instances suffice in science*'. But these remarks indicate a very one-sided theory of learning from experience. (Cf. chapter 1, p. 36, n. 2, and p. 38.)

[2] Duhem, as a staunch positivist within philosophy of science, would, no doubt, exclude most 'metaphysics' as unscientific and would not allow it to have any influence on science proper.

[3] Cf. volume 2, chapter 8, pp. 175–8, my [1968c], pp. 162–7, and this volume, pp. 31 ff and pp. 69 ff.

[4] Cf. Popper [1934], section 85.

It should be pointed out, however, that the methodology of scientific research programmes has more teeth than Duhem's conventionalism: instead of leaving it to Duhem's unarticulated common sense[1] to judge when a 'framework' is to be abandoned, I inject some hard Popperian elements into the appraisal of whether a programme progresses or degenerates or of whether one is overtaking another. That is, I give criteria of progress and stagnation within a programme and also rules for the 'elimination' of whole research programmes. A research programme is said to be *progressing* as long as its theoretical growth anticipates its empirical growth, that is, as long as it keeps predicting novel facts with some success ('*progressive problemshift*'); it is *stagnating* if its theoretical growth lags behind its empirical growth, that is, as long as it gives only *post hoc* explanations either of chance discoveries or of facts anticipated by, and discovered in, a rival programme ('*degenerating problemshift*').[2] If a research programme progressively explains more than a rival, it 'supersedes' it, and the rival can be eliminated (or, if you wish, 'shelved').[3]

(*Within* a research programme a theory can only be eliminated by a better theory, that is, by one which has excess empirical content over its predecessors, some of which is subsequently confirmed. And for this replacement of one theory by a better one, the first theory does not even have to be 'falsified' in Popper's sense of the term. Thus, progress is marked by instances verifying excess content rather

---

[1] Cf. Duhem [1906], part II, chapter VI, §10.

[2] In fact, I define a research programme as degenerating even if it anticipates novel facts but does so in a patched-up development rather than by a coherent, pre-planned positive heuristic. I distinguish three types of *ad hoc* auxiliary hypotheses: those which have no excess empirical content over their predecessor ('*ad hoc$_1$*'), those which do have such excess content but none of it is corroborated ('*ad hoc$_2$*') and finally those which are not *ad hoc* in these two senses but do not form an integral part of the positive heuristic ('*ad hoc$_3$*'). Examples of *ad hoc$_1$* hypotheses are provided by the linguistic prevarications of pseudosciences, or by the conventionalist stratagems discussed in my [1963–4], like 'monsterbarring', 'exceptionbarring', 'monsteradjustment', etc. A famous example of an *ad hoc$_2$* hypothesis is provided by the Lorentz–Fitzgerald contraction hypothesis; an example of an *ad hoc$_3$* hypothesis is Planck's first correction of the Lummer–Pringsheim formula (also cf. chapter 1, p. 79 ff). Some of the cancerous growth in contemporary social 'sciences' consists of a cobweb of such *ad hoc$_3$* hypotheses, as shown by Meehl and Lykken. (For references, cf. chapter 1, p. 88, n. 4).

[3] The rivalry of two research programmes is, of course, a protracted process during which it is rational to work in either (*or, if one can, in both*). The latter pattern becomes important, for instance, when one of the rival programmes is vague and its opponents wish to develop it in a sharper form in order to show up its weakness. Newton elaborated Cartesian vortex theory in order to show that it is inconsistent with Kepler's laws. (Simultaneous work on rival programmes, of course, undermines Kuhn's thesis of the psychological incommensurability of rival paradigms.)

The progress of one programme is a vital factor in the degeneration of its rival. If programme P_1 constantly produces 'novel facts' these, by definition, will be anomalies for the rival programme P_2. If P_2 accounts for these novel facts only in an *ad hoc* way, it is degenerating by definition. Thus the more P_1 progresses, the more difficult it is for P_2 to progress.

than by falsifying instances;[1] empirical 'falsification' and actual 'rejection' become independent.[2] Before a theory has been modified we can never know in what way it had been 'refuted', and some of the most interesting modifications are motivated by the 'positive heuristic' of the research programme rather than by anomalies. This difference alone has important consequences and leads to a rational reconstruction of scientific change very different from that of Popper's.[3])

It is very difficult to decide, especially since one must not demand progress at each single step, when a research programme has degenerated hopelessly or when one of two rival programmes has achieved a decisive advantage over the other. In this methodology, as in Duhem's conventionalism, there can be no instant – let alone mechanical – rationality. *Neither the logician's proof of inconsistency nor the experimental scientist's verdict of anomaly can defeat a research programme in one blow.* One can be 'wise' only after the event.[4]

In this code of scientific honour modesty plays a greater role than in other codes. One *must* realise that one's opponent, even if lagging badly behind, may still stage a comeback. No advantage for one side can ever be regarded as absolutely conclusive. There is never anything inevitable about the triumph of a programme. Also, there is never anything inevitable about its defeat. Thus pigheadedness, like modesty, has more 'rational' scope. *The scores of the rival sides, however, must be recorded*[5] *and publicly displayed at all times.*

(We should here at least refer to the main epistemological problem of the methodology of scientific research programmes. As it stands, like Popper's methodological falsificationism, it represents a very radical version of conventionalism. One needs to posit some extramethodological inductive principle to relate – even if tenuously – the scientific gambit of pragmatic acceptances and rejections to verisimilitude.[6] Only such an 'inductive principle' can turn science from a mere game into an epistemologically rational exercise; from a set of lighthearted sceptical gambits pursued for intellectual fun into a

---

[1] Cf. especially chapter 1, pp. 36–7.

[2] Cf. especially volume 2, chapter 8, p. 177 and this volume, p. 36.

[3] For instance, a rival theory, which acts as an *external* catalyst for the Popperian falsification of a theory, here becomes an *internal* factor. In Popper's (and Feyerabend's) reconstruction such a theory, after the falsification of the theory under test, can be removed from the rational reconstruction; in my reconstruction it has to stay within the internal history lest the falsification be undone. (Cf. p. 109, n. 5.)

  Another important consequence is the difference between Popper's discussion of the Duhem–Quine argument and mine; cf. on the one hand Popper [1934], last paragraph of section 18 and section 19, n. 1; Popper [1957b], pp. 131–3; Popper [1963a], p. 112, n. 26, pp. 238–9 and p. 243; and on the other hand, chapter 1, pp. 184–9.

[4] For the falsificationist this is a repulsive idea; cf. e.g. Agassi [1963], pp. 48ff.

[5] Feyerabend seems now to deny that even this is a possibility; cf. his [1970a] and especially [1970b] and [1974].

[6] I use 'verisimilitude' here in Popper's technical sense, as the difference between the truth content and falsity content of a theory. Cf. his [1963a], chapter 10.

– more serious – fallibilist venture of approximating the Truth about the Universe.[1])

The methodology of scientific research programmes constitutes, like any other methodology, a historiographical research programme. The historian who accepts this methodology as a guide will look in history for rival research programmes, for progressive and degenerating problemshifts. Where the Duhemian historian sees a revolution merely in simplicity (like that of Copernicus), he will look for a large scale progressive programme overtaking a degenerating one. When the falsificationist sees a crucial negative experiment, he will 'predict' that there was none, that behind any alleged crucial experiment, behind any alleged single battle between theory and experiment, there is a hidden war of attrition between two research programmes. The outcome of the war is only later linked in the falsificationist reconstruction with some alleged single 'crucial experiment'.

The methodology of research programmes – like any other theory of scientific rationality – must be supplemented by empirical–external history. No rationality theory will ever solve problems like why Mendelian genetics disappeared in Soviet Russia in the 1950s, or why certain schools of research into genetic racial differences or into the economics of foreign aid came into disrepute in the Anglo-Saxon countries in the 1960s. Moreover, to explain different speeds of development of different research programmes we may need to invoke external history. Rational reconstruction of science (in the sense in which I use the term) cannot be comprehensive since human beings are not *completely* rational animals; and even when they act rationally they may have a false theory of their own rational actions.[2]

But the methodology of research programmes draws a demarcation between internal and external history which is markedly different from that drawn by other rationality theories. For instance, what for the falsificationist looks like the (regrettably frequent) phenomenon of irrational adherence to a 'refuted' or to an inconsistent theory and which he therefore relegates to *external* history, may well be explained in terms of my methodology *internally* as a rational defence of a promising research programme. Or, the successful *predictions* of novel facts which constitute serious evidence for a research programme and therefore vital parts of internal history, are irrelevant both for the inductivist and for the falsificationist.[3] For the inductivist and the falsificationist it does not really matter whether the discovery of a fact preceded or followed a theory: only their logical relation is decisive. The 'irrational' impact of the historical coincidence that a theory

---

[1] For a more general discussion of this problem, cf. *below*, pp. 121–2.

[2] Also cf. pp. 105, 108, 110, 118, 122.

[3] The reader should remember that in this paper I discuss only naive falsificationism; cf. p. 108, n. 1.

happened to have *anticipated* a factual discovery, has no internal significance. Such anticipations constitute 'not proof but [mere] propaganda'.[1] Or again, take Planck's discontent with his own 1900 radiation formula, which he regarded as 'arbitrary'. For the falsificationist the formula was a bold, falsifiable hypothesis and Planck's dislike of it a non-rational mood, explicable only in terms of psychology. However, in my view, Planck's discontent can be explained internally: it was a rational condemnation of an '*ad hoc₃*' theory.[2] To mention yet another example: for falsificationism irrefutable 'metaphysics' is an external intellectual influence, in my approach it is a vital part of the rational reconstruction of science.

Most historians have hitherto tended to regard the solution of some problems as being the monopoly of externalists. One of these is the problem of the high frequency of *simultaneous discoveries*. For this problem vulgar-Marxists have an easy solution: a discovery is made by many people at the same time, once a social need for it arises.[3] Now what constitutes a 'discovery', and especially a major discovery, depends on one's methodology. For the inductivist, the most important discoveries are factual, and, indeed, such discoveries are frequently made simultaneously. For the falsificationist a *major* discovery consists in the discovery of a theory rather than of a fact. Once a theory is discovered (or rather invented), it becomes public property; and nothing is more obvious than that several people will test it simultaneously and make, simultaneously, (minor) factual discoveries. Also, a published theory is a challenge to devise higher-level, independently testable explanations. For example, given Kepler's ellipses and Galileo's rudimentary dynamics, simultaneous 'discovery' of an inverse square law is not so very surprising: a problem-situation being public, simultaneous solutions can be explained on *purely internal* grounds.[4] The discovery of a new problem, however, may not be so readily explicable. If one thinks of the history of science as composed of rival research programmes, then most simultaneous discoveries, theoretical or factual, are explained by the fact that research programmes being public property, many people work on them in different corners of the world, possibly not knowing of each other. However, really *novel, major, revolutionary* developments are rarely invented simultaneously. Some alleged simultaneous discoveries of novel programmes are seen as having been simultaneous discoveries only with false hindsight: in

---

[1] This is Kuhn's comment on Galileo's successful *pre*diction of the phases of Venus (Kuhn [1957], p. 224). Like Mill and Keynes before him, Kuhn cannot understand why the historic order of theory and evidence should count, and he cannot see the importance of the fact that Copernicans *pre*dicted the phases of Venus, while Tychonians only explained them by *post hoc* adjustments. Indeed, since he does not see the importance of the fact, he does not even care to mention it.

[2] Cf. p. 112, n. 2.

[3] For a statement of this position and an interesting critical discussion cf. Polanyi [1951], pp. 4ff. and pp. 78ff.

[5] Cf. Popper [1963b] and Musgrave [1969a].

fact they are *different* discoveries, merged only later into a single one.[1]

A favourite hunting ground of externalists has been the related problem of why so much importance is attached to – and energy spent on – *priority disputes*. This can be explained only *externally* by the inductivist, the naive falsificationist, or the conventionalist; but in the light of the methodology of research programmes some priority disputes are vital *internal* problems, since in this methodology *it becomes all-important for rational appraisal which programme was first in anticipating a novel fact and which fitted in the by now old fact only later*. Some priority disputes can be explained by rational interest and not simply by vanity and greed for fame. It then becomes important that Tychonian theory, for instance, succeeded in explaining – only *post hoc* – the observed phases of, and the distance to, Venus which were originally precisely anticipated by Copernicans;[2] or that Cartesians managed to explain everything that the Newtonians *pre*dicted – but only *post hoc*. Newtonian optical theory explained *post hoc* many phenomena which were anticipated and first observed by Huyghensians.[3]

All these examples show how the methodology of scientific research programmes turns many problems which had been *external* problems for other historiographies into internal ones. But occasionally the borderline is moved in the opposite direction. For instance there may have been an experiment which was accepted *instantly* – in the absence of a better theory – as a negative crucial experiment. For the falsificationist such acceptance is part of internal history; for me it is not rational and has to be explained in terms of external history.

Note. The methodology of research programmes was criticized both by Feyerabend and by Kuhn. According to Kuhn: '[Lakatos] must specify criteria which can be used *at the time* to distinguish a degenerate from a progressive research programme; and so on. Otherwise, *he has told us nothing at all*'.[4] Actually, I *do* specify such criteria. But Kuhn probably meant that '[my]

---

[1] This was illustrated convincingly, by Elkana, for the case of the so-called simultaneous discovery of the conservation of energy; cf. his [1971].

[2] Also cf. p. 115, n. 1.

[3] For the Mertonian brand of functionalism – as Alan Musgrave pointed out to me – priority disputes constitute a *prima facie* disfunction and therefore an anomaly for which Merton has been labouring to give a general socio-psychological explanation. (Cf. e.g. Merton [1957], [1963] and [1969].) According to Merton 'scientific *knowledge* is not the richer or the poorer for having credit given where credit is due: it is the social *institution* of science and individual men of science that would suffer from repeated failures to allocate credit justly' (Merton [1957], p. 648). But Merton overdoes his point: in important cases (like in some of Galileo's priority fights) there was more at stake than institutional interests: the problem was whether the Copernican research programme was progressive or not. (Of course, not all priority disputes have scientific relevance. For instance, the priority dispute between Adams and Leverrier about who was first to discover Neptune had no such relevance: whoever discovered it, the discovery strengthened the same (Newtonian) programme. In such cases Merton's external explanation may well be true.)

[4] Kuhn [1970*b*], p. 239, my italics.

standards have practical force only if they are combined with a *time limit* (what looks like a degenerating problemshift may be the beginning of a much longer period of advance)'.[1] Since I specify no such time limit, Feyerabend concludes that my standards are no more than '*verbal ornaments*'.[2] A related point was made by Musgrave in a letter containing some major constructive criticisms of an earlier draft, in which he demanded that I specify, for instance, at what point dogmatic adherence to a programme ought to be explained 'externally' rather than 'internally'.

Let me try to explain why such objections are beside the point. One may rationally stick to a degenerating programme until it is overtaken by a rival *and even after*. What one must *not* do is to deny its poor public record. Both Feyerabend and Kuhn conflate *methodological* appraisal of a programme with firm *heuristic* advice about what to do.[3] It is perfectly rational to play a risky game: what is irrational is to deceive oneself about the risk.

This does not mean as much licence as might appear for those who stick to a degenerating programme. For they can do this mostly only in private. Editors of scientific journals should refuse to publish their papers which will, in general, contain either solemn reassertions of their position or absorption of counterevidence (or even of rival programmes) by *ad hoc*, linguistic adjustments. Research foundations, too, should refuse money.[4]

These observations also answer Musgrave's objection by separating rational and irrational (or honest and dishonest) adherence to a degenerating programme. They also throw further light on the demarcation between internal and external history. They show that internal history is self-sufficient for the presentation of the history of disembodied science, including degenerating problemshifts. External history explains why some people have false beliefs about scientific progress, and how their scientific activity may be influenced by such beliefs.

[1] Feyerabend [1970a], p. 215.
[2] *Ibid.*
[3] Cf. p. 103, n. 1.
[4] I do, of course, *not* claim that such decisions are necessarily uncontroversial. In such decisions one has also to use one's *common sense*. Common sense (that is, judgment in *particular* cases which is not made according to mechanical rules but only follows general principles which leave some *Spielraum*) plays a role in all brands of non-mechanical methodologies. The Duhemian conventionalist needs common sense to decide when a theoretical framework has become sufficiently cumbersome to be replaced by a 'simpler' one. The Popperian falsificationist needs common sense to decide when a basic statement is to be 'accepted', or to which premise the *modus tollens* is to be directed. (Cf. chapter 1, p. 22 ff). But neither Duhem nor Popper gives a blank cheque to 'common sense'. They give very definite guidance. The Duhemian judge directs the jury of common sense to agree on comparative simplicity; the Popperian judge directs the jury to look out primarily for, and agree upon, accepted basic statements which clash with accepted theories. My judge directs the jury to agree on appraisals of progressive and degenerating research programmes. But, for example, there may be conflicting views about whether an accepted basic statement expresses a *novel* fact or not. Cf. chapter 1, p. 70.

Although it is important to reach agreement on such verdicts, there must also be the possibility of appeal. In such appeals inarticulated common sense is questioned, articulated and criticized. (The criticism may even turn from a criticism of law interpretation into a criticism of the law itself.)

(e) Internal and external history

Four theories of the rationality of scientific progress – or logics of scientific discovery – have been briefly discussed. It was shown how each of them provides a theoretical framework for the rational reconstruction of the history of science.

Thus the internal history of *inductivists* consists of alleged discoveries of hard facts and of so-called inductive generalizations. The internal history of *conventionalists* consists of factual discoveries and of the erection of pigeonhole systems and their replacement by allegedly simpler ones.[1] The internal history of *falsificationists* dramatizes bold conjectures, improvements which are said to be *always* content-increasing and, above all, triumphant 'negative crucial experiments'. The *methodology of research programmes*, finally, emphasizes long-extended theoretical and empirical rivalry of major research programmes, progressive and degenerating problemshifts, and the slowly emerging victory of one programme over the other.

Each rational reconstruction produces some characteristic pattern of rational growth of scientific knowledge. But all of these *normative* reconstructions may have to be supplemented by *empirical* external theories to explain the residual non-rational factors. The history of science is always richer than its rational reconstruction. *But rational reconstruction or internal history is primary, external history only secondary, since the most important problems of external history are defined by internal history.* External history either provides non-rational explanation of the speed, locality, selectiveness, etc. of historic events as *interpreted* in terms of internal history; or, when history differs from its rational reconstruction, it provides an empirical explanation of why it differs. But the *rational* aspect of scientific growth is fully accounted for by one's logic of scientific discovery.

Whatever problem the historian of science wishes to solve, he has first to reconstruct the relevant section of the growth of objective scientific knowledge, that is, the relevant section of 'internal history'. As it has been shown, what constitutes for him internal history, depends on his philosophy, whether he is aware of this fact or not. Most theories of the growth of knowledge are theories of the growth of disembodied knowledge: whether an experiment is crucial or not, whether a hypothesis is highly probable in the light of the available evidence or not, whether a problemshift is progressive or not, is not dependent in the slightest on the scientists' beliefs, personalities or authority. These subjective factors are of no interest for any internal history. For instance, the 'internal historian' records the Proutian programme with its hard core (that atomic weights of pure chemical elements are whole numbers) and its positive heuristic (to overthrow,

---

[1] Most conventionalists have also an intermediate inductive layer of 'laws' between facts and theories; cf. p. 106, n. 2.

and replace, the contemporary false observational theories applied in measuring atomic weights). This programme was later carried through.[1] The internal historian will waste little time on Prout's *belief* that if the 'experimental techniques' *of his time* were 'carefully' applied, and the experimental findings properly interpreted, the anomalies would *immediately* be seen as mere illusions. The internal historian will regard this historical fact as a fact in the second world which is only a caricature of its counterpart in the third world.[2] *Why* such caricatures come about is none of his business; he might – in a footnote – pass on to the externalist the problem of why certain scientists had 'false beliefs' about what they were doing.[3]

Thus, in constructing internal history the historian will be highly selective: he will omit everything that is irrational in the light of his rationality theory. But this normative selection still does not add up to a fully fledged rational reconstruction. For instance, Prout never articulated the 'Proutian programme': the Proutian programme is not Prout's programme. *It is not only the ('internal') success or the ('internal') defeat of a programme which can be judged only with hindsight: it is frequently also its content.* Internal history is not just a *selection* of methodologically interpreted facts: it may be, on occasions, their *radically improved version*. One may illustrate this using the Bohrian programme. Bohr, in 1913, may not have even thought of the possibility of electron spin. He had more than enough on his hands without the spin. Nevertheless, the historian, describing with hindsight the Bohrian programme, should include electron spin in it, since electron spin fits naturally in the original outline of the programme. Bohr might have referred to it in 1913. Why Bohr did not do so, is an interesting problem which deserves to be indicated in a footnote.[4]

---

[1] The proposition 'the Proutian programme was carried through' looks like a 'factual' proposition. But there are no 'factual' propositions: the phrase only came into ordinary language from dogmatic empiricism. *Scientific 'factual' propositions* are theory-laden: the theories involved are 'observational theories'. *Historiographical 'factual' propositions* are also theory-laden: the theories involved are methodological theories. In the decision about the truth-value of the 'factual' proposition, 'the Proutian programme was carried through', two methodological theories are involved. First, the theory that the units of scientific appraisal are research programmes; secondly, some *specific* theory of how to judge whether a programme was 'in fact' carried through. For all these considerations a Popperian internal historian will not need to take any interest whatsoever in the *persons* involved, or in their beliefs about their own activities.

[2] The 'first world' is that of matter, the 'second' the world of feelings, beliefs, consciousness, the 'third' the world of objective knowledge, articulated in propositions. This is an age-old and vitally important trichotomy; its leading contemporary proponent is Popper. Cf. Popper [1968a], [1968b] and Musgrave [1969] and [1974].

[3] Of course what, in this context, constitutes 'false belief' (or 'false consciousness'), depends on the rationality theory of the critic: cf. pp. 105, 107 and 109. But no rationality theory can ever succeed in leading to 'true consciousness'.

[4] If the publication of Bohr's programme had been delayed by a few years, further speculation might even have led to the spin problem without the previous observation of the anomalous Zeeman effect. Indeed, Compton raised the problem in the context of the Bohrian programme in his [1919].

(Such problems might then be solved either internally by pointing to rational reasons in the growth of objective, impersonal knowledge; or externally by pointing to psychological causes in the development of Bohr's personal beliefs.)

One way to indicate discrepancies between history and its rational reconstruction is to relate the internal history *in the text*, and indicate *in the footnotes* how actual history 'misbehaved' in the light of its rational reconstruction.[1]

Many historians will abhor the idea of *any* rational reconstruction. They will quote Lord Bolingbroke: 'History is philosophy teaching by example.' They will say that before philosophizing 'we need a lot more examples'.[2] But such an inductivist theory of historiography is utopian.[3] *History without some theoretical 'bias' is impossible.*[4] Some historians look for the discovery of hard facts, inductive generalizations, others for bold theories and crucial negative experiments, yet others for great simplifications, or for progressive and degenerating problemshifts; all of them have *some* theoretical 'bias'. This bias, of course, may be obscured by an eclectic variation of theories or by theoretical confusion: but neither eclecticism nor confusion amounts to an atheoretical outlook. What a historian regards as an external problem is often an excellent guide to his implicit methodology: some will ask why a 'hard fact' or a 'bold theory' was discovered exactly when and where it actually was discovered; others will ask why a 'degenerating problemshift' could have wide popular acclaim over an incredibly long period or why a 'progressive problemshift' was left 'unreasonably' unacknowledged.[5] Long texts have been devoted to the problem of

---

[1] I first applied this expositional device in my [1963–4]; I used it again in giving a detailed account of the Proutian and the Bohrian programmes; cf. chapter 1, pp. 51, 53, 58. This practice was criticized at the 1969 Minneapolis conference by some historians. McMullin, for instance, claimed that this presentation may illuminate a *methodology*, but certainly not real *history*: the text tells the reader what ought to have happened and the footnotes what in fact happened (cf. McMullin [1970]). Kuhn's criticism of my exposition ran essentially on the same lines: he thought that it was a specifically *philosophical* exposition: 'a *historian* would not include *in his narrative* a factual report which he knows to be false. If he had done so, he would be so sensitive to the offence that he could not conceivably compose a footnote calling attention to it.' (Cf. Kuhn [1970b], p. 256.)

[2] Cf. L. Pearce Williams [1970].

[3] Perhaps I should emphasize the difference between on the one hand, *inductivist historiography of science*, according to which *science* proceeds through discovery of hard facts (in nature) and (possibly) inductive generalizations, and, on the other hand, the *inductivist theory of historiography of science* according to which *historiography of science* proceeds through discovery of hard facts (in history of science) and (possibly) inductive generalizations. 'Bold conjectures', 'crucial negative experiments', and even 'progressive and degenerating research programmes' may be regarded as 'hard historical facts' by some inductivist historiographers. One of the weaknesses of Agassi's [1963] is that he omitted to emphasize this distinction between scientific and historiographical inductivism.

[4] Cf. Popper [1957b], section 31.

[5] This thesis implies that the work of those 'externalists' (mostly trendy 'sociologists of science') who claim to do social history of some scientific discipline without having

whether, and if so, why, the emergence of science was a purely European affair; but such an investigation is bound to remain a piece of confused rambling until one clearly defines 'science' according to some normative philosophy of science. One of the most interesting problems of external history is to specify the psychological, and indeed, social conditions which are necessary (but, of course, never sufficient) to make scientific progress possible; but in the very formulation of this 'external' problem *some* methodological theory, *some* definition of science is bound to enter. History of *science* is a history of events which are selected and interpreted in a normative way.[1] This being so, the hitherto neglected problem of appraising rival logics of scientific discovery and, hence, rival reconstructions of history, acquires paramount importance. I shall now turn to this problem.

2   CRITICAL COMPARISON OF METHODOLOGIES: HISTORY AS A TEST OF ITS RATIONAL RECONSTRUCTION

Theories of scientific rationality can be classified under two main heads.

(1) *Justificationist methodologies* set very high epistemological standards: for classical justificationists a proposition is 'scientific' only if it is *proven*, for neojustificationists, if it is *probable* (in the sense of the probability calculus) or *corroborated* (in the sense of Popper's third note on corroboration) to a proven degree.[2] Some philosophers of science gave up the idea of proving or of (provably) probabilifying scientific theories but remained dogmatic empiricists: whether inductivists, probabilists, conventionalists or falsificationists, they still stick to the provability of 'factual' propositions. By now, of course, all these different forms of justificationism have crumbled under the weight of *epistemological and logical criticism.*

(2) The only alternatives with which we are left are *pragmatic–conventionalist methodologies*, crowned by some global principle of induction. Conventionalist methodologies first lay down rules about

mastered the discipline itself, and its internal history, is worthless. Also cf. Musgrave [1974].

[1] Unfortunately there is only one single word in most languages to denote history$_1$ (the set of historical events) and history$_2$ (a set of historical propositions). Any history$_2$ is a theory- and value-laden reconstruction of history$_1$.

[2] That is, a hypothesis h is scientific only if there is a number q such that $p(h, e) = q$ where e is the available evidence and $p(h, e) = q$ can be *proved*. It is irrelevant whether p is a Carnapian confirmation function or a Popperian corroboration function as long as $p(h, e) = q$ is allegedly proved. (Popper's third note on corroboration, of course, is only a curious slip which is out of tune with his philosophy: cf. volume 2, chapter 8, pp. 194–200.)

Probabilism has never generated a programme of historiographical reconstruction; it has never emerged from grappling – unsuccessfully – with the very problems it created. As an epistemological programme it has been degenerating for a long time; as a historiographical programme it never even started.

'acceptance' and 'rejection' of factual and theoretical propositions – without yet laying down rules about proof and disproof, truth and falsehood. We then get *different systems of rules of the scientific game*. The inductivist game would consist of collecting 'acceptable' (not proven) data and drawing from them 'acceptable' (not proven) inductive generalizations. The conventionalist game would consist of collecting 'acceptable' data and ordering them into the simplest possible pigeonhole systems (or devising the simplest possible pigeonhole systems and filling them with acceptable data). Popper specified yet another game as 'scientific'.[1] Even methodologies which have been epistemologically and logically discredited, may go on functioning, in these emasculated versions, as guides for the rational reconstruction of history. But these *scientific games* are without any genuine epistemological relevance *unless* we superimpose on them some sort of metaphysical (or, if you wish, 'inductive') principle which will say that the game, as specified by the methodology, gives us the best chance of approaching the Truth. Such a principle then turns the pure conventions of the game into fallible conjectures; but without such a principle the scientific game is just like any other game.[2]

It is very difficult to criticize conventionalist methodologies like Duhem's and Popper's. There is no obvious way to criticize either a game or a metaphysical principle of induction. In order to overcome these difficulties I am going to propose a new theory of how to appraise such methodologies of science (the ones, which – at least in the first stage, before the introduction of an inductive principle – are conventionalist). I shall show that methodologies may be criticized without any direct reference to any epistemological (or even logical) theory, and without using directly any logico-epistemological criticism. The basic idea of this criticism is that *all methodologies function as historiographical (or meta-historical) theories (or research programmes) and can be criticized by criticizing the rational historical reconstructions to which they lead.*

I shall try to develop this historiographical method of criticism in a dialectical way. I start with a special case: I first 'refute' falsificationism by 'applying' falsificationism (on a normative historiographical meta-level) to itself. Then I shall apply falsificationism also to inductivism and conventionalism, and, indeed, argue that all methodologies are bound to end up 'falsified' with the help of this Pyrrhonian *machine de guerre*. Finally, I shall 'apply' not falsificationism but the methodology of scientific research programmes (again on a

---

[1] Popper [1934], sections 11 and 85. Also cf. the comment in chapter 3, p. 141, n. 8.

    The methodology of research programmes too is, in the first instance, defined as a game; cf. especially *above*, pp. 110–12.

[2] This whole problem area is the subject of chapter 8 of volume 2, pp. 181ff, but especially of chapter 3 of this volume.

normative–historiographical meta-level) to inductivism, conventional-ism, falsificationism and to itself, and show that – on this meta-criterion – methodologies can be constructively criticized and com-pared. This normative–historiographical version of the methodology of scientific research programmes supplies a general theory of how to compare rival logics of discovery in which (in a sense carefully to be specified) *history may be seen as a 'test' of its rational reconstructions.*

(a) Falsificationism as a meta-criterion:
history 'falsifies' falsificationism (and any other methodology)

In their purely 'methodological' versions scientific appraisals, as has already been said, are *conventions* and can always be formulated as a definition of science.[1] How can one criticize such a definition? If one interprets it nominalistically,[2] a definition is a mere abbreviation, a terminological suggestion, a tautology. How can one criticize a tau-tology? Popper, for one, claims that his definition of science is 'fruitful' because 'a great many points can be clarified and explained with its help'. He quotes Menger: 'Definitions are dogmas; only the conclu-sions drawn from them can afford us any new insight'.[3] But how can a definition have explanatory power or afford new insights? Popper's answer is this: 'It is only from the consequences of my definition of empirical science, and from the methodological decisions which depend upon this definition, that the scientist will be able to see how far it conforms to his intuitive idea of the goal of his endeavours'.[4]

The answer complies with Popper's general position that conven-tions can be criticized by discussing their 'suitability' relative to some purpose: 'As to the suitability of any convention opinions may differ; and a reasonable discussion of these questions is only possible between parties having some purpose in common. The choice of that purpose ...goes beyond rational argument'.[5] Indeed, Popper never offered a theory of rational criticism of consistent conventions. He does not raise, let alone answer, the question: ' *Under what conditions would you give up your demarcation criterion?*'[6]

---

[1] Cf. Popper [1934], sections 4 and 11. Popper's definition of science is, of course, his celebrated 'demarcation criterion'.

[2] For an excellent discussion of the distinction between nominalism and realism (or, as Popper prefers to call it, 'essentialism') in the theory of definitions, cf. Popper [1945], volume 2, chapter 11, and [1963a], p. 20.

[3] Popper [1934], section 11.

[4] *Ibid.*

[5] Popper [1934], section 4. But Popper, in his *Logik der Forschung*, never specifies a *purpose* of the game of science that would go beyond what is contained in its rules. The thesis that the *aim* of science is *truth*, occurs only in his writings since 1957. All that he says in his *Logik der Forschung* is that the quest for truth may be a psychological *motive* of scientists. For a detailed discussion cf. chapter 3.

[6] This flaw is the more serious since Popper himself has expressed qualifications about his criterion. For instance in his [1963a] he describes 'dogmatism', that is, treating

But the question can be answered. I give my answer in two stages: I propose first a naive and then a more sophisticated answer. I start by recalling how Popper, according to his own account,[1] arrived at his criterion. He thought, like the best scientists of his time, that Newton's theory, although refuted, was a wonderful scientific achievement; that Einstein's theory was still better; and that astrology, Freudianism and twentieth century Marxism were pseudoscientific. His problem was to find a definition of science which yielded these '*basic judgments*' concerning particular theories; and he offered a novel solution. Now let us consider the proposal that *a rationality theory – or demarcation criterion – is to be rejected if it is inconsistent with an accepted 'basic value judgment' of the scientific élite.* Indeed, this meta-methodological rule (*meta-falsificationism*) would seem to correspond to Popper's methodological rule (falsificationism) that a scientific theory is to be rejected if it is inconsistent with an ('empirical') basic statement unanimously accepted by the scientific community. Popper's whole methodology rests on the contention that there exist (relatively) singular statements on whose truth-value scientists can reach unanimous agreement; without such agreement there would be a new Babel and 'the soaring edifice of science would soon lie in ruins'.[2] But even if there were an agreement about 'basic' statements, if there were no agreement about how to appraise scientific achievement relative to this 'empirical basis', would not the soaring edifice of science equally soon lie in ruins? No doubt it would. While there has been little agreement concerning a *universal* criterion of the scientific character of theories, there has been considerable agreement over the last two centuries concerning *single* achievements. While there has been no *general* agreement concerning a theory of scientific rationality, there has been considerable agreement concerning whether a particular single step in the game was scientific or crankish, or whether a particular gambit was played correctly or not. A general definition of science thus must reconstruct the acknowledgedly best gambits as 'scientific': if it fails to do so, it has to be rejected.[3]

anomalies as a kind of 'background noise', as something that is 'to some extent necessary' (p. 49). But on the next page he identifies this 'dogmatism' with 'pseudo-science'. Is then pseudoscience 'to some extent necessary'? Also, cf. chapter 1, p. 89, n. 5.

[1] Cf. Popper [1963*a*], pp. 33–7.   [2] Popper [1934], section 29.

[3] This approach, of course, does not imply that we *believe* that the scientists 'basic judgments' are unfailingly rational; it only means that we *accept* them in order to criticize universal definitions of science. (If we were to add that no such *universal* definition has been found and no such *universal* definition will ever be found, the stage would be set for Polanyi's conception of the lawless closed autocracy of science.)

My meta-criterion may be seen as a 'quasi-empirical' self-application of Popperian falsificationism. I introduced this 'quasi-empiricalness' earlier in the context of mathematical philosophy. We may abstract from *what* flows in the logical channels of a deductive system, whether it is something certain or something fallible, whether it is truth and falsehood or probability and improbability, or even moral or scientific desirability and undesirability: it is the *how* of the flow which decides whether the

Then let us propose tentatively that *if a demarcation criterion is inconsistent with the 'basic' appraisals of the scientific élite, it should be rejected.*

Now *if* we apply this quasi-empirical meta-criterion (which I am going to reject later), Popper's demarcation criterion – that is, Popper's rules of the game of science – has to be rejected.[1]

Popper's basic rule is that the scientist must specify in advance under what experimental conditions he will give up even his most basic assumptions. For instance, he writes, when criticizing psychoanalysis: '*Criteria of refutation* have to be laid down beforehand: it must be agreed which observable situations, if actually observed, mean that the theory is refuted. But what kind of clinical responses would refute to the satisfaction of the analyst *not merely a particular analytic diagnosis but psychoanalysis itself?* And have such criteria ever been discussed or agreed upon by analysts?'[2] In the case of psychoanalysis Popper was right: no answer has been forthcoming. Freudians have been nonplussed by Popper's basic challenge concerning scientific honesty. Indeed, they have refused to specify experimental conditions under which they would give up their basic assumptions. For Popper this was the hallmark of their intellectual dishonesty. But what if we put Popper's question to the Newtonian scientist: 'What kind of observation would refute to the satisfaction of the Newtonian not merely a particular Newtonian explanation but Newtonian dynamics and gravitational theory itself? And have such criteria ever been discussed or agreed upon by Newtonians?' The Newtonian will, alas, scarcely be able to give a positive answer.[3] But then if analysts are to be condemned as dishonest by Popper's standards, Newtonians must also be condemned. Newtonian science, however, in spite of this sort of 'dogmatism', is highly regarded by the greatest scientists, and, indeed, by Popper himself. Newtonian 'dogmatism' then is a 'falsification' of Popper's definition: it defies Popper's rational reconstruction.

Popper may certainly withdraw his celebrated challenge and demand falsifiability – and rejection on falsification – only for systems of theories, including initial conditions and all sorts of auxiliary and observational theories.[4] This is a considerable withdrawal, for it allows

system is negativist, 'quasi-empirical', dominated by *modus tollens* or whether it is justificationist, 'quasi-Euclidean', dominated by *modus ponens*. (Cf. volume 2, chapter 2.) This 'quasi-empirical' approach may be applied to *any* kind of normative knowledge: Watkins has already applied it to ethics in his [1963] and [1967]. But now I prefer another approach: cf. p. 133, n. 4.

1 It may be noted that this meta-criterion does not have to be construed as psychological, or 'naturalistic' in Popper's sense. (Cf. his [1934], section 10.) The definition of the 'scientific élite' is not simply an empirical matter.

2 Popper [1963a], p. 38, n. 3, my italics. This, of course, is equivalent to his celebrated 'demarcation criterion' between (internal, rationally reconstructed) science and nonscience (or 'metaphysics'). The latter may be (externally) 'influential' and has to be branded as pseudoscience only if it declares itself to be science.

3 Cf. chapter 1, pp. 16–17.   4 Cf. e.g. his [1934], section 18.

the imaginative scientist to save his pet theory by suitable lucky alterations in some odd, obscure corner on the periphery of his theoretical maze. But even Popper's mitigated rule will show up even the most brilliant scientists as irrational dogmatists. For in large research programmes there are always known anomalies: normally the researcher puts them aside and follows the positive heuristic of the programme.[1] In general he rivets his attention on the positive heuristic rather than on the distracting anomalies, and hopes that the 'recalcitrant instances' will be turned into confirming instances as the programme progresses. On Popper's terms the greatest scientists in these situations used forbidden gambits, *ad hoc* stratagems: instead of regarding Mercury's anomalous perihelion as a falsification of the Newtonian theory of our planetary system and thus as a reason for its rejection, most physicists shelved it as a problematic instance to be solved at some later stage – or offered *ad hoc* solutions. This methodological attitude of treating as (mere) *anomalies* what Popper would regard as (dramatic) counterexamples is commonly accepted by the best scientists. Some of the research programmes now held in highest esteem by the scientific community progressed in an ocean of anomalies.[2] That in their choice of problems the greatest scientists 'uncritically' ignore anomalies (and that they isolate them with the help of *ad hoc* stratagems) offers, at least on our meta-criterion, a further falsification of Popper's methodology. He cannot interpret as rational some most important patterns in the growth of science.

Furthermore, for Popper, working on *an inconsistent system* must invariably be regarded as irrational: 'a self-contradictory system must be rejected...[because it] is uninformative...No statement is singled out...since all are derivable'.[3] But some of the greatest scientific research programmes progressed on inconsistent foundations.[4] Indeed in such cases the best scientists' rule is frequently: '*Allez en avant et la foi vous viendra*'. This anti-Popperian methodology secured a breathing space both for the infinitesimal calculus and for naive set theory when they were bedevilled by logical paradoxes.

Indeed, if the game of science had been played according to Popper's rule book, Bohr's 1913 paper would never have been published because it was inconsistently grafted on to Maxwell's theory, and Dirac's delta functions would have been suppressed until Schwartz. All these examples of research based on inconsistent foundations constitute further 'falsifications' of falsificationist methodology.[5]

---

[1] Cf. chapter 1, especially pp. 50 ff. [2] *Ibid.*, pp. 52 ff.
[3] Cf. Popper [1934], section 24.
[4] Cf. chapter 1, especially pp. 26 ff.
[5] In general Popper stubbornly overestimates the immediate striking force of purely negative criticism. 'Once a mistake, or a contradiction, is pinpointed, there can be no verbal evasion: it can be proved, and that is that' (Popper [1959a], p. 394). He adds: 'Frege did not try evasive manoeuvres when he received Russell's criticism.' But of course he did. (Cf. Frege's *Postscript* to the second edition of his *Grundgesetze*.)

Thus several of the 'basic' appraisals of the scientific *élite* 'falsify' Popper's definition of science and scientific ethics. The problem then arises, to what extent, given these considerations, can falsificationism function as a guide for the historian of science. The simple answer is, to a very small extent. Popper, the leading falsificationist, never wrote any history of science; possibly because he was too sensitive to the judgment of great scientists to pervert history in a falsificationist vein. One should remember that while in his autobiographical recollections he mentions Newtonian science as the paradigm of scientificness, that is, of falsifiability, in his classical *Logik der Forschung* the falsifiability of Newton's theory is nowhere discussed. The *Logik der Forschung*, on the whole, is dryly abstract and highly ahistorical.[1] Where Popper does venture to remark casually on the falsifiability of major scientific theories, he either plunges into some logical blunder,[2] or distorts history to fit his rationality theory. If a historian's methodology provides a poor rational reconstruction, he may either misread history in such a way that it coincides with his rational reconstruction, or he will find that the history of science is highly irrational. Popper's respect for great science made him choose the first option, while the disrespectful Feyerabend chose the second.[3] Thus Popper, in his historical asides, tends to turn anomalies into 'crucial experiments' and to exaggerate their immediate impact on the history of science. Through his spectacles, great scientists accept refutations readily and this is the primary source of their problems. For instance, in one place he claims that the Michelson–Morley experiment decisively overthrew classical ether theory; he also exaggerates the role of this experiment in the emergence of Einstein's relativity theory.[4] It takes a naive

[1] Interestingly, as Kuhn points out, 'a consistent interest in historical problems and a willingness to engage in original historical research distinguishes the men [Popper] has trained from the members of any other current school in the philosophy of science' (Kuhn [1970b], p. 236). For a hint at a possible explanation of the apparent discrepancy cf. p. 137, n. 1.

[2] For instance, he claims that a perpetual motion machine would 'refute' (on his terms) the first law of thermodynamics ([1934], section 15). But how can one interpret, on Popper's own terms, the statement that 'K is a perpetual motion machine' as a 'basic', that is, as a spatio-*temporally* singular statement?

[3] I am referring to Feyerabend's [1970b] and [1974].

[4] Cf. Popper [1934], section 30 and Popper [1945], volume 2, pp. 220–1. He stressed that Einstein's problem was how to account for experiments 'refuting' classical physics and he 'did not...set out to criticise our conceptions of space and time'. But Einstein certainly did. His Machian criticism of our concepts of space and time, and, in particular his operationalist criticism of the concept of simultaneity played an important role in his thinking.

I discussed the role of the Michelson–Morley experiments at some length in chapter 1.

Popper's competence in physics would never, of course, have allowed him to distort the history of relativity theory as much as Beveridge, who wanted to persuade economists to an empirical approach by setting them Einstein as an example. According to Beveridge's falsificationist reconstruction, Einstein 'started [in his work on gravitation] from facts [which refuted Newton's theory, that is,] from the movements

falsificationist's simplifying spectacles to see, with Popper, Lavoisier's classical experiments as refuting (or as 'tending to refute') the phlogiston theory; or to see the Bohr–Kramers–Slater theory as being knocked out with a single blow from Compton; or to see the parity principle 'rejected' by 'counterexample'.[1]

Furthermore, if Popper wants to reconstruct the provisional acceptance of theories as rational on *his* terms, he is bound to ignore the historical fact that most important theories are born refuted and that some laws are further explained, rather than rejected, in spite of the known counterexamples. He tends to turn a blind eye on all anomalies known before the one which later was enthroned as 'crucial counterevidence'. For instance, he mistakenly thinks that 'neither Galileo's nor Kepler's theories were refuted before Newton'.[2] The context is significant. Popper holds that the most important pattern of scientific progress is when a crucial experiment leaves one theory *unrefuted* while it refutes a rival one. But, as a matter of fact, in most, if not in all, cases where there are two rival theories, both are known to be simultaneously infected by anomalies. In such situations Popper succumbs to the temptation to simplify the situation into one to which his methodology is applicable.

Falsificationist historiography is then 'falsified'. But if we apply the

of the planet Mercury, the unexplained aberrancies of the moon' (Beveridge [1937]). Of course, Einstein's work on gravitation grew out from a 'creative shift' in the positive heuristic of his special relativity programme, and certainly not from pondering over Mercury's anomalous perihelion or the moon's devious, unexplained aberrancies.

[1] Popper [1963a], pp. 220, 239, 242–3 and [1963b], p. 965. Popper, of course, is left with the problem why 'counterexamples' (that is, anomalies) are not recognized immediately as causes for rejection. For instance, he points out that in the case of the breakdown of parity 'there had been many observations – that is, photographs of particle tracks – from which we might have read off the result, but the observations had been either ignored or misinterpreted' ([1963b], p. 965). Popper's – external – explanation seems to be that scientists have not yet learned to be sufficiently critical and revolutionary. But is it not a better – and internal – explanation that the anomalies *had* to be ignored until some progressive alternative theory was offered which turned the counterexamples into examples?

[2] Popper [1963a], p. 246.

[3] As I mentioned, one Popperian, Agassi, did write a book on the historiography of science (Agassi [1963]). The book has some incisive critical sections flogging inductivist historiography, but he ends up by replacing inductivist mythology by falsificationist mythology. For Agassi *only* those facts have scientific (internal) significance which can be expressed in propositions which conflict with some extant theory: only their discovery deserves the honorific title 'factual discovery'; factual propositions which *follow from* rather than *conflict with* known theories are irrelevant; so are factual propositions which are *independent* of them. If some valued factual discovery in the history of science is known as a confirming instance or chance discovery, Agassi boldly predicts that on *close* investigation they will turn out to be refuting instances, and he offers five case studies to support his claim (pp. 60–74). Alas, on *closer* investigation it turns out that Agassi got wrong all the five examples which he adduced as confirming instances of his historiographical theory. In fact all the five examples (in our normative meta-falsificationist sense) 'falsify' his historiography.

same meta-falsificationist method to inductivist and conventionalist historiographies, we shall 'falsify' them too.

The best logico-epistemological demolition of inductivism is, of course, Popper's; but even if we assumed that inductivism were philosophically (that is, epistemologically and logically) sound, Duhem's historiographical criticism falsifies it. Duhem took the most celebrated 'successes' of inductivist historiography: Newton's law of gravitation and Ampère's electromagnetic theory. These were said to be two most victorious applications of inductive method. But Duhem (and, following him, Popper and Agassi) showed that they were not. Their analyses illustrate how the inductivist, if he wants to show that the growth of actual science is rational, must falsify actual history out of all recognition.[1] Therefore, if the rationality of science is inductive, actual science is not rational; if it is rational, it is not inductive.[2]

Conventionalism – which, unlike inductivism, is no easy prey to logical or epistemological criticism[3] – can also be historiographically falsified. One can show that the clue to scientific revolutions is not the replacement of cumbersome frameworks by simpler ones.

The Copernican revolution was generally taken to be the *paradigm of conventionalist historiography*, and it is still so regarded in many quarters. For instance Polanyi tells us that Copernicus's 'simpler picture' had 'striking beauty' and '[justly] carried great powers of conviction'. [4] But modern study of primary sources, particularly by Kuhn,[5] has dispelled this myth and presented a clear-cut historiographical refutation of the conventionalist account. It is now agreed that the Copernican system was 'at least as complex as the Ptolemaic'.[6] But if this is so, then, if the acceptance of Copernican theory was rational, it was not for its superlative objective simplicity.[7]

Thus inductivism, falsificationism and conventionalism can be falsified as rational reconstructions of history with the help of the sort of historiographical criticism I have adduced.[8] Historiographical falsification of inductivism, as we have seen, was initiated already

[1] Cf. Duhem [1906], Popper [1948] and [1957a], Agassi [1963].
[2] Of course, an inductivist may have the temerity to claim that genuine science has not yet started and may write a history of extant science as a history of bias, superstition and false belief.
[3] Cf. Popper [1934], section 19.
[4] Cf. Polanyi [1951], p. 70.
[5] Kuhn [1957]. Also cf. Price [1959].
[6] Cohen [1960], p. 61. Bernal, in his [1954], says that '[Copernicus's] reasons for [his] revolutionary change were essentially philosophic and aesthetic [that is, in the light of conventionalism, scientific]; but in later editions he changes his mind: '[Copernicus's] reasons were mystical rather than scientific.'
[7] For a more detailed sketch cf. chapter 4.
[8] Other types of criticism of methodologies may, of course, be easily devised. We may, for instance, apply the standards of each methodology (not only falsificationism) to itself. The result, for most methodologies, will be equally destructive: inductivism cannot be proved inductively, simplicity will be seen as hopelessly complex. (For the latter cf. end of n. 2, p. 130.)

by Duhem and continued by Popper and Agassi. Historiographical criticisms of (naive) falsificationism have been offered by Polanyi, Kuhn, Feyerabend and Holton.[1] The most important historiographical criticism of conventionalism is to be found in Kuhn's – already quoted – masterpiece on the Copernican revolution.[2] The upshot of these criticisms is that all these rational reconstructions of history force history of science into the Procrustean bed of their hypocritical morality, thus creating fancy histories, which hinge on mythical 'inductive bases', 'valid inductive generalizations', 'crucial experiments', 'great revolutionary simplifications', etc. But critics of falsificationism and conventionalism drew very different conclusions from the falsification of these methodologies than Duhem, Popper and Agassi did from their own falsification of inductivism. Polanyi (and, seemingly, Holton) concluded that while proper, rational scientific appraisal can be made in *particular* cases, there can be no *general* theory of scientific rationality.[3] *All* methodologies, *all* rational reconstructions can be historiographically 'falsified': science *is* rational, but its rationality cannot be subsumed under the general laws of any methodology.[4] Feyerabend, on the other hand, concluded that not only can there be no general theory of scientific rationality but also that there is no such thing as scientific rationality.[5] Thus Polanyi swung towards conservative authoritarianism, while Feyerabend swung towards sceptical

---

[1] Cf. Polanyi [1958], Kuhn [1962], Holton [1969], Feyerabend [1970b] and [1971]. I should also add Lakatos [1963–4], [1968c], and chapter 1 *above*.

[2] Kuhn [1957]. Such historiographical criticism can easily drive some rationalists into an irrational defence of their favourite falsified rationality theory. Kuhn's historiographical criticism of the simplicity theory of the Copernican revolution shocked the conventionalist historian Richard Hall so much that he published a polemic article in which he singled out and reasserted those aspects of Copernican theory which Kuhn himself had mentioned as possibly having a claim to higher simplicity, and ignored the rest of Kuhn's – valid – argument (Hall [1970]). No doubt, simplicity can always be defined for *any* pair of theories T_1 and T_2 in such a way that the simplicity of T_1 is greater than that of T_2.
For further discussion of conventionalist historiography cf. chapter 4.

[3] Thus Polanyi is a conservative rationalist concerning science, and an 'irrationalist' concerning the philosophy of science. But, of course, this meta-'irrationalism' is a perfectly respectable brand of rationalism: to claim that the concept of 'scientifically acceptable' cannot be further defined, but only transmitted by the channels of 'personal knowledge', does not make one an outright irrationalist, only an outright conservative. Polanyi's position in the philosophy of natural science corresponds closely to Oakeshott's ultra-conservative philosophy of political science. (For references and an excellent criticism of the latter cf. Watkins [1952]. Also cf. pp. 35–6.)

[4] Of course, none of the critics were aware of the exact logical character of meta-methodological falsificationism as explained in this section and none of them applied it completely consistently. One of them writes: 'At this stage we have not yet developed a general theory of criticism even for scientific theories, let alone for theories of rationality: therefore if we want to falsify methodological falsificationism, we have to do it before having a theory of how to do it' (*above*, chapter 1, p. 30).

[5] I use the critical machinery developed in this paper against Feyerabend's epistemological anarchism in chapter 4.

anarchism. Kuhn came up with a highly original vision of irrationally changing rational authority.[1]

Although, as it transpires from this section, I have high regard for Polanyi's, Feyerabend's and Kuhn's criticisms, of extant ('internalist') theories of method, I drew a conclusion completely different from theirs. I decided to look for an improved methodology which offers a better *rational* reconstruction of science.

Feyerabend and Kuhn immediately tried to 'falsify' my improved methodology in turn.[2] I soon had to discover that, at least in the sense described in the present section, my methodology too – and any methodology whatsoever – *can* be 'falsified', for the simple reason that no set of human judgments is completely rational and thus no rational reconstruction can ever coincide with actual history.[3]

This recognition led me to propose a new *constructive* criterion by which methodologies *qua* rational reconstructions of history might be appraised.

(b) The methodology of historiographical research programmes.
History – to varying degrees – corroborates its rational reconstructions

I should like to present my proposal in two stages. First, I shall amend slightly the falsificationist historiographical meta-criterion just discussed, and then replace it altogether with a better one.

First, the slight amendment. If a universal rule clashes with a particular 'normative basic judgment', one should allow the scientific community time to ponder the clash: they may give up their particular judgment and submit to the general rule. 'Second-order' – historiographical – falsifications must not be rushed any more than 'first order' – scientific – ones.[4]

Secondly, since we have abandoned naive falsificationism in *method*, why should we stick to it in *meta-method*? We can easily replace it with

[1] Kuhn's vision was criticized from many quarters; cf. Shapere [1964] and [1967], Scheffler [1967] and especially the critical comments by Popper, Watkins, Toulmin, Feyerabend and Lakatos – and Kuhn's reply – in Lakatos and Musgrave [1970]. But none of these critics applied a systematic *historiographical* criticism to his work. One should also consult Kuhn's 1970 *Postscript* to the second edition of his [1962] and its review by Musgrave (Musgrave [1971]).

[2] Cf. Feyerabend [1970a], [1970b], and [1974]; and Kuhn [1970b].

[3] For instance, one may refer to the actual immediate impact of at least *some* 'great' negative crucial experiments, like that of the falsification of the parity principle. Or one may quote the high respect for at least *some* long, pedestrian, trial-and-error procedures which occasionally precede the announcement of a major research programme, which in the light of my methodology is, at best, 'immature science'. (Cf. chapter 1, p. 87; also cf. L. P. Williams's reference to the history of spectroscopy between 1870 and 1900 in his [1970].) Thus the judgment of the scientific élite, on occasions, goes also against *my* universal rules too.

[4] There is a certain analogy between this pattern and the occasional appeal procedure of the theoretical scientist against the verdict of the experimental jury; cf. chapter 1, pp. 42–6.

a methodology of scientific research programmes of second order, or if you wish, a methodology of historiographical research programmes.

While maintaining that a theory of rationality has to try to organize basic value judgments in universal, coherent frameworks, we do not have to reject such a framework immediately merely because of some anomalies or other inconsistencies. We should, of course, insist that a good rationality theory must anticipate further basic value judgments unexpected in the light of its predecessors or that it must even lead to the revision of previously held basic value judgments.[1] We then reject a rationality theory only for a better one, for one which, in this 'quasi-empirical' sense, represents a *progressive shift* in the sequence of research programmes of rational reconstructions. Thus this new – more lenient – meta-criterion enables us to compare rival logics of discovery and discern growth in 'meta-scientific' – methodological – knowledge.

For instance, Popper's theory of scientific rationality need not be rejected simply because it is 'falsified' by some actual 'basic judgments' of leading scientists. Moreover, on our new criterion, Popper's demarcation criterion clearly represents progress over its justificationist predecessors, and in particular, over inductivism. For, contrary to these predecessors, it rehabilitated the scientific status of falsified theories like phlogiston theory, thus reversing a value judgment which had expelled the latter from the history of science proper into the history of irrational beliefs.[2] Also, it successfully rehabilitated the Bohr–Kramers–Slater theory.[3] In the light of most justificationist theories of rationality the history of science is, at its best, a history of *pre*scientific preludes to some *future* history of science.[4] Popper's methodology enabled the historian to interpret more of the *actual* basic value judgments in the history of science as rational: in *this* normative–historiographical sense Popper's theory constituted progress. In the light of better rational reconstructions of science one can always reconstruct more of actual great science as rational.[5]

I hope that my modification of Popper's logic of discovery will be seen, in turn – on the criterion I specified – as yet a further step

---

[1] This latter criterion is analogous to the exceptional 'depth' of a theory which clashes with some basic statements available at the time and, at the end, emerges from the clash victoriously. (Cf. Popper's [1957a].) Popper's example was the inconsistency between Kepler's laws and the Newtonian theory which set out to explain them.

[2] Conventionalism, of course, had performed this historic role to a great extent before Popper's version of falsificationism.

[3] van der Waerden had thought that the Bohr–Kramers–Slater theory was bad: Popper's theory showed it to be good. Cf. van der Waerden [1967], p. 13 and Popper [1963a], pp. 242 ff; for a critical discussion cf. chapter 1, p. 82, nn. 1 and 2.

[4] The attitude of some modern logicians to the history of mathematics is a typical example; cf. my [1963–4], p. 3.

[5] This formulation was suggested to me by my friend Michael Sukale.

forward. For it seems to offer a coherent account of *more* old, isolated basic value judgments; moreover, it has led to new and, at least for the justificationist or naive falsificationist, surprising basic value judgments. For instance, according to Popper's theory, it was irrational to retain and further elaborate Newton's gravitational theory after the discovery of Mercury's anomalous perihelion; or again, it was irrational to develop Bohr's old quantum theory based on inconsistent foundations. From my point of view these were perfectly rational developments: some rearguard actions in the defence of defeated programmes – even after the so-called 'crucial experiments' – are perfectly rational. Thus my methodology leads to the reversal of those historiographical judgments which deleted these rearguard actions both from inductivist and from falsificationist party histories.[1]

Indeed, this methodology confidently predicts that where the falsificationist sees the instant defeat of a theory through a simple battle with some fact, the historian will detect a complicated war of attrition, starting long before, and ending after, the alleged 'crucial experiment'; and where the falsificationist sees consistent and unrefuted theories, it predicts the existence of hordes of known anomalies in research programmes progressing on possibly inconsistent foundations.[2] Where the conventionalist sees the clue to the victory of a theory over its predecessor in the former's intuitive simplicity, this methodology predicts that it will be found that victory was due to empirical degeneration in the old and empirical progress in the new programme.[3] Where Kuhn and Feyerabend see irrational change, I predict that the historian will be able to show that there has been rational change. The methodology of research programmes thus predicts (or, if you wish, 'postdicts') novel historical facts, unexpected in the light of extant (internal and external) historiographies and these predictions will, I hope, be corroborated by historical research. If they are, then the methodology of scientific research programmes will itself constitute a progressive problemshift.

Thus progress in the theory of scientific rationality is marked by discoveries of novel historical facts, by the reconstruction of a growing bulk of value-impregnated history as rational.[4] In other words, the theory of scientific

[1] Cf. chapter 1, section 3(c).

[2] Cf. chapter 1, pp. 52–86.

[3] Duhem himself gives only one explicit example: the victory of wave optics over Newtonian optics [1906], chapter VI, §10 (also see chapter IV, §4). But where Duhem relies on intuitive 'common sense', I rely on an analysis of rival problemshifts.

[4] One may introduce the notion of *'degree of correctness'* into the meta-theory of methodologies, which would be analogous to Popper's empirical content. Popper's empirical 'basic statements' would have to be replaced by quasi-empirical 'normative basic statements' (like the statement that 'Planck's radiation formula is arbitrary').

Let me point out here that the methodology of research programmes may be applied not only to norm-impregnated historical knowledge but to any normative knowledge, including even ethics and aesthetics. This would then supersede the naive falsificationist 'quasi-empirical' approach as outlined in n. 3, p. 124.

rationality progresses if it constitutes a 'progressive' historiographical research programme. I need not say that no such historiographical research programme can or should explain *all* history of science as rational: even the greatest scientists make false steps and fail in their judgment. Because of this *rational reconstructions remain for ever submerged in an ocean of anomalies. These anomalies will eventually have to be explained either by some better rational reconstruction or by some 'external' empirical theory.*

This approach does not advocate a cavalier attitude to the 'basic normative judgments' of the scientist. 'Anomalies' may be rightly ignored by the internalist *qua* internalist and relegated to external history only as long as the internalist historiographical research programme is *progressing*; or if a supplementary empirical externalist historiographical programme absorbs them *progressively*. But if in the light of a rational reconstruction the history of science is seen as increasingly irrational *without* a progressive externalist explanation (such as an explanation of the degeneration of science in terms of political or religious terror, or of an antiscientific ideological climate, or of the rise of a new parasitic class of pseudoscientists with vested interests in rapid 'university expansion'), then historiographical innovation, proliferation of historiographical theories, is vital. Just as scientific progress is possible even if one never gets rid of scientific anomalies, progress in rational historiography is also possible even if one never gets rid of historiographical anomalies. The rationalist historian need not be disturbed by the fact that actual history is more than, and, on occasions, even different from, internal history, and that he may have to relegate the explanation of such anomalies to external history. But this unfalsifiability of internal history does not render it immune to constructive, but only to negative, criticism – just as the unfalsifiability of a scientific research programme does not render it immune to constructive, but only to negative, criticism.

Of course, one can criticize internal history only by making the historian's (usually latent) methodology explicit, showing how it functions as a historiographical research programme. Historiographical criticism frequently succeeds in destroying much of fashionable externalism. An 'impressive', 'sweeping', 'far-reaching' external explanation is usually the hallmark of a weak methodological substructure; and, in turn, the hallmark of a relatively weak internal history (in terms of which most actual history is either inexplicable or anomalous) is that it leaves too much to be explained by external history. When a better rationality theory is produced, internal history may expand and reclaim ground from external history. The competition, however, is not as open in such cases as when two rival scientific research programmes compete. Externalist historiographical programmes which supplement internal histories based on naive methodologies (whether aware or unaware of the fact) are likely either to degenerate quickly or never even to get off the ground, for the simple reason that they set out to offer psychological or sociological 'explanations' of methodologically induced fantasies rather than of (more rationally interpreted) historical facts. Once an externalist account uses, whether consciously or not, a naive methodology (which can so easily creep into its 'descriptive' language), it turns into a fairy tale which, for all its apparent scholarly sophistication, will collapse under historiographical scrutiny.

Agassi has already indicated how the poverty of inductivist history opened the door to the wild speculations of vulgar-Marxists.[1] His falsificationist historiography, in turn, flings the door wide open to those trendy 'sociologists of knowledge' who try to explain the further (possibly unsuccessful) development of a theory 'falsified' by a 'crucial experiment' as the manifestation of the irrational, wicked, reactionary resistance by established authority to enlightened revolutionary innovation.[2] But in the light of the methodology of scientific research programmes such rearguard skirmishes are perfectly explicable *internally*: where some externalists see power struggle, sordid personal controversy, the rationalist historian will frequently find rational discussion.[3]

An interesting example of how a poor theory of rationality may impoverish history is the treatment of degenerating problemshifts by historiographical positivists.[4] Let us imagine for instance that in spite of the objectively progressing astronomical research programmes, the astronomers are suddenly all gripped by a feeling of Kuhnian 'crisis'; and then they all are converted, by an irresistible *Gestalt*-switch, to

---

[1] Cf. text to n. 1, p. 105. (The terminology 'wild speculation' is, of course, inherited from inductivist methodology. It should now be reinterpreted as 'degenerating programme'.)

[2] The fact that even degenerating externalist theories have been able to achieve some respectability was to a considerable extent due to the weakness of their previous internalist rivals. Utopian Victorian morality either creates false, hypocritical accounts of bourgeois decency, or adds fuel to the view that mankind is totally depraved; utopian scientific standards either create false, hypocritical accounts of scientific perfection, or add fuel to the view that scientific theories are no more than mere beliefs bolstered by some vested interests. This explains the 'revolutionary' aura which surrounds some of the absurd ideas of contemporary sociology of knowledge: some of its practitioners claim to have unmasked the bogus rationality of science, while, at best, they exploit the weakness of outdated theories of scientific rationality.

[3] For examples cf. Cantor [1971] and the Forman–Ewald debate (Forman [1969] and Ewald [1969]).

[4] I call '*historiographical positivism*' the position that history can be written as a completely *external* history. For historiographical positivists history is a purely empirical discipline. They deny the existence of objective standards as opposed to mere beliefs about standards. (Of course, they too hold beliefs about standards which determine the choice and formulation of their historical problems.) This position is typically Hegelian. It is a special case of *normative positivism*, of the theory that sets up might as the criterion of right. (For a criticism of Hegel's ethical positivism cf. Popper [1945], volume 1, pp. 71–2, volume 2, pp. 305–6 and Popper [1962].) Reactionary Hegelian obscurantism pushed values back completely into the world of facts; thus reversing their separation by Kantian philosophical enlightenment.

astrology. I would regard this catastrophe as a horrifying *problem*, to be accounted for by some empirical externalist explanation. But not a Kuhnian. All he sees is a 'crisis' followed by a mass conversion effect in the scientific community: an ordinary revolution. Nothing is left as problematic and unexplained.[1] The Kuhnian psychological epiphenomena of 'crisis' and 'conversion' can accompany either objectively progressive or objectively degenerating changes, either revolutions or counterrevolutions. But this fact falls outside Kuhn's framework. Such historiographical anomalies cannot be formulated, let alone be progressively absorbed, by his historiographical research programme, in which there is no way of distinguishing between, say, a 'crisis' and 'degenerating problemshift'. But such anomalies might even be predicted by an externalist historiographical theory based on the methodology of scientific research programmes that would specify social conditions under which degenerating research programmes may achieve socio-psychological victory.

(c) Against aprioristic and anti-theoretical approaches to methodology

Finally, let us contrast the theory of rationality here discussed with the strictly aprioristic (or, more precisely, 'Euclidean') and with the anti-theoretical approaches.[2]

'Euclidean' methodologies lay down *a priori general rules* for scientific appraisal. This approach is most powerfully represented today by Popper. In Popper's view there must be the constitutional authority of an *immutable statute law* (laid down in his demarcation criterion) to distinguish between good and bad science.

Some eminent philosophers, however, ridicule the idea of statute law, the possibility of any valid demarcation. According to Oakeshott and Polanyi there must be – and can be – no statute law at all: only case law. They may also argue that even if one mistakenly allowed for statute law, statute law too would need authoritative interpreters. I think that Oakeshott's and Polanyi's position has a great deal of truth in it. After all, one must admit (*pace* Popper) that until now all the

---

[1] Kuhn seems to be in two minds about objective scientific progress. I have no doubt that, being a devoted scholar and scientist, he *personally* detests relativism. But his *theory* can either be interpreted as denying scientific progress and recognizing only scientific change; or, as recognizing scientific progress but as 'progress' marked solely by the march of actual history. Indeed, on his criterion, he would have to describe the catastrophe mentioned in the text as a proper 'revolution'. I am afraid this might be one clue to the unintended popularity of his theory among the New Left busily preparing the 1984 'revolution'.

[2] The technical term 'Euclidean' (or rather 'quasi-Euclidean') means that one starts with universal, high level propositions ('axioms') rather than singular ones. I suggested that the 'quasi-Euclidean' versus 'quasi-empirical' distinction is more useful than the '*a priori*' versus '*a posteriori*' distinction (see volume 2, chapters 1 and 2).

Some of the 'apriorists' are, of course, empiricists. But empiricists may well be apriorists (or, rather, 'Euclideans') on the meta-level here discussed.

'laws' proposed by the apriorist philosophers of science have turned out to be wrong in the light of the verdicts of the best scientists. Up to the present day it has been the scientific standards, as applied 'instinctively' by the scientific *élite* in *particular* cases, which have constituted the main – although not the exclusive – yardstick of the philosopher's *universal* laws. But if so, methodological progress, at least as far as the most advanced sciences are concerned, still lags behind common scientific wisdom. Is it not then *hubris* to try to impose some *a priori* philosophy of science on the most advanced sciences? Is it not *hubris* to demand that if, say, Newtonian or Einsteinian science turns out to have violated Bacon's, Carnap's or Popper's *a priori* rules of the game, the business of science should be started anew?

I think it is. And, indeed, the methodology of historiographical research programmes implies a pluralistic system of authority, partly because the wisdom of the scientific jury and its case law has not been, and cannot be, fully articulated by the philosopher's statute law, and partly because the philosopher's statute law may occasionally be right when the scientists' judgment fails. I disagree, therefore, both with those philosophers of science who have taken it for granted that general scientific standards are immutable and reason can recognize them *a priori*,[1] and with those who have thought that the light of reason illuminates only particular cases. The methodology of historiographical research programmes specifies ways both for the philosopher of science to learn from the historian of science and *vice versa*.

But this two-way traffic need not always be balanced. The statute law approach should become much more important when a tradition degenerates[2] or a new bad tradition is founded.[3] In such cases statute law may thwart the authority of the corrupted case law, and slow down or even reverse the process of degeneration.[4] When a scientific school degenerates into pseudoscience, it may be worthwhile to force a methodological debate in the hope that working scientists will learn more from it than philosophers (just as when ordinary language degenerates into, say, journalese, it may be worthwhile to invoke the rules of grammar).[5]

---

[1] Some might claim that Popper does *not* fall into this category. After all, Popper defined 'science' in such a way that it should include the refuted Newtonian theory and exclude unrefuted astrology, Marxism and Freudianism.

[2] This seems to be the case in modern particle physics; or according to some philosophers and physicists even in the Copenhagen school of quantum physics.

[3] This is the case with some of the main schools of modern sociology, psychology and social psychology.

[4] This, of course, explains why a good methodology – 'distilled' from the mature sciences – may play an important role for immature and, indeed, dubious disciplines. While Polanyiite academic autonomy should be defended for departments of theoretical physics, it must not be tolerated, say, in institutes for computerized social astrology, science planning or social imagistics. (For an authoritative study of the latter, cf. Priestley [1968].)

[5] Of course, a critical discussion of scientific standards, possibly leading even to their improvement, is impossible without articulating them in general terms; just as if one

(d) Conclusion

In this paper I have proposed a 'historical' method for the evaluation of rival methodologies. The arguments were primarily addressed to the philosopher of science and aimed at showing how he can – and should – learn from the history of science. But the same arguments also imply that the historian of science must, in turn, pay serious attention to the philosophy of science and decide upon which methodology he will base his internal history. I hope to have offered some strong arguments for the following theses. First, each methodology of science determines a characteristic (and sharp) demarcation between (primary) internal history and (secondary) external history and, secondly, both historians and philosophers of science must make the best of the critical interplay between internal and external factors.

Let me finally remind the reader of my favourite – and by now well-worn – joke that history of science is frequently a caricature of its rational reconstructions; that rational reconstructions are frequently caricatures of actual history; and that some histories of science are caricatures both of actual history and of its rational reconstructions.[1] This paper, I think, enables me to add: *Quod erat demonstrandum.*

wants to challenge a language, one has to articulate its grammar. Neither the conservative Polanyi nor the conservative Oakeshott seem to have grasped (or to have been inclined to grasp) the *critical* function of language – Popper has. (Cf. especially Popper [1963*a*], p. 135.)

[1] Cf. e.g. volume 2, chapter 1, p. 4 or volume 2, chapter 8, p. 178, n. 3.

3

Popper on demarcation and induction*

INTRODUCTION

Popper's ideas represent the most important development in the philosophy of the twentieth century; an achievement in the tradition – and on the level – of Hume, Kant, or Whewell. Personally, my debt to him is immeasurable: more than anyone else, he changed my life. I was nearly forty when I got into the magnetic field of his intellect. His philosophy helped me to make a final break with the Hegelian outlook which I had held for nearly twenty years.[1] And, more important, it provided me with an immensely fertile range of problems, indeed, with a veritable research programme. Work on a research programme is, of course, a critical affair, and it is no wonder that my work on Popperian problems has frequently led me into conflict with Popper's own solutions.[2]

In the present note I shall sketch my position on what Popper himself frequently referred to as the two main problems of his now classical *Logik der Forschung*: the problem of demarcation and the problem of induction. Popper first gave a solution of the problem of demarcation and then, having claimed that 'the problem of induction is only an instance or facet of the problem of demarcation', he applied

---

* This paper was written in 1970–1 and first appeared in English as Lakatos [1974]. Lakatos's acknowledgments read: 'I should like to thank my friends Colin Howson, Alan Musgrave, Helmut Spinner, John Worrall, Elie Zahar and especially John Watkins for their critical scrutiny of previous versions. Their comments and objections are referred to throughout the paper.' (*Eds.*)

[1] Since Hegel each generation has unfortunately needed – and has fortunately had – philosophers to break Hegel's spell on young thinkers who so frequently fall into the trap of 'impressive and all-explanatory theories [like Hegel's or Freud's] which act upon weak minds like revelations' (cf. Popper [1963a] p. 39). Moore was the liberator in Cambridge before the first war, Popper in the London School of Economics after the second.

[2] Cf. my [1968b], [1968c], [1970a] and [1971b] (see this volume, chapters 1 and 2, and volume 2, chapter 8). In these papers I tried to explain why I think Popper's philosophy is so immensely important. The reason why I continue to criticize various aspects of Popper's philosophy is my conviction that it represents the most advanced philosophy of our time, and that philosophical progress can only be based – even if 'dialectically' – on its achievements. Although the present paper is meant to be self-contained, some of its formulations had, for the sake of brevity, to be crude. The reader will find it helpful, and perhaps, at times necessary, to compare it primarily with chapter 1 for more detailed expositions of some issues.

his demarcation criterion to solve the problem of induction.[1] In my view, Popper's solution of *the problem of demarcation* is a great achievement but can be improved upon, and even in its improved form opens up large problems hitherto unsolved. But I think that *the problem of induction* is definitely more than merely an 'instance or facet' of the problem of demarcation. Popper, in his early philosophy, offered decisive criticisms of earlier solutions of the problem – or rather problems – of induction, and suggested a purely *negative* solution. His later philosophy (based on the idea of truth-content and verisimilitude) involved a shift of the problem and also a *positive* solution of the shifted problem; but, to my mind, he has not yet realized the *full* implications of his own achievement.

I   POPPER ON DEMARCATION
(a) *Popper's game of science*

Popper's 'logic of scientific discovery' (or 'methodology', or 'system of appraisals' or 'demarcation criterion' or 'definition of science')[2] is a theory of scientific rationality; more specifically, a set of standards for scientific theories. Originally people had hoped that a 'logic of discovery' would provide them with a mechanical book of rules for solving problems. This hope was given up: for Popper the logic of discovery or 'methodology' consists merely of a set of (tentative and far from mechanical) rules for the appraisal of ready articulated theories. All the rest he sees as a matter for an empirical psychology of discovery, outside the normative realm of the logic of discovery. This represents an all-important shift in the problem of normative philosophy of science. *The term 'normative' no longer means rules for arriving at solutions, but merely directions for the appraisal of solutions already there.* Some philosophers are still not aware of this problem-shift.[3]

Popper's logic of discovery contains 'proposals', 'conventions', about when a theory should be taken seriously (when a crucial experiment could, and indeed has been, devised against it) and about when a theory should be rejected (when it has failed a crucial experiment). Popper's logic of discovery gives, for the first time in the context of a major epistemological research programme, a new role to experience

[1] Cf. e.g. chapter 1 of his [1934]; and also chapter 1 of his [1963a], esp. pp. 52ff and 58. (The phrase quoted is on p. 54.)

[2] This profusion of synonyms has proved to be rather confusing.

[3] I should like to say here that I always had *doubts* about whether this (no doubt progressive) problemshift had not gone a bit too far. This shift had been even more pronounced in the philosophy of mathematics than in the philosophy of science. Following Pólya, I have held that there might well be a *limbo* for a 'genuine' heuristic which is rational and non-psychologistic; it was in this vein that I expressed some reservations concerning Tarski's novel use of the term 'methodology'; cf. my [1963–4], p. 4, n. 4. But I cannot here pursue this matter further.

in science: scientific theories are not based on, established or 'pro-babilified' by, 'facts' but rather eliminated by them. For Popper, pro-gress consists of an incessant, ruthless, revolutionary confrontation of bold, speculative theories and repeatable observations, and of the subsequent fast elimination of the defeated theories: 'The method of trial and error is a *method of eliminating false theories* by observation statements'.[1] 'Conjectures [are] boldly put forward for trial, to be eliminated if they clash with observations'.[2] Thus, the history of science is seen as a series of *duels* between theory and experiment, duels in which only experiments can score decisive victories. The theoretician proposes a scientific theory; some basic statements con-tradict it; if one of these becomes 'accepted'[3] the theory is 'refuted' and must be rejected and a new one has to take its place. 'What ultimately decides the fate of a theory is the result of a test, *i.e.*, an agreement about basic statements.'[4] Popper realizes, of course, that we always test large systems of theories rather than isolated ones. But he does not regard this as an insurmountable difficulty: he suggests that we should *guess* – and, indeed, *agree* – which part of such a system is responsible for the refutation (that is, which part is to be regarded as false), perhaps helped by independent tests of some portions of the system. Within Popper's philosophy this kind of guessing is absolutely indispensable: if one were allowed to blame refutations upon the initial conditions *all the time*, no major theory need ever be rejected. He is not content with tests which are designed to test large systems: he calls on the scientist to specify, beforehand, those experiments which will, if their outcome is negative, lead to the falsification of the very heart of the system.[5] He demands of the scientist that he specify in advance under what experimental conditions he would give up his *most basic* assumptions.[6] This, indeed, is the gist of Popper's 'demar-cation criterion' or, to use a better term, of his definition of science.[7]

Popper's definition of science can best be put in terms of 'conven-tions' or 'rules' governing the '*game of science*'.[8]

---

[1] Popper [1963a], p. 56 (Popper's italics). Cf. *below*, p. 155, n. 3.
[2] Popper [1963a], p. 46.
[3] For the conditions of acceptance of basic statements, cf. Popper [1934], section 22 and chapter 1, pp. 23–4.   [4] Popper, *ibid.*, section 30.
[5] For references, cf. *below*, p. 147, n. 1 and p. 150, n. 2.
[6] Cf. p. 146, text to n. 3. Also chapter 1, p. 24.
[7] Cf. chapter 1, p. 25. For an interesting discussion cf. also Musgrave [1968].
[8] Popper [1934], sections 11 and 85. The first paragraph in section 11 explains why he gave the title *The Logic of Scientific Discovery* to his book and is worth quoting: 'Methodological rules are here regarded as *conventions*. They might be described as the rules of the game of empirical science. They differ from the rules of pure logic rather as do the rules of chess, which few would regard as part of *pure* logic: seeing that the rules of pure logic govern transformations of linguistic formulae, the result of an inquiry into the rules of chess could perhaps be entitled "The Logic of Chess", but hardly "Logic" pure and simple. (Similarly, the result of an inquiry into the rules of the game of science – that is, of scientific discovery – may be entitled "The Logic of Scientific Discovery".)'

The opening move must be a *consistent, falsifiable hypothesis*; that is, a consistent hypothesis which has agreed-on potential falsifiers. A potential falsifier is a 'basic statement' whose truth-value is decidable with the help of the experimental techniques of the time. The scientific jury must agree *unanimously* that there is an experimental technique which will enable them to assign a truth-value to the 'basic statement'. (Unanimity can, of course, be reached by expelling the minority as pseudoscientists or cranks.[1])

The next move is the repeated performance of the test in a con-trolled experiment,[2] and the second decision of the jury on what actual truth-value (truth or falsehood) to attribute to the potential falsifier. (If this second decision is not unanimous there are two possible moves: either the status of 'potential falsifier' must be withdrawn and, unless a replacement is found, the opening move cancelled; *or, alternatively, the dissenting minority must be declared cranks and excluded from the jury.*[3])

If the second verdict is *negative,* and the potential falsifier is rejected, then the hypothesis is declared 'corroborated', which only means that it invites *further* challenges. If the second verdict is *positive,* and the potential falsifier accepted, then the hypothesis is declared 'falsified', which means that it is *rejected,* 'overthrown', 'dropped', buried with military honours.[4] (In 1960 Popper introduced a new rule: military pomp can only be awarded to an eliminated hypothesis, if, before it was falsified, it was at least once – in a different experiment – corroborated.[5])

After the burial a new hypothesis is invited. This new hypothesis must, however, explain the partial success, if any, of its predecessor, and also something *more.* A hypothesis, however novel in its intuitive aspects, will not be allowed to be proposed, unless it has novel empirical content in excess of its predecessor. If it has no such excess content,

---

[1] I am afraid Popper did not spell out this implication; although he mentions, as if it were a matter of fact, that cranks do not 'seriously disturb the working of the various social institutions which have been designed to further scientific objectivity' (Popper [1945], volume II, p. 218). Then he goes on: '*Only political power*...can impair [their] functioning.' (Also cf. his [1957b], p. 32.) I wonder.

[2] For the concept of 'controlled experiment', cf. chapter 1, p. 27, n. 4.

[3] Cf. *above*, n. 1.

[4] Popper [1934], sections 3 and 4. Also cf. section 22 ('Falsifiability and Falsification') for 'special rules...which determine under what conditions a system is to be regarded as falsified'. It is intriguing that *at least in this particular section* (section 22) there is not a word about the identification of 'falsification', in the sense here described, with 'overthrow' or 'elimination'. Some of my friends used this omission as evidence that Popper did not advocate such an identification, but left the problem of elimination – as opposed to 'falsification' – open. But, in other passages, especially in works related to social sciences (cf. e.g., his [1957b], pp. 133–4), Popper clearly identifies 'falsification' with 'rejection' and 'elimination'. And if falsification does *not* mean rejection, what *does* it mean? Popper tells us nothing about how we can continue to play the game of science with a *falsified* hypothesis.

[5] Cf. his [1963a], pp. 242–5.

the referee will declare it '*ad hoc*' and make the proposer withdraw it. If the new hypothesis is not *ad hoc*, the standard procedure for falsifiable hypotheses, as described above, is followed for the new hypothesis.[1]

This 'scientific game', if properly played, will 'progress' in the sense that the theories subsequently proposed will have increasing generality (or 'empirical content'); they will pose ever deeper *questions* about the universe.[2]

Just as the rules of chess do not explain why some people should play the game and, indeed, devote their life to it, the rules of science do not explain why some people should play the game of science and, indeed, devote their life to it. The rules decide whether a particular *move* is 'proper' (or 'scientific') or not, but they remain silent about whether the *game as a whole* is 'proper' (or 'rational') or not. The rules say nothing either about the (psychological) motives of the players or about the (rational) purpose of the game. One can of course play the game as a genuine game and enjoy it for itself, without caring for its purpose or being aware of one's motives.

Note. I had endless discussions with some of my Popperian friends about the identification of Popper and 'Popper$_1$' (the naive methodological falsificationist) in my [1968*b*] and [1970] and in this section. I should like to say that never in my life have I experienced more sharply the pains of the historian than in this analysis. My [1968*c*], especially pp. 384f. shows that then I identified Popper with my 'Popper$_2$', the sophisticated methodological falsificationist. In my [1968*b*] I shifted my position and then suggested that *Popper conflated the two positions* (see volume 2, chapter 8). I held the same position in the text of my [1970], but in the Appendix, *I identified Popper essentially with Popper$_1$*, the naive methodological falsificationist (see this volume, chapter 1). I maintain this position in my present paper, but with the grave suspicion that I might have missed some vital ingredient in the whole analysis. Could it be that the problem of *The Logic of Scientific Discovery* was a different one from the one I reconstructed? Is *my* split of Popper into Popper$_1$ and Popper$_2$ a result of *my* problemshift? No doubt, the most characteristic Popperian$_1$ quotations occur in Popper's *Poverty of Historicism* and *Open Society*. Are these no more than occasional exaggerations which occur only in his passionate condemnation of social pseudosciences? But surely Popper himself describes his original problem as how to demarcate science from pseudoscience! I confess that I am now at a loss as an exegetic and only hope that Popper's reply will dissolve my puzzlement.*

[1] Following Popper's new rule referred to in the previous footnote, the anti-*adhoc*ness rules may also be tightened; and we have to distinguish between *adhoc*$_1$ and *adhoc*$_2$, cf. volume 2, chapter 8, pp. 170–81, esp. p. 180, n. 1.

[2] Popper [1934], section 85, last sentence.

* For Popper's reply, see his [1974]. (*Eds.*)

(b) How can one criticize the rules of the scientific game?

The rules of the game are *conventions*, and can be formulated in terms of a *definition*.[1] How can one criticize a definition, in particular, if one interprets it nominalistically?[2] A definition is then a mere abbreviation, a tautology. What can one criticize about a tautology? Popper claims that his definition of science is 'fruitful': 'that a great many points can be clarified and explained with its help'. He quotes Menger: 'Definitions are dogmas; only the conclusions drawn from them can afford us any new insight'.[3] But how can a definition have explanatory power or afford new insights? Popper's answer is this: 'It is only from the consequences of my definition of empirical science, and from the methodological decisions which depend upon this definition, that the scientist will be able to see how far it conforms to his intuitive idea of the goal of his endeavours'.[4]

This answer complies with Popper's general position that conventions can be criticized by discussing their 'suitability' relative to some purpose: 'As to the suitability of any convention opinions may differ; and a reasonable discussion of these questions is only possible between parties having some purpose in common. The choice of that purpose ...goes beyond rational argument'.[5] But Popper, in his *Logik der Forschung* never specifies a *purpose* of the game of science that would go beyond what is contained in its rules. The idea that the *aim* of science is *truth*, occurs in his writings for the first time in 1957.[6] In his *Logik der Forschung* the quest for truth may be a psychological *motive* of scientists – it is not a rational *purpose* of science.[7]

Even in Popper's later writings we find no suggestion of how to appraise one consistent set of rules (or demarcation criterion) as leading more successfully towards Truth than another.[8] Indeed, the thesis that any such argument connecting method and success is impossible, has been a cornerstone of Popper's philosophy from 1920 to 1970. Thus I conclude that Popper never offered a theory of rational criticism of consistent conventions.[9] He does not answer the

---

[1] Cf. Popper [1934], sections 4 and 11.

[2] For an excellent discussion of the distinction between nominalism and realism (or, as Popper prefers to call it, 'essentialism') in the theory of definitions, cf. Popper [1945c], chapter 11, and [1963a], p. 20. Also cf. chapter 1, p. 41.

[3] Popper [1934], section 11.   [4] *Ibid.*

[5] *Ibid.*, section 4.   [6] Cf. his [1957a].

[7] He calls the search for truth 'the strongest [unscientific] motive' ([1934], section 85). Also cf. *above*, pp. 160–1.

[8] Popper's crucial arguments against various inductivist theories of science show that they are *inconsistent*. On the other hand, he admits that the conventionalist theory is consistent, 'self-contained and defensible', and concludes: 'My conflict with the conventionalists is not one that can be settled by a detached theoretical discussion' (Popper [1934], section 19). Is then the choice between *consistent* sets of rules a matter of subjective taste?

[9] In the early 1960s Popper adopted Bartley's comprehensively critical rationalism. According to this theory all propositions accepted by a rational person must be open

question: '*Under what conditions would you give up your demarcation criterion?*'.[1]

But the question can be answered. I shall give my answer in two stages: first a naive and then a more sophisticated answer. I start by recalling how Popper, according to his own account, had arrived at his criterion. He thought, like the best scientists of his time, that Newton's theory, although refuted, was a wonderful scientific achievement: that Einstein's theory was still better; and that astrology, Freudianism and twentieth-century Marxism were pseudoscientific. His problem was to find a definition of science from which these '*basic judgments*' concerning each of these theories followed; and he offered a novel solution. Now let us agree *provisionally* on the meta-criterion that *a rationality theory – or demarcation criterion – is to be rejected if it is inconsistent with accepted 'basic value judgments' of the scientific community.*[2] Indeed, this meta-methodological rule would seem to correspond to Popper's methodological rule that a scientific theory is to be rejected if it is inconsistent with an ('empirical') basic statement unanimously accepted by the scientific community. Popper's whole methodology rests on the contention that there exist (relatively) singular statements on whose truth-value scientists can reach unanimous agreement; without such agreement there would be a 'new Babel' and 'the soaring edifice of science would soon lie in ruins'.[3] But, even if there is agreement about 'basic' statements, if there were no agreement whatsoever about how to appraise scientific achievement relative to this 'empirical basis', would not the soaring edifice of science equally soon lie in ruins? No doubt it would. Although there has been little agreement concerning a *universal* criterion of the scientific character of theories, there has been considerable agreement over the last two centuries concerning *single* achievements. While there has been no *general* agreement concerning a theory of scientific rationality, there has been considerable agreement concerning the rationality of a *particular* step in the game – was it scientific or crankish? A general definition of science thus must reconstruct the acknowledgedly best games and the most esteemed gambits as 'scientific'; if it fails to do so, it has to be rejected.

Then let us propose tentatively that, *if a demarcation criterion is*

to criticism. But the basic weakness of this position is its emptiness. There is not much point in affirming the criticizability of any position we hold without concretely specifying the forms such criticism might take. (For an interesting criticism of Bartley's position cf. Watkins [1971].)

[1] This flaw is the more serious since Popper himself has expressed qualifications about his criterion. For instance in his [1963a] he describes 'dogmatism', that is, treating anomalies as a kind of 'background noise', as something that is 'to some extent necessary' (p. 49). But on the next page he identifies this 'dogmatism' with 'pseudoscience'. Is then pseudoscience 'to some extent necessary'? Also, cf. chapter 1, p. 89, n. 5.

[2] 'Basic value judgments' would sound better in German: '*normative Basissätze*'.

[3] Popper [1934], section 29.

inconsistent with the basic appraisals of the scientific élite, it should be given up.[1] This meta-criterion was suggested to me by Popper's own description of his original problem, and by his own brand of method-ological falsificationism (but, I should stress, one can accept Popper's falsificationism and yet reject this meta-falsificationism). However, *if* we apply this meta-criterion (which I am going to reject later), Popper's demarcation criterion – that is, Popper's rules of the game of science – has to be rejected.[2]

(c) A quasi-Polanyiite 'falsification' of Popper's demarcation criterion

Popper's demarcation criterion can indeed be easily 'falsified' by using the meta-criterion proposed in the last section; that is, by showing that in its light the best scientific achievements were unscientific and that the best scientists, in their greatest moments, broke the rules of Popper's game of science.

Popper's basic rule is that *the scientist must specify in advance under what experimental conditions he will give up even his most basic assumptions:*

Criteria of refutation have to be laid down beforehand: it must be agreed which observable situations, if actually observed, mean that the theory is refuted. But what kind of clinical responses would refute to the satisfaction of the analyst *not merely a particular clinical diagnosis but psychoanalysis itself?* And have such criteria even been discussed or agreed upon by analysts?[3]

In the case of psychoanalysis Popper was right: no answer has been forthcoming. Freudians have been nonplussed by Popper's basic chal-lenge concerning scientific honesty. Indeed, they have refused to specify experimental conditions under which they would give up their basic assumptions. For Popper this is the hallmark of their intellectual

---

[1] This approach, of course, does not mean that we *believe* that the scientists' 'basic judgments' are unfailingly rational; it only means that we *accept* them in order to criticize universal definitions of science. (If we add that no such *universal* criterion has been found and no such *universal* criterion will ever be found, the stage is set for Polanyi's conception of the lawless closed autocracy of science.)

The idea of this meta-criterion may be seen as a 'quasi-empirical' self-application of Popperian falsificationism. I had introduced this 'quasi-empiricalness' earlier in the context of mathematical philosophy. We may abstract from *what* flows in the logical channels of a deductive system, whether it is something certain or something fallible, whether it is truth and falsehood or probability and improbability, or even moral or scientific desirability and undesirability: it is the *how* of the flow which decides whether the system is negativist, 'quasi-empirical', dominated by *modus tollens* or whether it is justificationist, 'quasi-Euclidean', dominated by *modus ponens*. (Cf. volume 2, chapter 2.) This 'quasi-empirical' approach may be applied to *any* kind of normative knowledge like ethical or aesthetic, as has already been done by Watkins in his [1963] and [1967]. But now I prefer another approach: cf. *below*, p. 152, n. 2.

[2] It may be noted that this meta-criterion does not have to be construed as psycho-logical, or 'naturalistic' in Popper's sense. (Cf. his [1934], section 10.) The definition of the 'scientific élite' is not an empirical matter.

[3] Popper [1963a], p. 38, n. 3 (my italics). This, of course, is equivalent to his celebrated 'demarcation criterion' between science and pseudoscience – or, as he put it, 'meta-physics'. (For this point, also cf. Agassi [1964], section VI.)

dishonesty. But what if we put Popper's question to the Newtonian scientist: 'What kind of observation would refute to the satisfaction of the Newtonian not merely a particular Newtonian explanation but Newtonian dynamics and gravitational theory itself? And have such criteria even been discussed or agreed upon by Newtonians?' The Newtonian will, alas, scarcely be able to give a positive answer.[1] But then, if psychoanalysts are to be condemned as dishonest by Popper's standards, must not Newtonians be similarly condemned?

Popper may certainly withdraw his celebrated challenge and demand falsifiability – and rejection on falsification – only for *systems* of theories, including initial conditions and all sorts of auxiliary and observational theories. This is a considerable withdrawal, for it allows the imaginative scientist to save his pet theory by suitable lucky alterations in some odd corner of the theoretical maze. But even Popper's mitigated rule will make life impossible for the most brilliant scientist. For, in large research programmes there are always known anomalies: normally the researcher puts them aside and follows the positive heuristic of the programme.[2] In general he rivets his attention on the positive heuristic rather than on the distracting anomalies, and hopes that the 'recalcitrant instances' will be turned into confirming instances as the programme progresses. On Popper's terms, even great scientists use forbidden gambits, *ad hoc* stratagems: instead of regarding Mercury's anomalous perihelion as a falsification of the Newtonian theory of our planetary system and thus as a reason for its rejection, most of them shelved it as a problematic instance to be solved at some later stage – or offered *ad hoc* solutions. This methodological attitude of treating as *anomalies* what Popper would regard as counterexamples is commonly accepted by the best scientists. Some of the research programmes now held in highest esteem by the scientific community progressed in an ocean of anomalies.[3] Rejection of such work by Popper as irrational ('uncritical') implies – at least on our quasi-Polanyiite meta-criterion – a falsification of his definition.

Moreover, for Popper, *an inconsistent system* does not forbid any observable state of affairs and working on it must be invariably regarded as irrational: 'a self-contradictory system must be rejected... [because it] is uninformative... No statement is singled out... since all are derivable'.[4] But some of the greatest scientific research programmes progressed on inconsistent foundations.[5] Indeed, in such cases the best scientists' rule is frequently: 'Allez en avant et la foi vous viendra'. This anti-Popperian rule secured a sanctuary for the infinitesimal calculus hounded by Bishop Berkeley, and for naive set theory in the period of the first paradoxes. Indeed, if the game of science had been played according to Popper's rule book, Bohr's 1913 paper would

[1] Cf. chapter 1, pp. 16–17.
[2] *Ibid.*, exp. p. 50 ff.
[3] *Ibid.*, p. 52 ff.
[4] Cf. Popper [1934], section 24.
[5] Cf. chapter 1, esp. p. 55 ff.

never have been published, inasmuch as it was inconsistently grafted on to Maxwell's theory, and Dirac's delta functions would have been suppressed until Schwartz.

In general, Popper stubbornly overestimates the immediate striking force of purely negative criticism. 'Once a mistake, or a contradiction, is pinpointed, there can be no verbal evasion: it can be proved, and that is that.'[1]

This is how some of the 'basic' appraisals of the scientific *élite* 'falsify' Popper's definition of science and scientific ethics.

Note. I do not actually hold the meta-criterion described in section (b) and applied in section (c). I shall negate the theses of both sections in what follows. I only chose this Socratic–Popperian dialectical way of developing my position because I think this is the best way of developing a complex argument: by asking a simple question, giving a simple answer and then by criticizing the answer (and possibly the question), thus being led to more sophisticated questions and to more sophisticated solutions. This approach also suggests that the dialectic does not end in some 'final solution'.

(d) An amended demarcation criterion

One can easily amend Popper's definition of science so that it no longer rules out essential gambits of actual science. I tried to bring about such an amendment, primarily by shifting the problem of appraising theories to the problem of appraising historical series of theories, or, rather, of 'research programmes', and by changing Popper's rules of theory rejection.[2]

First, *one may 'accept' not only basic but also universal statements as conventions: indeed, this is the most important clue to the continuity of scientific growth.*[3] The basic unit of appraisal must be not an isolated theory or conjunction of theories but rather a *research programme*, with a conventionally accepted (and thus by provisional decision 'irrefut-

---

[1] Popper [1959a], p. 394. He adds: 'Frege did not try evasive manoeuvres when he received Russell's criticism'. But of course he did. (Cf. Frege's *Postscript* to the second edition of his *Grundgesetze*.) This historiographical mistake may also be related to Popper's earlier overconfidence in the unambiguity of mathematical reasoning. Also cf. volume 2, chapter 8, p. 157, n. 3.

[2] Cf. chapter 1 and volume 2, chapter 8. Popper always held, and in his later philosophy particularly emphasized, that 'the influence of [some] nontestable metaphysical theories upon science exceeded that of many testable theories', and even started to talk about 'metaphysical research programmes'. (Cf. chapter 1, p. 95, n. 6.) But whereas Popper acknowledged the *influence* of metaphysics *upon* science, I see metaphysics as an integral part of science. For Popper – and for Agassi and Watkins – metaphysics is *merely* 'influential'; I specify concrete patterns of appraisal. And these *conflict* with Popper's earlier appraisals of 'falsifiable' theories which he has not yet abandoned.

[3] Popper does not permit this: 'There is a vast difference between my views and conventionalism. I hold that what characterises the empirical method is just this: our conventions determine the acceptance of the *singular*, not of *universal* statements'. (Popper [1934], section 30.)

able') *hard core* and with a *positive heuristic* which defines problems, foresees anomalies and turns them victoriously into examples according to a preconceived plan. The scientist lists anomalies, but, as long as his research programme sustains its momentum, ignores them. *It is primarily the positive heuristic of his programme, not the anomalies, which dictate the choice of his problems.*[1] Only when the driving force of the positive heuristic weakens, may more attention be given to anomalies. (The methodology of research programmes can explain in this way *the relative autonomy of theoretical science*; Popper's disconnected chains of conjectures and refutations cannot.)

The *appraisal* of large units like research programmes is in one sense much more liberal and in another much more strict than Popper's appraisal of theories. This new appraisal is *more tolerant* in the sense that it allows a research programme to outgrow infantile diseases, such as inconsistent foundations and occasional *ad hoc* moves. Anomalies, inconsistencies, *ad hoc* stratagems can be consistent with progress. The old rationalist dream of a mechanical, semi-mechanical or at least fast-acting method for showing up falsehood, unprovenness, meaningless rubbish or even irrational choice has to be given up. It takes a long time to appraise a research programme: Minerva's owl flies at dusk. But this new appraisal is also *more strict* in that it demands not only that a research programme should successfully predict novel facts, but also that the protective belt of its auxiliary hypotheses should be largely built according to a preconceived unifying idea, laid down in advance in the positive heuristic of the research programme.[2]

It is very difficult to decide, especially if one does not demand progress at each single step, when a research programme has degenerated hopelessly; or when one of two rival programmes has achieved a decisive advantage over the other. There can be no 'instant rationality'. *Neither the logician's proof of inconsistency nor the experimental scientist's verdict of anomaly can defeat a research programme at one blow.* One can be 'wise' only after the event. Nature may shout NO, but human ingenuity – contrary to Weyl and Popper[3] – may always be able

---

[1] Agassi, in some passages, seems to deny this: 'Learning from experience is learning from a refuting instance. The refuting instance then becomes a problematic instance'. (Agassi [1964b], p. 201.) In his [1969] he attributes to Popper the statement that 'we learn from experience by refutations' (p. 169), and adds that, according to Popper, one can learn *only* from refutation but not from corroboration (p. 167). But this is a very one-sided theory of learning from experience. (Cf. chapter 1, p. 36, n. 2, and p. 38.)

Feyerabend, in his [1969b], says that '*negative instances suffice in science*'. (He adds in a footnote that he omits Popper's 'somewhat strange theory of corroboration'.) These problems of demarcation are, of course, closely connected with the problem of induction; also cf. *below*, p. 157, n. 6.

[2] In my [1970a] I called patched-up developments which did not meet such criteria *ad hoc*$_3$ stratagems (see chapter 1). Planck's first correction of the Lummer–Pringsheim formula was *ad hoc* in *this* sense. A particularly good example is Meehl's anomaly. (Cf. chapter 1, p. 88, nn. 2 and 4.)

[3] Popper [1934], section 85.

to shout louder. With sufficient brilliance, and some luck, any theory, even if it is false, can be defended 'progressively' for a long time.

But when should a particular theory, or a whole research programme, be rejected? I claim, only if there is a better one to replace it.[1] Thus I separate Popperian 'falsification' and 'rejection', the conflation of which turned out to be the main weakness of his 'naive falsificationism'.[2]

My modification, then, presents a very different picture of the game of science from Popper's. The best opening gambit is not a falsifiable (and therefore consistent) hypothesis, but a research programme. Mere 'falsifications' (that is, anomalies) are recorded but not acted upon. 'Crucial experiments' in Popper's sense do not exist: at best they are honorific titles conferred on certain anomalies *long after the event*, when one programme has been defeated by another one. For Popper, a crucial experiment is described by an accepted basic statement which is inconsistent with a theory. I, for one, hold that no accepted basic statement *alone* entitles us to reject a theory. Such a clash may present a problem (major or minor), but in no circumstance a 'victory'. No experiment is crucial at the time it is performed (except perhaps psychologically). The Popperian pattern of 'conjectures and refutations', that is, the pattern of trial-by-hypothesis followed by error-shown-by-experiment breaks down.[3] A theory can only be eliminated by a *better* theory, that is, by one which has excess empirical content over its predecessors, some of which is subsequently confirmed. And for this replacement of one theory by a better one, the first theory does not even have to be 'falsified' in Popper's sense of the term.[4] Thus progress is marked by instances verifying excess content rather than by falsifying instances,[5] and 'falsification' and 'rejection' become logically independent.[6] Popper says explicitly that 'before a theory has been refuted we can never know in what way it may have to be modified'.[7] In my view it is rather the opposite way around: before the modifi-

---

[1] Cf. volume 2, chapter 8, pp. 175–8, my [1968c], pp. 162–7, and this volume, chapter 1, p. 31 ff. and p. 69 ff.

[2] One important consequence is the difference between Popper's discussion of the 'Duhem–Quine argument' and mine; cf. on the one hand Popper [1934], last paragraph of section 18 and section 19, n. 1; Popper [1957b], pp. 131–3; Popper [1963a], p. 112, n. 26, pp. 238–9 and p. 243; and on the other hand, chapter 1, pp. 96–101.

[3] Popper, in one interesting passage, tries to define the difference between the amoeba's and Einstein's method; they both seem to pursue the method of conjectures and refutations ([1963a], p. 52). Popper thinks that Einstein has a '*more* critical and constructive attitude' than the amoeba (my italics). I think that a better solution is that the amoeba has *no (articulated) research programmes*.

[4] Popper occasionally – and Feyerabend systematically – stressed the *catalytic* role of alternative theories in devising so-called 'crucial experiments'. But alternatives are not merely catalysts, which can be later removed in the rational reconstruction, but are *necessary* parts of the falsifying process. (Cf. chapter 1, p. 37, n. 1.)

[5] Cf. esp. chapter 1, pp. 36–7.

[6] Cf. esp. volume 2, chapter 8, p. 177 and this volume, chapter 1, p. 37.

[7] Popper [1963a], p. 51.

cation we do not know in what way, if at all, the theory had been 'refuted', and some of the most interesting modifications are motivated by the 'positive heuristic' of the research programme rather than by anomalies.[1]

(e) An amended meta-criterion

An opponent could claim that the falsification of my criterion is not much more difficult than Popper's. What about the immediate impact of great crucial experiments, like that of the falsification of the parity principle? Or the long, pedestrian, trial-and-error procedures which occasionally precede the announcement of a major research programme? Will not the judgment of the scientific *élite* go against my universal rules?

I should like to present my answer in two stages. First, I should like to amend slightly my previously announced provisional meta-criterion,[2] and then replace it altogether with a better one.

First, the slight amendment. If a universal rule clashes with a particular 'normative basic judgment' one should allow some time to the scientific community to ponder about the clash: they may give up their particular judgment and submit to the general rule.[3] These 'second-order' falsifications must not be rushed.

Secondly, if we abandon naive falsificationism in *method*, why stick to it in *meta-method*? We can easily have a second-order methodology of scientific research programmes.

While maintaining that a theory of rationality has to try to organize basic value judgments in universal, coherent frameworks, we do not have to reject such a framework immediately merely because of some anomalies or other inconsistencies. On the other hand, a good rationality theory must anticipate further basic value judgments unexpected in the light of their predecessors or even lead to the revision of previously held basic value judgments.[4] We reject a rationality theory only for a better one, for one which, in this quasi-empirical sense, represents a *progressive shift*. Thus this new – more lenient – meta-criterion enables us to compare rival logics of discovery and discern growth in 'meta-scientific' knowledge.

For instance, Popper's theory of scientific rationality need not be seen as 'falsified' simply because it clashes with some actual basic judgments of leading scientists. On the contrary, on our new criterion

---

[1] Cf. esp. chapter 1, pp. 50–2.

[2] Cf. *above*, p. 146.

[3] There is a certain analogy between this pattern and the occasional appeal procedure of the theoretical scientist against the verdict of the experimental jury; cf. chapter 1, pp. 42–5.

[4] This latter criterion is analogous to the exceptional 'depth' of a theory which clashes with some basic statements available at the time and, at the end, emerges victorious. (Cf. Popper's [1957a].) Popper's example was the inconsistency between Kepler's laws and the Newtonian theory which set out to explain them.

it represents progress over its justificationist predecessors. For, contrary to these predecessors, it rehabilitated the scientific status of falsified theories like phlogiston theory, thus reversing a value judgment which expelled the latter from the history of science proper into the history of irrational beliefs. Also, it reversed the appraisal of the falling star of the 1920s: the Bohr–Kramers–Slater theory.[1] In the light of most justificationist theories of rationality the history of science is, at its best, a history of *prescientific* preludes to some *future* history of science.[2] Popper's methodology enabled the historian to interpret more of the *actual* basic value judgments in the history of science as rational: it constituted progress.

On the other hand, I hope that my modification of Popper's logic of discovery will be seen, in turn – on the criterion I specified – as a further step forward. For it seems to offer a coherent account of *more* old, isolated basic value judgments as rational; indeed, it has led to new and, at least for the justificationist or naive falsificationist, *surprising* basic value judgments. For instance, on Popper's theory, it becomes *irrational* to retain and further elaborate Newton's gravitational theory after the discovery of Mercury's anomalous perihelion; or it becomes *irrational* to develop Bohr's old quantum theory based on inconsistent foundations. From my point of view, these were perfectly *rational* developments. My theory, unlike Popper's, explains some rearguard skirmishes for defeated programmes as perfectly rational, and thus leads to the reversal of those standard historiographical judgments which led to the disappearance of many of these skirmishes from history of science textbooks.[3] These rearguard skirmishes were previously deleted both by the inductivist and by the naive falsificationist party histories.

Progress in the theory of rationality is thus marked by historical discoveries: by the reconstruction of a growing bulk of value-impregnated history as rational.[4] This idea may be seen as a self-application of my theory of scientific research programmes to a (non-scientific) research programme concerning scientific appraisals.[5]

I, of course, can easily answer the question when I would give up my criterion of demarcation: when another one is proposed which is better on my meta-criterion. (I have not yet answered the question

---

[1] Van der Waerden thought that the Bohr–Kramers–Slater theory was bad; Popper's theory showed it to be good. Cf. van der Waerden [1967], p. 13 and Popper [1963a], pp. 242 ff.; for a critical discussion cf. chapter 1, p. 82, nn. 1 and 2.

[2] The attitude of some modern logicians to the history of mathematics is a typical example; cf. my [1963–4], p. 3.

[3] Cf. chapter 1, section 3(c).

[4] I need not say that no rationality theory can or should explain *all* history of science as rational: even the greatest scientists make wrong steps and fail in their judgment.

[5] The methodology of research programmes may thus be applied to normative knowledge, including even ethics and aesthetics; this would then supersede the (naive falsificationist) 'quasi-empirical' approach as outlined *above*, p. 146, n. 1.

under what circumstances I shall give up my meta-criterion; but one must always stop somewhere.[1])

Finally let me elaborate two characteristics of my methodology and meta-methodology somewhat further.

First, I advocate a primarily quasi-empirical approach instead of Popper's aprioristic approach for law-giving to science.[2] I do not lay down general rules of the game *a priori,* so that, if history of science turns out to violate the rules, I would have to call the business of science to start anew. The law must take into account, if not be based upon, the verdict of the scientific jury. According to the conservative doctrine of Oakeshott and Polanyi, there must be only the jury, unfettered by written law. According to Popper merely the jury – even with common law – is not enough. There must be the authority of statute law to distinguish between good and bad science and to direct the jury in periods when a good tradition is in danger of degeneration or when new bad traditions emerge.[3] But in my view, there must be a dual system of authority, because the wisdom of the scientific *jury* has not been, and cannot be, fully articulated by the philosopher's *law.* Laws need authoritative interpreters. This is why, in matters of academic autonomy and the authority of tradition, I stand, even if only slightly, to the 'right' of the more 'liberal' Popper, who, to my mind, has a rather naive trust in the power of his (right!) law of scientific behaviour, and forgets that until now all the 'laws' proposed by the philosophers of science have turned out to be false generalizing interpretations of the verdicts of the best scientists. Up to the present day it has been the scientific norms, as applied instinctively by the scientific *élite* in *particular* cases, which have constituted the main yardstick of the philosopher's *universal* laws. Methodological progress still lags behind instinctive scientific verdicts in the sense that the main problem is to find, if possible, a theory of rationality which would explain *actual* scientific rationality rather than to bring legislative

---

[1] For an interesting discussion cf. Naess [1964].

[2] Alternatively, one might claim that this quasi-empirical approach is already implicit in Popper's meta-method, and I only make it *explicit.* After all, Popper's starting point was to define 'science' in such a way that it should include the refuted Newtonian theory and exclude unrefuted astrology, Marxism and Freudianism. Indeed, he says in the Preface of his [1959a] that 'since we possess many detailed reports of the discussions pertaining to the problem whether a theory such as Newton's or Maxwell's or Einstein's should be accepted or rejected, we may look at these discussions as if through a microscope that allows us to study in detail, and objectively some of the more important problems of "reasonable belief"'. Thus one might argue that Popper's meta-method was in my sense 'quasi-empiricist', even though he was not aware of it.

Kraft is very near to my quasi-empirical methodological approach. (Cf. Kraft [1925], esp. pp. 28–31.) Popper's description of Kraft's position as 'naturalistic' (Popper [1934b], section 10, n. 5) seems to be based on a misreading of some ambiguous passages. Kraft, in fact, advocates a meta-methodology which *learns* primarily from historical case studies, but in a normative–critical way.

[3] The former seems to apply to modern particle physics; the latter to some of the main schools of modern sociology, psychology and social psychology.

153

interference by the philosophy of science to the most advanced sciences.[1]

Secondly, I hold that philosophy of science is more of a guide to the historian of science than to the scientist. Since I think that philosophies of rationality lag behind scientific rationality even today, I find it difficult fully to share Popper's optimism that a better philosophy of science will be of *considerable* help to the scientist;[2] although no doubt it may help – and Popper's philosophy *has* helped – those great scientists whose scientific judgment was warped by the influence of previous, worse philosophies.

All this raises a host of problems about age-old problems of the role of authority, the right balance between the law and the jury, the mechanism of constitutional change, as applied to science. Institutionalized science is not participatory democracy (as some students, American senators and British MPs seem to think).[3] Scientific decision cannot be based on majority vote. But should it then be guided by enlightened despotism? Is the scientific community an 'open' society, as Popper sees it, or a 'closed' one, as Polanyi and Kuhn do? And which *ought* it to be?[4]

Instead of going any further into this field of problems, where Kuhn's theory is now the centre of discussion, I shall turn to the problem of induction and its relation to the problem of demarcation.

2 NEGATIVE AND POSITIVE SOLUTIONS TO THE PROBLEM OF INDUCTION: SCEPTICISM AND FALLIBILISM

(a) The game of science and the search for truth

A 'logic of discovery' in the Popperian sense, that is a system of appraisal of scientific theories, defines 'rules of the scientific game'.[5] These rules demarcate science from non-science and in particular from pseudoscience, and thus offer a *demarcation criterion*. But, in one respect, this demarcation criterion is poorer than most previous criteria. Most previous criteria laid down the aim of science as the discovery of the blueprint of the universe. Each 'discovery' discerns a piece of this blueprint: thus each step of the 'game' is seen as a step towards the goal. But what is the aim of *Popper*'s 'scientific game'? In inductivism the game was strictly connected with, and subordinated to, the Aim. In Popper's philosophy this link seems to be severed. The rules of the game, the methodology, stand on their own feet; but these feet dangle in the air without philosophical support.

The problem of induction, as Popper rightly pointed out, was originally identical with the problem of demarcation. Justificationists

---

[1] The situation may be changing now: cf. the previous footnote.
[2] Cf. Popper [1959a], p. 19.   [3] Cf. volume 2, chapter 12.
[4] Cf. Watkins [1970], p. 26.   [5] Popper [1934], section 85.

rigorously subordinated the rules of the game to the aim of science, to the finding of the Blueprint of the Universe: a step in the scientific game was proper only if it was *proved* to be a step in reconstructing this blueprint or, as they later more modestly claimed, if it was *proved* to be a likely (or 'probable') step towards it. But Popper, in the early stage of his philosophy, shifted the centre of gravity to the problem of demarcation and separated it from the problem of induction. He solved the problem of demarcation without justifying the game by subordinating it to a final *aim*; and then he claimed to have negatively solved (or, rather, dissolved) the problem of induction. He supported this latter claim with the courageous assertion that the game is *autonomous*, that one cannot – and need not – *prove* that the game actually progresses towards its aim; one may only piously *hope* that it does.

Popper's classical *Logik der Forschung* is consistent with the game of science being pursued simply for its own sake.[1] Of course it is abundantly clear that Popper's *instinctive* answer was that the aim of the science *was* indeed the pursuit of Truth; but, inasmuch as in 1934 the correspondence theory was in eclipse, he thought he could do nothing but adopt a cautious position, which, in its formulation if not in its spirit, was entirely sceptical: science could at best – tentatively – detect error. He proudly noted that 'in [his] logic of science it is possible to avoid using the concepts "true" and "false"'.[2] If science was victorious, it was victorious in *rejecting* refuted and provisionally *accepting* corroborated theories.[3] The 'success' of science was nothing but unmasking alleged successes; indeed, 'those who are unwilling to expose their ideas to the hazard of refutation do not take part in the scientific game'.[4] If a theory stands up to severe tests, it is awarded the honorific title 'corroborated'. But the *only* function of high corroboration is to challenge the ambitious scientist to overthrow the theory.[5] Scientific 'progress' is increased awareness of ignorance rather than growth of knowledge. It is '*learning*' without ever *knowing*.

[1] Some of my friends objected that this is not so; that the *aim* of science, according to Popper's [1934], is clearly to discover ever deeper questions, and that Popper's methodology follows from this presupposition. I reject this objection: asking 'ever deeper questions' is *synonymous* with the ban on 'conventionalist stratagems', that is, 'asking deeper questions' is a *rule* of the game; if it is also its purpose, then the game has its purpose in itself.

[2] Popper [1934], section 84.

[3] The whole of *Logik der Forschung* is in an important sense a *pragmatic* treatise; it is about *acceptance* and *rejection* and not about *truth* and *falsehood*. (But it is not *pragmatist*: it does not identify acceptance with truth and rejection with falsehood.) Popper occasionally deviates from his pragmatic–methodological terminology, and he slips, no doubt unintentionally, into the language of 'dogmatic falsificationism'. (For this concept cf. chapter 1, p. 12 ff.) For instance, in his *Open Society* he describes the '*main point*' of his *Logik der Forschung* in these words: 'We can never rationally establish the truth of scientific laws; all we can do is...to eliminate the *false* ones' (volume II, p. 363, my horrified italics).

[4] Popper [1934], section 85.   [5] Popper [1959a], p. 419.

(Popper does not seem fully to have realized that, within the framework of his *Logik der Forschung*, he cannot even answer the question 'what *can* one *learn* in the game of science?' One cannot learn about the world even from one's '*mistakes*', one cannot detect genuine epistemological error unless one has a theory of truth and a theory of how one may recognize increasing or decreasing truth-content. A 'dogmatic falsificationist', of course, *can* learn about the world from his mistakes; a 'methodological falsificationist' *can not*, as I shall later argue, without invoking some principle of induction.[1])

To put it more sharply: *Popper's demarcation criterion has nothing to do with epistemology*. It says nothing about the epistemological value of the scientific game.[2] One may, of course, *independently* of one's logic of discovery, *believe* that the external world exists, that there are natural laws and even that the scientific game produces propositions ever nearer to Truth; but there is nothing *rational* about these metaphysical beliefs; they are mere animal beliefs. There is nothing in the *Logik der Forschung* with which the most radical sceptic need disagree.

Tarski's rehabilitation of the correspondence theory of truth came to Popper's attention only after the publication of the *Logik der Forschung*. But, when it did, it changed radically the general tone of Popper's philosophy of science. It stimulated Popper to complement his logic of discovery with his own theory of verisimilitude and of approximation to the Truth, an achievement marvellous both in its simplicity and in its problem-solving power.[3] It became possible, for the first time, to define *progress* even for a sequence of false theories: such a sequence constitutes progress, if its truth-content, or, as Popper proposed, its verisimilitude (truth-content minus falsity-content) increases. But this is not enough: we have to *recognize* progress. This can be done easily by an inductive principle which connects realist metaphysics with methodological appraisals, verisimilitude with corroboration, which reinterprets the rules of the 'scientific game' as a – conjectural – theory about the *signs* of the *growth of knowledge*, that is, about the signs of *growing verisimilitude of our scientific theories*.[4] Popper's 'rules' are then no longer pursued for their own sake;

---

[1] For the terms 'dogmatic' and 'methodological' falsificationism cf. my [1968c] and chapter 1.

[2] This is characteristic of the demarcation criterion of 'methodological falsificationism'. The demarcation criterion of 'dogmatic falsificationism' on the other hand, is genuinely epistemological. (For the two criteria cf. chapter 1, pp. 10–31.)

[3] Cf. 'Truth, Rationality and the Growth of Scientific Knowledge', forming chapter 10 of his [1963a].

[4] The expression '*growth of scientific knowledge*' appears characteristically as the subtitle of the *chef d'oeuvre* of his later philosophy. In his [1934] he claimed that 'that main problem of philosophy is the critical analysis of the appeal to the authority of experience' (section 10). But in the new Preface to the 1959 English edition he says that 'the central problem of epistemology has always been and still is the growth of knowledge'. There is a marked shift from the negativist 1934 text to the optimistic 1958 Preface.

victories of science are then no longer victories merely in a game; they are even more than mere detections of error and replacements of erroneous theories by ever more comprehensive errors: they become instead putative milestones in approximating the Truth. (Popper's famous 'third requirement', introduced in this very paper, may also be seen against this background: *corroborations* of major theories, rather than perpetual detections of failure, become signposts of success.[1])

As a consequence, the tone of Popper's discussion of *scepticism* has changed markedly since 1960. Before 1960 he never said anything against scepticism nor did he distinguish scepticism from fallibilism. But, since 1960, Popper has shifted *towards* epistemological optimism. He now consistently separates scepticism and fallibilism; and, indeed, his celebrated first *Addendum* to the fourth edition of his *Open Society* consists almost entirely of a sermon against scepticism. Even though in his methodology *decisions* play a vital role,[2] he is now firmly and explicitly against interpreting them as 'leaps in the dark'. Such an interpretation would be 'an exaggeration as well as an over-dramatization',[3] it would be 'nihilist ado about nothing'.[4] 'Philosophical despair is not called for', he writes, for we *can* cope with the task of 'getting to know the beautiful world we live in and ourselves; and, fallible though we are, we nevertheless find that our powers of understanding, surprisingly, are almost adequate for the task – more so than we ever dreamt in our wildest dreams.'[5]

To some of Popper's students all this looked like the betrayal of everything that Popper had stood for; it seemed to be a break with the very essence of his *Logik der Forschung*.[6]

But it is only in the light of Popper's Tarskian turn that his *Logik der Forschung* can be properly understood. For we now understand why Popper had not offered a positive solution of induction in 1934. The main achievement of his *Logik der Forschung* was to show that the problem of demarcation can be solved without any 'inductive principle' being involved, which in turn could only rest on some satisfactory theory of truth. This was a most important achievement. But, after the problem of demarcation has been solved *in this autonomous way*, the link has to be re-established between the game of science on the one hand and the growth of knowledge on the other. If once one

---

[1] For a detailed critical discussion and references, cf. volume 2, chapter 8, pp. 173–81.

[2] This is why I called it 'revolutionary conventionalism'; cf. chapter 1, p. 21.

[3] Popper [1962], pp. 380–1.   [4] *Ibid.*, p. 383.

[5] *Ibid.*, p. 382.

[6] Agassi accused Popper of a 'verificationist' turn. (Cf. Agassi [1959]; for Popper's reply see Popper [1963a], p. 248, n. 31.) Later Agassi tried to attribute to Popper the strange view that corroboration may guide us in our 'choice', but we can 'learn' only from refutations (Agassi [1969]). Feyerabend too seems to think that corroboration plays no real role in science or learning from experience (cf. Feyerabend [1969b]). Also, cf. *above*, p. 149, n. 1.

accepts Popper's problemshift, demarcation and 'induction' become *separate* problems, the solution of the second becoming a possibly trivial corollary to the solution of the first. But the remainder must not be forgotten. The positive solution of the problem of induction is that the scientific game, as played by the greatest scientists, is the best extant way of increasing the verisimilitude of our knowledge, of approaching Truth; the *sign* of increasing verisimilitude is increasing degree of corroboration. I have little doubt that Popper would have started his *Logik der Forschung* with this *positive* solution of the problem of induction, had Tarski's theory of truth come in 1925 (and had Popper arrived at his idea of truth-content and verisimilitude by 1930). But, since the idea of truth was in disarray in the 1920s and, since he did not know at the time of Tarski's results, he formulated the 'rules' of science in the pragmatic terms of rejection and acceptance *alone*. He did this so ingeniously that those who tried to show that his hidden, instinctive guiding idea must actually be *there* as a hidden inductive principle, were foiled.[1] In the terminology of my 'Changing Problem of Inductive Logic', Popper managed to put acceptability$_1$ and acceptability$_2$ (his methodological appraisals) on their own feet and to make them *logically* independent from acceptability$_3$.[2] But philosophically, as I said before, these feet dangled in the air without the support of an underlying conjectural 'inductive' metaphysics. Popper's methodological appraisals are interesting primarily because of the hidden *inductive assumption* that, if one lives up to them, one has a better chance to get nearer to the Truth than otherwise. The value of excess corroboration is that it indicates that the scientists *might* be approaching truth, just as the value of the birds above Columbus's ship was that they indicated that the discoverers *might* be approaching land.[3]

Thus, once we have the theory of verisimilitude, we can correlate methodological appraisals with genuine epistemological appraisals. Methodological appraisals are *analytic*;[4] but without a *synthetic* interpretation they remain devoid of any genuine epistemological significance, they remain part of a pure game. A new, *synthetic* interpretation must be given to Popper's methodological appraisals with the help of an inductive principle: there must be an 'acceptance$_3$' based on 'acceptance$_1$' and 'acceptance$_2$'.[5]

Only such a positive solution of the problem of induction can

---

[1] For example, J. O. Wisdom and Ayer argued that only an inductive principle can prevent upholding refuted theories in the hope that the refutations will come to an end; only an inductive principle can explain why we hold that refuted theories stay refuted. I have shown that they were wrong. Cf. volume 2, chapter 8, p. 182.

[2] This is the message of section 79 of Popper's [1934].

[3] The analogy must be taken with a pinch of salt. Columbus's inference from the sighting of birds to the nearness of land was easily refutable; my 'inductive principle' is not.

[4] For references cf. *below*, p. 163, n. 3.

[5] Cf. volume 2, chapter 8, pp. 170–81.

separate constructive fallibilism from scepticism and from all its evil consequences, like relativism, irrationalism, mysticism. Popper, however, after having provided the tools for such a positive solution in the form of his theory of verisimilitude, shrank back from stating clearly and explicitly a positive solution of the (Popperian) problem of induction, that is, of the problem of the epistemological value of his logic of discovery.

(b) A plea to Popper for a whiff of 'inductivism'

Popper has not fully exploited the possibilities opened up by his Tarskian turn. While he now talks freely about the metaphysical ideas of truth and falsity, he still will not say unequivocally that the positive appraisals in his scientific game may be seen as a – conjectural – sign of the growth of conjectural knowledge; that corroboration is a *synthetic* – albeit conjectural – measure of verisimilitude. He still emphasizes that 'science often errs and that pseudoscience may happen to stumble on the truth'.[1] Although making strongly optimistic sermons in praise of human knowledge,[2] when it comes to making a precise statement he restricts his 'optimism' to a classical *sceptical* thesis: 'I am a metaphysical realist, and an epistemological optimist in the sense that I hold that the truthlikeness ("versimilitude") of our scientific theories can increase: this is how our knowledge grows'.[3] A sceptic, of course, may hold realist *beliefs*; but, from the statement that 'the verisimilitude of our scientific theories *can* increase', it only follows that 'our knowledge *can* grow – but without our knowing it'. *If so, even Popper's newly found fallibilism is nothing more than scepticism together with a eulogy of the game of science.* Popper's theory of verisimilitude remains a metaphysical–logical theory which has nothing to do with epistemology.

No wonder, then, as Watkins put it, that 'in critical discussion of Popper's epistemology [we usually find] the suspicion that, far from solving the problem of rational choice between competing hypotheses, his methodology really leads to thorough-going scepticism'.[4]

Watkins's reply is exceptionally lucid. It is worth quoting a passage in full:

Many philosophers who have given up the hope that any of our empirical statements about the external world are certain, cling all the more tenaciously to the hope that some of them are at least less uncertain than others. Such philosophers tend to characterize as *scepticism* the thesis that *all empirical statements about the external world are equally uncertain*. I will use ST_1 as an abbreviation for this (first) 'sceptical' thesis. Now Popper's philosophy is 'sceptical' in the sense of ST_1; but then 'scepticism' in this sense seems to me to be unavoidable.[5]

---

[1] Popper [1968c], p. 91.
[2] Cf. *above*, p. 157.
[3] Popper [1968c], p. 93.
[4] Watkins [1968], pp. 277–8.
[5] *Ibid.*

Then Watkins goes on:

Philosophers who place their hopes, not on certainties, whether absolute or relative, but on rational argument and criticism, will prefer to characterize as *scepticism* the thesis that *we never have any good reason for preferring one empirical statement about the external world to another*. I will use ST_2 as an abbreviation for this second sceptical thesis. ST_1 and ST_2 are by no means equivalent. ST_2 implies ST_1 (on the assumption that, if one hypothesis *were* less uncertain than another, that *would*, other things being equal, be a reason for preferring it). But ST_1 does not imply ST_2: there may be reasons having nothing to do with relative certainty for preferring one hypothesis to another. Empirical scientists cannot expect to have good reasons for preferring a particular explanatory hypothesis to all the (infinitely many) *possible* alternatives to it. But they often do have good reasons for preferring one out of several competing hypotheses which have actually been proposed. How one hypothesis may be rationally appraised as better than the other hypotheses under discussion, and what a future hypothesis would need to do for it to be even better than this one – this is what Popper's methodology is about.[1]

But the '*good reasons* for preferring one empirical statement about the external world to another' are laid down in Popper's demarcation criterion, in his rules of the game of science. Preference is only a pragmatic concept *within the context* of this game. This preference can only assume epistemological significance with the help of an additional, synthetic, *inductive (or, if you wish, quasi-inductive) principle* which would somehow assert the epistemological superiority of science over pseudoscience. Such an inductive principle must be based on some sort of correlation between 'degree of corroboration' and 'degree of verisimilitude'. But, both Popper's and Watkins's positions are ambiguous on whether the degree of corroboration can be interpreted *synthetically*. For instance, Watkins claims: 'We may have good reasons for claiming that a particular hypothesis h_2 is closer to the truth than a rival hypothesis h_1'. [2] But this contradicts his previous assertion that h_1 and h_2 are *equally* uncertain, unless he uses the terms 'equally uncertain' and 'closer to the truth' in the Pickwickian sense that we can have good reasons for holding that h_2 is closer to the truth than h_1, even though they are equally uncertain.[3] But such paradoxes are inevitable for philosophers who want the impossible: to fight pseudoscience from a sceptical position.

Indeed, Popper recently tends to complain that some of his critics believe that he is a mere 'negativist', that he is 'flippant about the search for truth, and addicted to barren and destructive criticism and to the propounding of views which are clearly paradoxical'.[4] Popper's answer is as beautiful as it is unconvincing:

[1] *Ibid.*, p. 279.
[2] *Ibid.*, p. 280.
[3] This inconsistency also occurs in the celebrated chapter 10 of Popper's [1963a]. I quote Watkins only because his exposition is so clear.
[4] Popper [1963a], p. 229.

This mistaken picture of our views seems to result largely from the adoption of a justificationist programme, and of the mistaken subjectivist approach to truth which I have described. For the fact is that we too see science as the search for truth, and that, at least since Tarski, we are no longer afraid to say so. Indeed, it is only with respect to this aim, the discovery of truth, that we can say that though we are fallible, we hope to learn from our mistakes. It is only the idea of truth which allows us to speak sensibly of mistakes and of rational criticism, and which makes rational discussion possible – that is to say, critical discussion in search of mistakes with the serious purpose of eliminating as many of these mistakes as we can, in order to get nearer to the truth. Thus the very idea of error – and of fallibility – involves the idea of an objective truth as the standard of which we may fall short. (It is in this sense that the idea of truth is a *regulative* idea.)

There is not a word in this passage about how to recognize the *signs* of being nearer to the Truth, nothing which amounts to more than the assertion that we must play the scientific game *seriously*, in the *hope* of getting nearer to the Truth. But did Pyrrho or Hume have anything against being 'serious' or entertaining 'hopes'?

In order further to clarify this whole issue, I shall analyse briefly Popper's criticism of induction.

Popper's reputation is rightly that of the scourge of induction. But, as I have pointed out before,[1] in Popper's anti-inductivist campaign (at least) three logically independent issues have to be carefully distinguished.

(i) First, there is the campaign against the *inductivist logic of discovery*. This is the Baconian doctrine according to which a discovery is scientific only if it is *guided* by facts and not *misguided* by theory. The scientist must start by purging his mind of theories (or rather *bias*); nature will then become for him an open book.[2] This doctrine was already opposed by rationalists, like Descartes and Kant; but even they demarcated misguiding bad theories from good *a priori* principles which intuition can recognize as true. The method of free, creative conjectures and empirical tests developed only in stages from Whewell, Bernard, through Peirce and finally the Bergsonians, to achieve unique clarity and force in Popper's 'demarcation criterion', which demarcated this method of discovery and scientific progress both from inductive fact-collecting and from 'metaphysical' speculation. In this campaign Popper achieved a decisive success, not only intellectually but socio-psychologically; at least among philosophers of science Baconian method is now only taken seriously by the most provincial and illiterate. In this line he also proposed a *positive* theory about the

---

[1] Cf. volume 2, chapter 8, pp. 190 ff.

[2] This method may be associated – as in Descartes' case – with an intuitive–psychologistic theory of content-increasing ('inductive') logic. But one may try to dispense with such logic and search for some universal inductive principles which would turn inductive logic into a deductive system. For this programme of deductive reconstruction of induction cf. Max Black [1967], pp. 174 ff.

role of speculation and experience in the growth of science;[1] but this was not the last word on the subject, and I hope to have developed it one step further.[2]

(ii) The second prong of Popper's attack was directed against the programme of an *a priori* probabilistic inductive logic or confirmation theory. This programme postulates that it is possible to assign – with the certainty of logic – to any pair of propositions a 'degree of confirmation', which characterizes the evidential support that the second proposition lends to the first. The function obeys the axioms of the probability calculus. The heart of this programme is the construction of an *a priori* meta-science (by defining a distribution function over a finite or enumerably infinite number of possible states of the universe) that enables one to compute confirmation functions. Thus certainty is shifted from the science of the actual to the meta-science of the possible, which, in turn, provides a proven confirmation theory for science. This programme was initiated by Cambridge philosophers (Johnson, Broad, Keynes) and its most persistent and influential proponents became Hans Reichenbach and then Rudolf Carnap.[3] In this campaign too Popper achieved a complete victory, although 'inductive logic', displaying all the characteristics of a degenerating research programme, is still – sociologically – a booming industry.[4]

(A weakness of this second part of Popper's anti-inductivist campaign was his determination to achieve an ultimate, clear-cut victory with one single blow; either by showing that Carnap's approach was inconsistent, or by showing that, *if* inductive logic was possible, *then* the virtue of a theory was its improbability rather than its probability, given the evidence. He did not realize that fighting a research programme – in this case a non-empirical one – by showing up its degen-

---

[1] Cf. his [1963a], pp. 42–6. However, Popper fails to emphasize that there can be no such thing as a purely empirical theory of learning. Before studying the psychology of learning, we must agree on a *normative* demarcation between learning and being indoctrinated. Cf. chapter 1, p. 38, text to n. 2.

[2] Cf. chapter 2, esp. section 2(*b*).

[3] Carnap confused the philosophical issue by his conviction that all *a priori* true propositions are bound to be analytic; therefore the inductive principle is analytic. This confusion was exposed by Nagel and Popper. (For references cf. volume 2, chapter 8, p. 160, n. 2.)

[4] It is important to realize that the introduction of an inductive principle gives 'induction' a deductive structure. (Cf. *above*, p. 157, n. 2.) Victor Kraft, for instance, proposed, in 1925, such a 'deductivist' approach. To claim that this is the view that Popper adopted later (as Feyerabend puts it in his [1963]) is incorrect. Victor Kraft, in his undeservedly neglected [1925], may have anticipated Popper on many points, but *not* in his radical anti-inductivism. Kraft, in this work, contrary to Feyerabend's false account, proposed that an inductive assumption may provide a 'logically justified' expectation for the future (p. 253), and pointed out that therefore his position differed significantly from Hume's (pp. 254–5). (Incidentally, according to Feyerabend, 'Popper himself refers to Kraft as one of his predecessors'. This is untrue: there are two references to Victor Kraft in the *Logik der Forschung*, both critical.) Today Kraft still advocates an inductive principle, which, once introduced, would make science completely 'deductive' (Kraft [1966]).

eration and developing a rival programme, cannot be a fast process; I hope that my development also of this prong of his campaign contributed to the clarification of some of his points.)

But the second prong of Popper's anti-inductivist campaign can be interpreted in an even stronger sense. It can be said to have been directed against *any* infallible *a priori* metaphysical inductive principle, whether probabilistic or non-probabilistic, which would serve to assign a *proven* metric to the field of scientific statements.[1]

Non-probabilistic logics of confirmation are still being produced – some with great ingenuity – by philosophers of science who understood Popper's arguments against probability logic, but not this more general message.[2]

(iii) The third prong of Popper's anti-inductivist campaign is less easily discernible. It consists of a tacit but stubborn refusal to accept any *synthetic* inductive principle connecting Popperian *analytic* theory-appraisals (like content and corroboration) with verisimilitude.[3] But why should we exclude a *conjectural* inductive principle from rationality? Why relegate the *application* of science to its 'animal', 'biological' function?[4] Popper's master argument against a justificationist principle of induction (namely that it leads either to infinite regress or apriorism[5]) is, *in this case*, invalid; Popper's powerful argument only applies to a principle which would serve as a premise to a *proven* measure function of (spatio-temporally local)[6] verisimilitude (one like Popper's degrees of corroboration). A conjectural inductive principle would be abhorrent only to the sceptico-dogmatist,[7] for whom the combination of total lack of proof and strong assent indicated mere animal belief. For the Humean sceptical pessimist this is the end of the road; for the Kantian dogmatic optimist this is a 'scandal of philosophy' to be ironed out. But to the Popperian fallibilist, for whom conjectural metaphysics can be, at least in principle, rationally appraised, it *should* be a cause neither for sceptical resignation nor for apriorism.[8] Only some such conjectural metaphysics connecting

---

[1] Popper was so much preoccupied with his fight against *a priori probabilistic* measures of confirmation that he, at least for a brief moment, seems to have faltered in his stand against *a priori nonprobabilistic* measures; cf. volume 2, chapter 8, pp. 193 ff.

[2] Hintikka, L. J. Cohen and, perhaps, Levi, could be mentioned here.

[3] Popper, and following him, Agassi and Watkins have interpreted 'degree of corroboration' as a strictly tautologous appraisal. (For references, cf. volume 2, chapter 8, pp. 188–9, esp. p. 189, n. 2, and p. 190, n. 4.) This interpretation bears out my analysis of Popper's 'third anti-inductivist campaign'.

[4] Popper [1934], section 85.

[5] Popper [1934], section 1.

[6] Cf. volume 2, chapter 8, p. 187.

[7] For the 'dialectical unity' of dogmatism and scepticism as two poles of justificationism, cf. Popper [1963a], p. 228; and also my [1970b] and volume 2, chapter 8.

[8] Victor Kraft seems to have come very near to such a position. He abhorred Humean scepticism which 'denies rationality to empirical science and characterises it as being as irrational a phenomenon as the belief in paradise or in demons'. ([1925], p. 208.) He abhorred the idea that 'general knowledge about reality had no more validity than

corroboration and verisimilitude would separate Popper from the sceptics and establish his point of view, in Feigl's words, 'as a *tertium quid* between Hume's and Kant's epistemologies'.[1]

I had long discussions with Popper in 1966–7 about these issues; I profited immensely from them. But I was left with the impression that on what I called the 'third prong of his anti-inductivist campaign' we may never see eye to eye. *The reason is not that our disagreement is too big; but that it is so very small.* The difference between total scepticism and humble fallibilism is so small that one frequently feels that one is engaged in a mere verbal quibble: should the 'inductive principle' I advocate[2] be referred to as a 'rationally entertained speculation', which even might be seen as very weakly 'vindicated'; or should it be referred to as stark 'animal belief', conditioned in the Darwinian struggle for survival? I inserted at the end of my 'Changes in the problem of inductive logic' a brief section of three pages on 'Popper's opposition to "acceptability₃"' (see volume 2, chapter 8). This, I am afraid, is rather a trivial section. For, although in my lengthy and pedantic discussion of 'acceptability₃', I thought to offer a new, *positive* solution to the old problem of induction, the 'solution' was very thin. Alas, a solution is interesting only if it is embedded in, or leads to, a major research programme; if it creates new problems – and solutions – in turn. *But this would be the case only if such an inductive principle could be sufficiently richly formulated so that one may, say, criticize our scientific game from its point of view.* My inductive principle tries to explain why we 'play' the game of science. But it does so in an *ad hoc*, not in a 'fact-correcting' (or, if you wish, 'basic value judgment correcting') way. *Ad hoc* explanations are very near to mere linguistic transformations; although they may *also* be happy phrases suggesting and protecting later development. Such metaphysical developments were barred by Popper when he sternly announced that: 'As for inductive logic, I do not believe that it exists. There is, of course, a logic of

conjectural' (p. 255). On the other hand, he rejected Kantian apriorism and pointed out that Kant's very question ('how is [infallible] science possible?') assumed the existence of an infallible science. In fact, he points out, science is fallible and thus the question disappears. 'Then one can go on to reconstruct science as free, basisless – as completely arbitrary' (p. 31).

This is, of course, the step from Kant to LeRoy (cf. chapter 1, pp. 21 ff). But then Kraft, disappointingly, introduces 'simplicity' as a *validating* criterion ([1925], pp. 257–8); and he even asserts the *absolute validity* of basic statements (p. 253).

[1] Feigl [1964], p. 47.

[2] In my [1968b] I contrasted my fallible 'metaphysical principle' to 'inductive principles', which then I took to be *by definition* infallible. I chose this terminology in order not to offend Popper on a purely semantic point and to uphold the claim that he destroyed *all* possible kinds of inductive principle (see volume 2, chapter 8, p. 186). Now I have changed my terminology, inasmuch as Popper himself has started to talk about a 'positive solution' of the problem of induction (cf. *below*, p. 166, n. 3); and, indeed, there is nothing wrong with preserving old time-honoured terms (like 'inductive principle') even after a problem has been as radically shifted as the problem of induction was by Popper.

science, but that is part of an applied deductive logic; the logic of testing theories, or the logic of the growth of knowledge'.[1] I, on the contrary, hold that the 'logic of the growth of knowledge' *must* include – in addition to Popper's *logico-metaphysical* theory of verisimilitude – *some* speculative *genuinely epistemological* theory connecting scientific standards with verisimilitude.

I think it is the present thinness of a conjectural inductive metaphysics that makes Popper reluctant to see anything in it, and I appreciate his point.[2] Yet, although both 'tautological' appraisals and metaphysical inductive principles are equally irrefutable, there is an immense philosophical difference between interpreting an appraisal as tautologous and interpreting it as metaphysical. For *this choice – as I already indicated – is the choice between scepticism with a purely negative solution of the problem of induction and fallibilism with a – momentarily very weak – positive solution.* By refusing to accept a 'thin' metaphysical principle of induction Popper fails to separate rationalism from irrationalism, weak light from total darkness. Without this principle Popper's 'corroborations' or 'refutations' and my 'progress' or 'degeneration', would remain mere honorific titles awarded in a pure game.[3] With a *positive* solution of the problem of induction, however thin, methodological theories of demarcation can be turned from arbitrary conventions into rational metaphysics.

Popper, of course, might well retort that this 'positive solution' itself is merely an arbitrary convention. The rationalist wants a positive solution of the problem of induction, therefore he postulates one. But, as Russell put it: 'The method of postulating what we want has many advantages; they are the same as the advantages of theft over honest toil'.[4]

Yet, why should we be more sceptical about some such metaphysical postulates than we are about 'accepted' basic statements? Why not extend Popperian hardheaded conventionalism from the acceptance (without belief) of some spatio-temporally singular statements to granting similar acceptance to some universal statements (in my 'hard cores') and even beyond that, to some conjectural weak 'inductive

---

[1] Popper [1968c], p. 139.

[2] 'Inductive principles' which use methodological appraisals (like Popper's corroboration or my problemshift appraisals) as tentative measures of verisimilitude are, I admit, sadly irrefutable. Only God can see the discrepancy between the verisimilitude and the scientific appraisal of our best theories. This is the crucial support for Popper's scepticism.

(The actual principle, as posited in the discussion of 'acceptance₃' in vol. 2, ch. 8, is rather complicated. Now I would prefer to state it in the form that – roughly speaking – the methodology of scientific research programmes is better suited for approximating the truth in our actual universe than any other methodology; cf. this volume, chapter 2.)

[3] As Feigl put it: 'The problem is precisely to show what entitles us to use honorific descriptions' (Feigl [1964], p. 49).

[4] Russell [1919], p. 71.

principle'? Why should Popper attribute high rational–scientific (although, as I have mentioned, not genuinely epistemological) status to absurd statements like 'nothing can assume higher velocity than the velocity of light', or 'there is attraction between two distant masses', but classify a plausible statement like 'physics has higher verisimilitude than astrology' as 'animal belief'? Why should only a 'basic', but not a 'metaphysical', statement be accepted as long as there is no serious alternative offered?

Thus the third prong of Popper's anti-inductivist campaign leads into a Humean irrationalist theory of practical human action and of applied science.[1] Indeed, only a positive solution of the problem of induction can save Popperian rationalism from Feyerabend's epistemological anarchism.[2]

Finally, let me say that, although I do think that my criticism of Popper's solution of the *problem of demarcation* is a genuine further development in the very tradition he himself set for the 'logic of scientific discovery', I do not think that my 'criticism' of Popper's 'solution' of the *problem of induction* is more than an attempt to make explicit the *full* implications of his own theory of verisimilitude for the problem of induction, and thus make the epistemological difference between classical scepticism and his fallibilism sharp and explicit. I hope he will be able to accept my modifications on both issues.[3]

[1] There is, of course, an alternative: to elaborate a rational theory of practical action which is *independent* of scientific rationality. There are traces of this approach in Popper and it was explicitly advocated by Watkins. This is how Popper and Watkins, leading proponents of the scientific weltanschauung landed in a position for which science as a guide of life is an anomaly. (Cf. volume 2, chapter 8, pp. 189 ff.)

[2] I think that Feyerabend's transformation from the Popperian Feyerabend$_1$ into the anarchist darling of the New Left (Feyerabend$_2$) was due to his change to a radically sceptical interpretation of Popper's own philosophy of science. Also, my discussion explains Popkin's puzzlement as to whether or not Popper is a sceptic. (Cf. Popkin [1967], p. 458.)

[3] Indeed, I was pleased to learn from Popper that, in response to my [1968b], he has now inserted a short *Addendum* on p. 226 of his [1969]. In this he says: 'The logico-methodological problem of induction is not unsolvable, but has been (negatively) solved in my book: (a) *Negative solution.* We cannot justify our theories, either as true or as probable. This solution and the following solution are compatible; (b) *Positive solution. We can justify* the choice of certain theories in the light of their corroboration, that is, in the light of the present state of rational discussion of the rival theories from the point of view of their verisimilitude.'

This is the first time Popper mentions a 'positive' solution to the problem of induction. This 'positive solution' is, then, simply that we base our guess concerning which theory has higher verisimilitude on the comparison of their degrees of corroboration. (Of course, Popper here would need my *corrected* version of degree of corroboration which assigns positive degrees of corroboration or of 'acceptability$_2$' even to refuted theories: volume 2, chapter 8, pp. 176–7). Moreover, he says that this also solves the '*practical problem of induction*': we choose the hypothesis which is estimated to have higher verisimilitude. He calls this a risky but rational choice.

But even Popper's *Addendum* does not fully clarify the queries I raised. On a careful reading of the text it transpires that Popper still has not realized that the 'positive solution' which he now proposes implies the existence of a synthetic inductive principle. He still has not withdrawn his claim that his degree of corroboration is

[*Added in 1971:*] Popper now published a major paper on induction in order to clarify his position on this subject. Large sections of Popper [1971] consist of responses to my [1968*b*] (reprinted as volume 2, chapter 8) and to my present contribution.

I was interested to see that *on some minor points* Popper has now adopted some of my earlier suggestions. For instance, he now equates boldness with non-*adhoc*ness, that is, with excess content rather than content.[1] Also, he now gave up his long held and tenaciously defended doctrine that the degree of corroboration of an unrefuted theory cannot be smaller than the degree of corroboration of any of its consequences;[2] instead of this he has now radically moved towards the position outlined in my 'Theoretical support for predictions versus evidential support for theories'.[3] Unfortunately, the one point which Popper explicitly refers to in my work he misquotes: he claims that I 'suspect that the actual attribution of numbers to [his] "degree of corroboration", if possible, would render [his theory] inductivist in the sense of a probabilistic theory of induction'. Popper 'sees no reason whatever why this should be so'.[4] Nor do I; and I said no such thing on pp. 410–12 of my paper to which he refers the reader – nor did I say anything like this anywhere else.

On the major issue – on induction – Popper's [1971] does not contain anything new.[5] His 'criticism' of a plea for an inductive principle[6] leaves my argument for such a principle completely intact.

analytic. But, if so, then he needs an additional *synthetic* principle which will turn this analytic measure function into a synthetic function estimating verisimilitude. There remains an unresolved inconsistency between a genuine (that is, metaphysical) 'positive solution' of the problem of induction and the 'third prong' of his anti-inductivist campaign.

1   Popper [1971], p. 181; cf. volume 2, chapter 8, p. 170.
2   Cf. e.g. Popper [1959*a*], p. 270 and Watkins [1964], p. 98.
3   Volume 2, chapter 8, section 7 (pp. 192–3).
4   Popper [1971], p. 184, n. 23.
5   He repeats his well-worn tautology that 'in so far as we *have* to choose, it will be "rational" to choose the best tested theory. This will be "rational" in the most obvious sense of the word known to me: the best tested theory is the one which, in the light of our *critical discussion* appears to be the best so far, and I do not know of anything more "rational" than a well-conducted critical discussion' (p. 188). This insistence that the game of science is in no need of an extra-methodological rationale, leads him to discourage epistemologists: 'No theory of knowledge should attempt to explain why we are successful in our attempts to explain things' (p. 189). What then should a theory of knowledge attempt to explain?
6   Cf. especially the last two paragraphs of section 12 of his [1971], p. 195.

4

Why did Copernicus's research programme supersede Ptolemy's?*

INTRODUCTION

I first should like to offer an apology for imposing a philosophical talk upon you on the occasion of the quincentenary of Copernicus's birth. My excuse is that a few years ago I suggested a specific method for using history of science as an arbiter of some authority when it comes to debates in philosophy of science and I thought that the Copernican revolution might in particular serve as an important test case between some contemporary philosophies of science.

I am afraid that first I have to explain – very roughly – what philosophical issues I have in mind and how historiographical criticism may help in deciding some of them.

The central problem in philosophy of science is the problem of normative appraisal of scientific theories; and, in particular, the problem of stating *universal* conditions under which a theory is scientific. This latter limiting case of the *appraisal problem* is known in philosophy as the *demarcation problem* and it was dramatized by the Vienna Circle and especially by Karl Popper who wanted to show that some *allegedly* scientific theories, like Marxism and Freudianism, are pseudoscientific and hence that they are no better than, say, astrology. The problem is not an unimportant one and much is still to be done towards its solution. To mention a minor example, the Velikovsky affair revealed that scientists cannot readily articulate standards which are understandable to the layman (or, as my friend Paul Feyerabend reminds me, to themselves), and in the light of which one can defend as rational the rejection of a theory which *claims* to constitute a revolutionary scientific achievement.

This problem of appraisal is completely different from the problem of why and how new theories emerge. Appraisal of change is a normative problem and thus a matter for philosophy; explanation of change (of actual acceptance and rejection of theories) is a psycho-

---

* This paper was written with Elie Zahar in 1972–3. It was first published as Lakatos and Zahar [1976a]. Lakatos gives the following account of the paper's history: 'This talk was first given at the Quincentenary Symposium on Copernicus of the British Society for the History of Science, on 5th January 1973. The paper is the result of joint efforts by the co-authors, but it is narrated in the first person by Imre Lakatos. Previous versions were criticized by Paul Feyerabend and John Worrall.' (*Eds.*)

logical problem. I take this Kantian demarcation between the 'logic of appraisal' and the 'psychology of discovery' for granted. Attempts to blur it have only yielded empty rhetoric.[1]

The generalized demarcation problem is closely linked with the problem of the rationality of science. Its solution ought to give us guidance as to when the acceptance of a scientific theory is rational or irrational. There is still no agreed universal criterion on the basis of which we can say whether the rejection of the Copernican theory by the Church in 1616 was rational or not, or whether or not the rejection of Mendelian genetics by the Soviet Communist Party in 1949 was rational. (Of course, we, hopefully, all agree that both the *banning* of *De revolutionibus* and the *murder* of Mendelians were deplorable.) Or to mention a contemporary example, whether or not the present rejection by so-called American liberals of the application of genetics to intelligence by Jensen and others is rational, is a highly controversial question.[2] (We may nevertheless agree that even if it were decided that a theory ought to be rejected, this decision should not carry with it physical threats to its tenacious proponents; and that '[nothing] be condemned without understanding it, without learning it, without even hearing it'.[3])

I   EMPIRICIST ACCOUNTS OF THE 'COPERNICAN REVOLUTION'

Let me first define the term 'Copernican Revolution'. Even in the descriptive sense, this term has been ambiguously applied. It is frequently interpreted as the acceptance by the 'general public' of the belief that the Sun, and not the Earth is the centre of our planetary system. But neither Copernicus nor Newton held this belief.[4] Anyway, *changes* from one popular belief to another fall outside the province of the history of *science* proper. Let us, for the time being, forget about beliefs and states of mind and consider only *statements* and their objective (in Frege's and Popper's sense, 'third-world'[5]) contents. In particular, let us regard the Copernican Revolution as the hypothesis that it is the Earth that is moving around the Sun *rather* than *vice versa*, or, more precisely, that the fixed frame of reference for planetary motion is the fixed stars and not the Earth. This interpretation is held

---

[1] This draft is concerned only with the normative aspect of the problem indicated in the title of the paper. It does not attempt to go into the socio-psychological study of the Copernican Revolution.

[2] According to Urbach (Urbach [1974]) it is irrational. But whether Urbach is right or wrong, the decision of Stanford University not to allow Nobel prize winner Shockley to lecture on race and intelligence is as shocking as the decision of Leeds University to refuse him his honorary doctorate in engineering because Lord Boyle and Jerry Ravetz (a brilliant Copernican scholar!) found that he held a theory which was contrary to so-called 'liberal' doctrine.

[3] Galileo [1615].   [4] Cf. e.g. Price [1959], pp. 204–5.

[5] Cf. e.g. Popper [1972], especially chapters 3 and 4.

mostly by those who hold that isolated hypotheses are the proper units of appraisal (rather than research programmes or 'paradigms').[1] Let us take different versions of this approach in turn, and show how each version fails.

I first discuss the views of those people who attribute the superiority of the Copernican hypothesis to *straightforward empirical considerations*. These 'positivists' are either inductivists or probabilists or falsificationists.

According to the *strict inductivists* one theory is better than another if it was deduced from the facts while its rival was not (otherwise the two theories are both mere speculations and rank equal). But even the most committed inductivist has been wary of applying this criterion to the Copernican Revolution. One can hardly claim that Copernicus deduced his heliocentrism from the facts. Indeed, now it is acknowledged that both Ptolemy's and Copernicus's theories were inconsistent with known observational results.[2] Yet many distinguished scholars, like Kepler, claimed that Copernicus derived his results 'from the phenomena, from effects, from the consequences, like a blind man who secures his steps by means of a stick'.[3]

Strict inductivism was taken seriously and criticized by many people from Bellarmine to Whewell and was finally demolished by Duhem and Popper,[4] although some scientists and some philosophers of science, like Born, Achinstein and Dorling, still believe in the possibility of deduction or valid induction of theories from (selected?) facts.[5] But the downfall of Cartesian and, in general, psychologistic logic and the rise of Bolzano–Tarski logic sealed the fate of 'deduction from the phenomena'. *If scientific revolution lies in the discovery of new facts and in valid generalizations from them, then there was no Copernican (Scientific) Revolution.*

Let us turn then to the *probabilistic inductivists*. Can *they* explain why Copernicus's theory of celestial motions was better than Ptolemy's? According to probabilistic inductivists one theory is better than another if it has a higher probability relative to the total available evidence at the time. I know of several (unpublished) efforts to calculate

---

[1] Cf. *below*, sections 3, 4, and 5.

[2] Let me quote on this point an authoritative source: 'Ptolemy's theory was not very accurate. The positions for Mars, for example, were sometimes wrong by nearly 5°. But...the planetary positions predicted by Copernicus...were nearly as bad' (Gingerich [1972]). This error was known to Kepler and he complained about it in the preface to his *Rudolphine Tables*. It was even known to Adam Smith as it transpires from his [1799]. (Smith's essay was written some time before 1773, when he mentioned it in a letter to David Hume.) Gingerich also reminds us that 'in Tycho's observation books, we can see occasional examples where the older scheme based on the *Alfonsine Tables* yielded better predictions than could be obtained from the Copernican *Prutenic Tables*' (Gingerich [1973]; cf. especially his n. 6 in the same paper).

[3] Kepler [1604]. Jeans describes the idea of the moving Earth as Copernicus's 'theorem' [1948], p. 359 and claims that 'Copernicus had proved his case' (*ibid.*, p. 133).

[4] Cf. volume 2, chapter 8 and this volume, chapter 3.

[5] Cf. Born [1949], pp. 129–34; Achinstein [1970] and Dorling [1971].

the probabilities of the two theories, given the data available in the sixteenth century, and show that Copernicus's was the more probable. All these efforts failed. I understand that Jon Dorling is now trying to elaborate a new Bayesian theory of the Copernican Revolution. He will not succeed. *If scientific revolution lies in proposing a theory which is much more probable given the available evidence than its predecessor, then there was no Copernican (Scientific) Revolution.*

Falsificationist philosophy of science can give two independent grounds on which the superiority of Copernicus's theory of celestial motions might rest.[1] According to one version, Ptolemy's theory was irrefutable (that is, pseudoscientific) and Copernicus's theory refutable (that is, scientific). If this were true, we really should have a case for identifying the Copernican revolution with the Great Scientific Revolution: it constitutes the switch from irrefutable speculation to refutable science. In this interpretation Ptolemaic heuristic was inherently *ad hoc*: it could accommodate *any* new fact by increasing the incoherent mess of epicycles and equants. Copernican theory, on the other hand, is interpreted as empirically refutable (at least 'in principle'). This is a somewhat dubious reconstruction of history: Copernican theory might well use any number of epicyclets with no difficulty. The myth that the Ptolemaic theory included an indefinite number of epicycles which could be manipulated to fit any planetary observations, is anyway a myth invented after the discovery of Fourier series. But, as Gingerich recently discovered, this parallel between epicycles-on-epicycles and Fourier analysis was not seen either by Ptolemy or by his followers. Indeed, the recomputation of the Alfonsine Tables by Gingerich shows that for actual computations Alfonso's Jewish astronomers used only a single-epicycle theory.

Another version of falsificationism claims that both theories were for a long time equally refutable. They were mutually incompatible rivals, both unrefuted; *finally*, however, some later crucial experiment refuted Ptolemy while corroborating Copernicus. As Popper put it: 'Ptolemy's system was not refuted when Copernicus produced his... It is in these cases that crucial experiments become decisively important.'[2] But Ptolemy's system (any given version of it) was commonly known to be refuted and anomaly-ridden long before Copernicus. Popper cooks up his history to fit his naive falsificationism. (Of course, he might *now* (in 1974) distinguish between mere anomalies which do not refute, and crucial experiments which do. But this general *ad hoc* manoeuvre which he produced in response to my criticisms[3] will not help him to specify in general terms the alleged 'crucial

---

[1] For a third, cf. *below*, p. 175.
[2] Popper [1963a], p. 246. Popper, ignoring Tycho, thinks that the phases of Venus decided the issue for Copernicus.
[3] Cf. section 6 and my [1974d], n. 49.

experiment'.[1]) As we have seen, the alleged superiority of Reinhold's Prutenic tables over the Alfonsine ones could not provide the crucial test. But what about the phases of Venus discovered by Galileo in 1616? Could *this* have formed the crucial test which showed Copernicus's superiority? I think that this might be a quite reasonable answer if not for the ocean of anomalies in which both rivals were equally engulfed. The phases of Venus may have established the superiority of Copernicus's theory over Ptolemy's, and if they did, would make the Catholic decision to ban Copernicus's work in the very moment of its victory all the more horrifying. But if we apply the falsificationist criterion to the question of when Copernicus's theory superseded not only Ptolemy's but also Tycho Brahe's (which was very well known in 1616), then falsificationism has only an absurd reply: that it did so *only in 1838*.[2] The discovery of stellar parallax by Bessel was the crucial experiment between the two. But surely we cannot uphold the view that the abandonment of geocentric astronomy by the whole scientific community could only be defended *rationally* after 1838. This approach requires strong – and implausible – socio-psychological premises in order to explain the rash switch away from Ptolemy. Indeed, the late discovery of stellar parallax had very little effect. The discovery was made a few years *after* Copernicus's work had been removed from the *Index* on the grounds that Copernicus's theory had already been proved to be true.[3] Johnson surely must be wrong when he writes:

The fact that should be emphasized and re-emphasized is that there were no means whereby the validity of the Copernican planetary system could be verified by observation until instruments were developed, nearly three centuries later, capable of measuring the parallax of the nearest fixed star. For that length of time the truth or falsity of the Copernican hypothesis had to remain an open question in science.[4]

Something must be wrong with the falsificationist account. This is a typical example of how history of science can undermine a philosophy of science – too much of the actual history of science was irrational if scientific rationality is falsificationist rationality.[5] *If a scientific*

---

[1] Indeed, once a Popperian 'potential falsifier' can be interpreted as either serious or unserious depending on the great scientists' authority, Popper's whole philosophy of science collapses.

[2] *Not* in 1723, when there occurred a 'crucial experiment' on the aberration of light.

[3] This is very reminiscent of the story of the role in the optical revolution of the determination of the speed of light in media optically denser than air. Prior to Fresnel's work it was agreed both by the corpuscular and the wave theorists that the discovery of the speed of light in, say, water, would be the decisive factor in the debate. But when Foucault's and Fizeau's results in the 1850s eventually came out in favour of the wave theory, they had little effect – the issue had already been decided. (Cf. Worrall [1976b].)

[4] Johnson [1959], p. 220. Johnson's mistake is made even worse by conflating verification and truth. Watkins too seems to have held, in his otherwise excellent criticism of Kuhn, that the rivalry between the Copernicans and their adversaries was decided by the crucial experiment of 1838. (Watkins [1970], p. 36.)

[5] For the outlines of a general theory of how history of science can be a test of its philosophical 'rational reconstructions' cf. chapters 2 and 3.

revolution lies in the refutation of a major theory and in its replacement by an unrefuted rival, the Copernican Revolution took place (at best) in 1838.

2 SIMPLICISM

According to conventionalism, theories are accepted by convention. Indeed we can, given sufficient ingenuity, force the facts into *any* conceptual framework. This Bergsonian position is logically impeccable,[1] but it leads to cultural relativism (a position assumed both by Bergson and Feyerabend) *unless* a criterion for when one theory is better than another (even though the two theories may be observationally equivalent) is added to it. Most conventionalists try to avoid relativism by adopting some form of *simplicism*. I use this rather ugly term for methodologies according to which one cannot decide between theories on empirical grounds: a theory is better than another if it is simpler, more 'coherent', more 'economical' than its rival.[2]

The first man to claim that the chief merit of Copernicus's achievement was to produce a simpler, and *therefore* better, system than Ptolemy's was, of course, Copernicus himself. If his theory at the time had been observationally equivalent (if restricted to celestial kinematics) to Ptolemy's, this would have been understandable.[3] He was followed by Rheticus and Osiander; and Brahe too judged there was something in the claim. The superior simplicity of Copernicus's theory of celestial 'orbs' became an unchallenged *fact* in the history of science from Galileo to Duhem: all that Bellarmine questioned was the *further* inference from impressive simplicity to Truth. Adam Smith, for example, in his beautiful *History of Astronomy*, argued for the superiority of the Copernican hypothesis on the basis of its superlative 'beauty of simplicity'.[4] He disclaimed the inductivist idea that the Copernican tables were more accurate than their Ptolemaic predecessors and that therefore, Copernican theory was superior. According to Adam Smith the new, accurate observations were equally compatible with Ptolemy's system. The advantage of the Copernican system lay in the 'superior degree of coherence, which it bestowed upon the celestial appearances, the simplicity and uniformity which it introduced into the real directions and velocities of the Planets'.[5]

But the superior simplicity of Copernican theory was just as much of a myth as its superior accuracy. The myth of superior simplicity was dispelled by the careful and professional work of modern his-

[1] Cf. chapter 1, pp. 21–2 and p. 100.
[2] Cf. chapter 1, p. 22.
[3] This 'observational equivalence' is actually a great simplicist myth; cf. *below*, p. 175. It should, however, be remembered that Copernicus thought that this greater simplicity would also provide, *eo ipso*, better astronomical tables, that is, it would lead to saving *more* phenomena. Thus he did not believe in the 'observational equivalence' of his theory with Ptolemy's.
[4] Smith [1773], p. 72.   [5] *Ibid.*, p. 75.

torians. They reminded us that while Copernican theory solves certain problems in a simpler way than does the Ptolemaic one, the price of the simplifications is unexpected complications in the solution of other problems.[1] The Copernican system is certainly simpler since it dispenses with equants and with some eccentrics; but each equant and eccentric removed has to be replaced by new epicycles and epicyclets. The system is simpler in so far as it leaves the eighth sphere of fixed stars immobile and removes its two Ptolemaic motions; but Copernicus has to pay for the immobile eighth sphere by transferring its irregular Ptolemaic movements to the already corrupt earth which Copernicus sets spinning with a rather complicated wobble; he also has to put the centre of the universe, not at the Sun, as he originally intended, but at an empty point fairly near to it.

I think it is fair to say that the 'simplicity-balance' between Ptolemy's and Copernicus's system is roughly even. This is reflected in de Solla Price's remark that Copernicus's system was 'more complicated but more economical';[2] and also in Pannekoek's view that 'the new world structure, notwithstanding its simplicity in broad outline, was still extremely complicated in the details'.[3] According to Kuhn, Copernicus's account of the *qualitative* aspect of the major problems of planetary motion (e.g. the retrograde motion) is much neater, much 'more economical', than Ptolemy's, 'but this apparent economy...is [only] a propaganda victory...[and in fact] is largely an illusion'.[4] When it comes to details, '[Copernicus's] full system was little if any less cumbersome than Ptolemy's had been'. As he succinctly puts it: Copernicus introduced a 'great, and yet strangely small' change.[5] While the Copernican theory has more 'aesthetic harmony', gives a more 'natural' account of the *basic* features of the heavens, has 'fewer *ad hoc* assumptions', it is in the end 'a failure...neither more accurate nor significantly simpler than its Ptolemaic predecessors'.[6] According to Ravetz, the 'irregularly moving stellar sphere' in Ptolemy's system brought with it a 'fundamental measure of time [as] a motion along an irregularly moving orbit'. In Ravetz's judgment this is '*strictly incoherent*', but, if this irregularity in the motion of the stars is transferred to the motion of the Earth, as it is in Copernicus's system, we get a '*coherent*' astronomy.[7] But if so, coherence seems to be in the eye of the beholder. Simplicity seems to be relative to one's subjective taste.[8] *If dramatic increase in simplicity of observationally equivalent theories is the hallmark of scientific revolution, the Copernican*

---

[1] Cf. e.g. Kuhn [1957] and Ravetz [1966a].

[2] Price [1959], p. 216. According to Price, Copernicus '*increased* the complexity of the (Ptolemaic) system without increasing the accuracy' (my italics).

[3] Pannekoek [1961], p. 193.   [4] Kuhn [1957], p. 169.

[5] Kuhn [1957], p. 133.   [6] *Ibid.*, p. 174.

[7] Ravetz [1966b].

[8] The most beautiful argument for this statement is on pp. xvi–xvii of Santillana [1953]. One glance suffices to demonstrate the point.

Revolution cannot be regarded as one (even if some people like Kepler thought that its superiority was due to the beautiful harmony which it introduced)[1].

Let us now return to Popperian falsificationism. Popper lays great stress on crucial experiments and, in this respect, he is, on my terms, an empiricist. Man proposes and Nature disposes. But at the same time he proposes a new brand of simplicism: he claims that even *before* Nature disposes we should already regard a theory as better than its rival if it has more falsifiable content, more potential falsifiers.[2] Since Popper offered his 1934 falsifiability criterion as an explication of 'simplicity',[3] his *Logic of Scientific Discovery* should be regarded as a new, original brand of simplicism. In this sense then, especially in its realist interpretation,[4] the Copernican theory may have been better than Ptolemy's already in 1543, even had they been *observationally* equivalent at the time.

But the two theories were not observationally equivalent. Simplicists usually take it *too easily* for granted that the rival theories which they appraise are either logically or in some other strict sense equivalent so that the claim that only simplicity, and not facts, can decide should sound more plausible. The conventionalist idea that Ptolemy's and Copernicus's theories are *bound to be in some strong sense equivalent* is common currency among 'simplicists': after all, they accept conventionalism but want to find a way out of its relativist implications. The idea has been propounded by Dreyer, the Halls, Price, Kuhn and others.[5] Hanson is right in saying, in his criticism of their views, that 'in no ordinary sense of "simplicity" is the Copernican theory simpler than the Ptolemaic'; but he still preserves their 'Line of Sight Equivalence'.[6]

[1] For why Kepler *thought* he preferred Copernicus to Ptolemy and to Brahe cf. Westman [1972]. Why he did prefer it is more difficult to say.

[2] He strengthened his empiricism in his 'third requirement' (I called it 'acceptability$_2$'; cf. volume 2, chapter 8, pp. 173 ff.).

[3] Popper [1953], chapter VII.

[4] Cf. Feyerabend [1964]; an excellent paper from his almost-Popperian period. Agassi holds that Copernicus' theory had no *empirical* superiority: indeed, Agassi claims that Copernicus 'did not succeed in showing that his system is better than Ptolemy's, let alone in refuting him'(Agassi [1963], p. 5).

[5] For a criticism of Dreyer's, the Halls's, Price's, Kuhn's overstatements, cf. Hanson [1973], pp. 200–20. That he himself overemphasizes simplicity ('systemacity') transpires from his arguments and from absurd statements like: '(Copernicus), like Newton after him, and Aristotle before, revealed no new data, nor did he seek any' (*ibid.*, p. 87).

[6] Hanson [1973], pp. 212 and 233. Ironically, on p. 233, Hanson absentmindedly exchanged in his manuscript 'Ptolemaic' and 'Copernican' and the editor of the posthumous work did not notice or correct the slip of pen.

3 POLANYIITE AND FEYERABENDIAN ACCOUNTS OF THE COPERNICAN REVOLUTION

All the philosophies so far discussed are based on universal demarcation criteria. According to them *all* major changes in science can be explained using the *same* single criterion of scientific merit. But none of these philosophies has been able to give a clear and acceptable account of any rational grounds on which geocentric theories were inferior to Copernicus's *De revolutionibus*. The failure of 'demarcationists' to solve this problem (and other similar problems) has led to a situation in which some, if not most, scientists and quite a few philosophers of science *deny* that there can ever be any valid universal demarcation criterion or system of appraisals for judging scientific theories. The most influential contemporary proponent of this view is Polanyi, according to whom the search for a universal rationality criterion is utopian. There can be only a *case law*, no *statute law* for deciding what is scientific and pseudoscientific, what is a better and what is a worse theory. It is the jury of scientists which decides in each separate case and as long as scientific autonomy – and *eo ipso* the independence of this jury – is upheld, nothing will go very wrong. If Polanyi is right, the Royal Society's refusal to sponsor philosophy of science is quite reasonable: ignorant *philosophers* of science should not be allowed to *judge* scientific theories, that is the scientist's own business. The Royal Society is of course, willing to finance *historians* of science who *describe* their activities as constituting triumphant progress.[1]

In the Polanyiite view, in each individual case of rivalry between two scientific theories, one has to leave it to the inarticulate *Fingerspitzengefühl* (Holton's favourite expression) of the great scientists to decide which theory is better. The great scientists are the ones who have 'tacit knowledge' of the way things will go. Polanyi writes about the

> foreknowledge the Copernicans must have meant to affirm when they passionately maintained, against heavy pressure, during one hundred and forty years before Newton proved the point, that the heliocentric theory was not merely a convenient way of computing the paths of planets, but was really true.[2]

But of course, this '*foreknowledge*' – unlike a simple conjecture – cannot be articulated and made available to the layman outsider. Toulmin seems to have a similar view of the Copernican Revolution.[3] So does Kuhn. Kuhn claims that

---

[1] The Royal Society gives financial support to history of science, but none to philosophy of science.

[2] Polanyi [1966], p. 23. Also cf. his [1958], *passim*.

[3] I take it that the following passage bears this claim out: 'If Kepler and Galileo preferred Copernicus' new heliostatic system, their reasons for doing so were far more specific, varied, and sophisticated than are hinted at by such vague terms as "simplicity" and "convenience": especially at the outset, indeed the Copernican theory was by many tests substantially less simple or convenient than, the traditional

to astronomers the initial choice between Copernicus' system and Ptolemy's could only be a matter of taste, and matters of taste are the most difficult of all to define or debate. Yet as the Copernican Revolution itself indicates, matters of taste are not negligible. *The ear equipped to discern geometric harmony* could detect a new neatness and coherence in the sun-centered astronomy of Copernicus, and if that neatness and coherence had not been recognized, there might have been no Revolution.[1]

According to a *later* account of Kuhn's,[2] Ptolemaic astronomy was by 1543 in a state of 'paradigm-crisis' which is the inevitable prelude to any scientific 'revolution', i.e. mass-conversion: 'The state of Ptolemaic astronomy was a recognized scandal before Copernicus proposed a basic change in astronomical theory, and the preface in which Copernicus described his reasons for innovation provides a classic description of the crisis state'.[3] But how many apart from Copernicus felt this communal 'crisis'? After all there was not much of a 'scientific community' in Copernicus's time. And if Kuhn thinks that his full analysis of scientific revolutions applies to the Copernican case, why did so few scholars join the Copernican 'bandwagon' before Kepler and Galileo?

In Kuhn's judgment there is no *explicit* criterion on the basis of which Copernicus's system can be judged superior to Ptolemy's. But the scientific élite with an inarticulable and esoteric 'ear for geometric harmony' or crisis-sensitive psyche could tell which theory was better. It seems, however, that once it comes to details, Kuhn's account is no more trouble-free than the accounts of the demarcationists. He has to invent a socio-intellectual 'crisis' in the scientific élite working in the Ptolemaic paradigm in the sixteenth century and then a sudden switch to Copernicanism. *If these are necessary conditions for a scientific revolution, then the Copernican Revolution was not a scientific revolution.*

For Feyerabend, the failure both of demarcationists and élitists is

---

Ptolemaic analysis. When we consider the conceptual changes between successive physical theories, therefore, the rationality we are concerned with is neither a merely formal matter, like the internal articulation of a mathematical system, nor a merely pragmatic matter, of simple utility or convenience. Rather, we can understand on what foundation it rests, only if we look and see how, in practice, successive theories and sets of concepts are first applied, and later modified within the historical development of the relevant intellectual activity'. (Toulmin [1972], p. 65.)

[1] Kuhn [1957], p. 177, my italics. For a general criticism of this Polanyiite position cf. chapter 3, p. 153 and my [1974d], p. 372.

[2] Kuhn's position concerning the Copernican Revolution changed radically from the essentially internalist simplicism of his [1957] to his radically sociologistic [1962] and [1963].

[3] Kuhn [1963] p. 367. For Kuhn a 'crisis' *must* precede a 'revolution' exactly as for a naive falsificationist a refutation *must* precede a new conjecture. No surprise that Kuhn writes that there is 'unequivocal historical evidence' that 'the state of Ptolemaic astronomy was a scandal before Copernicus's announcement' (Kuhn [1962], pp. 67–8). Gingerich [1973] showed that Kuhn conjures up a scandal where there was none. (Of course, a progressive 'research programme' (in my sense) need not be preceded by the degeneration of its rival.)

only to be expected. For him, our brilliant leading cultural relativist, the Ptolemaic system was just one system of belief, the Copernican system another. The Ptolemaists did their thing and the Copernicans did theirs and at the end the Copernicans scored a propaganda victory. To quote Westman's summary of his position:

We are given two theories, the Copernican and the Ptolemaic, both of which provide reliable predictions, but where the former contradicts the accepted laws and facts of the contemporary terrestrial physics. Belief in the success of the new theory cannot be based upon methodological assumptions for no such principles can ever certify the correctness of a theory at its inception; nor, at the start, does there exist any new factual support. Therefore, the acceptance of the Copernican theory becomes a matter of metaphysical belief.[1]

According to Feyerabend *nothing more can be said.* Feyerabend's account is much more difficult to rebut than anybody else's. Indeed, we may in the end have to admit that Copernicus's and Kepler's and Galileo's adoption of the heliocentric theory and its victory is not rationally explicable, that it was largely a matter of taste, a *Gestalt-switch*, or a propaganda victory. But even if this *did* turn out to be the case we need not allow ourselves to be steamrollered by Feyerabend into *general* cultural relativism or by Kuhn into *general* élitism. Fresnel's wave theory of light, for example, was by 1830 clearly better than Newton's corpuscular theory on explicit objective criteria, but Fresnel's first adoption of the old wave idea was clearly a question of taste.[2] If it were irrational to *work on* a theory whose superiority was not yet established then almost all of the history of science would indeed be rationally inexplicable. But, as it happens, the Copernican Revolution can be explained as rational on the basis of the methodology of scientific research programmes.

4 THE COPERNICAN REVOLUTION IN THE LIGHT OF THE METHODOLOGY OF SCIENTIFIC RESEARCH PROGRAMMES

The methodology of scientific research programmes is a new demarcationist methodology (i.e. a *universal* definition of progress) which I have been advocating now for some years and which, it seems to me, improves on previous demarcationist methodologies and at the same time escapes at least some of the criticisms which élitists and relativists have levelled against inductivism, falsificationism and the rest.

Let me first roughly explain the central features of this methodology.[3]

First of all my unit of appraisal is not an isolated hypothesis (or a

---

[1] Westman [1972], p. 234. In his [1972], Feyerabend slips into a Polanyiite view: he thinks that Copernicans achieved a victory of *Reason* with the help of their '*Lebendigheit des Geistes*'.

[2] Cf. Worrall [1976b].

[3] For my use of the technical term 'methodology' cf. chapter 3, p. 153 and chapter 2, p. 103, n. 1.

conjunction of hypotheses): a research programme is rather a special kind of 'problemshift'.[1] It consists of a developing series of theories. Moreover, this developing series has a structure. It has a tenacious *hard core*, like the three laws of motion and the law of gravitation in Newton's research programme, and it has a *heuristic*, which includes a set of problem-solving techniques. (This, in Newton's case, consisted of the programme's mathematical apparatus, involving the differential calculus, the theory of convergence, differential and integral equations.) Finally, a research programme has a vast belt of auxiliary hypotheses on the basis of which we establish initial conditions. The protective belt of the Newtonian programme included geometrical optics, Newton's theory of atmospheric refraction, and so on. I call this belt a *protective belt* because it protects the hard core from refutations: anomalies are not taken as refutations of the hard core but of some hypothesis in the protective belt. Partly under empirical pressure (but partly *planned* according to its heuristic) the protective belt is constantly modified, increased, complicated, while the hard core remains intact.

Having specified that the unit of mature science is a research programme, I now lay down rules for appraising programmes. A research programme is either progressive or degenerating. It is *theoretically progressive* if each modification leads to new unexpected predictions and it is *empirically progressive* if at least some of these novel predictions are corroborated. It is always easy for a scientist to deal with a *given* anomaly by making suitable adjustments to his programme (e.g. by adding a new epicycle). Such manoeuvres are *ad hoc*, and the programme is *degenerating*, unless they not only explain the given facts they were intended to explain but also predict some new fact as well. The supreme example of a progressive programme is Newton's. It successfully anticipated novel facts like the return of Halley's comet, the existence and the course of Neptune and the bulge of the Earth.

A research programme never solves all its anomalies. 'Refutations' always abound. What matters is a few dramatic signs of empirical progress. This methodology also contains a notion of *heuristic progress*: the successive modifications of the protective belt must be in the spirit of the heuristic. Scientists rightly dislike artificial *ad hoc* devices for countering anomalies.

One research programme *supersedes* another if it has excess truth content over its rival, in the sense that it predicts progressively all that its rival truly predicts and some more besides.[2]

Before we apply this new and perhaps a bit too elaborate philoso-

---

[1] Cf. volume 2, chapter 8, p. 178, Lakatos [1968c] and this volume, chapter 1, p. 33 ff.

[2] For an interesting discussion of 'superseding' versus 'incommensurability' cf. Feyerabend [1974].

phical framework[1] to appraising the rival theories, or, rather, rival *programmes,* of Ptolemy and Copernicus, one important remark has to be made.

Any two rival research programmes can be made observationally equivalent by producing observationally equivalent falsifiable versions of the two with the help of suitable *ad hoc* auxiliary hypotheses. But such equivalence is uninteresting. Two rival research programmes are only equivalent if they are identical. Otherwise the two different heuristics proceed at different speeds. Even if two rival programmes explain the same range of evidence, the same evidence will give more support to the one than to the other depending on whether the evidence was, as it were, 'produced' by the theory or explained in an *ad hoc* way. The weight of evidence is not merely a function of a falsifiable hypothesis and the evidence; it is also function of temporal and heuristic factors.[2] The starting point of the methodology of scientific research programmes is the normative problem posed by 'revolutionary conventionalism'.[3] But if revolutionary conventionalism is correct, observational equivalence can always be produced between two rival theories. Simplicism concluded that empirical evidence loses its weight: only the degree of simplicity counts. Popper's falsifiability and Lakatos's and Zahar's degree of progressiveness discard the ambiguity and pitfalls of degrees of coherence and rehabilitate, in radically new ways, a 'positivistic' respect for facts.

The descriptive aspect of the methodology of scientific research programmes is clearly superior to the descriptive aspect of the methodologies previously discussed. Both Ptolemy and Copernicus worked on *research programmes*: they did not simply test conjectures or try to harmonize a vast conjunction of observational results, nor did they *commit* themselves to any community based 'paradigms'. I shall offer a description of the two research programmes (this, I take it, will be fairly uncontroversial) and I shall also offer an appraisal of their respective progress and degeneration.

Both *programmes* branched off from the Pythagorean–Platonic programme whose basic principle was that since heavenly bodies are perfect, all astronomical phenomena should be saved by a combination of as few uniform circular motions (or uniform spherical rotations about an axis) as possible. This principle remained the cornerstone of the heuristic of both programmes. This proto-programme contained no directives as to where the centre of the universe lies. The heuristic in this case was primary, the 'hard-core'

---

[1] For more careful formulations the reader has to be referred to my [1968c], chapter 1, chapter 2 and chapter 3. Also cf. my [1974d].

[2] Zahar's achievement lies primarily in producing an improved notion of 'weight of evidence', cf. *below*, pp. 184–5.

[3] Cf. chapter 1, p. 21.

secondary.[1] Some people, like Pythagoras, believed the centre was a fireball invisible from the inhabited regions of the earth; others, like some Platonists, that it was the sun; and still others, like Eudoxus, that it was the Earth itself. The geocentric hypothesis 'hardened' into a real hard core assumption only with the development of an elaborate Aristotelian terrestrial physics, with its natural and violent motion and its separation of the terrestrial or sublunar chemistry of four elements from the pure and eternal celestial *quinta essentia.*

The first rudimentary geocentric theory of the heavens consisted of concentric orbs around the Earth, one for the stars and one for each other celestial body. But this was known to be a false 'ideal model' and, as Eudoxus already realized, even if the rudimentary scheme worked for the stars, it definitely did not work for the planets. As is well known, Eudoxus devised a system of rotating spheres in order to account for planetary motion. He introduced twenty-six such spheres in order to explain – or rather to save – the stations and retrogressions of the planets. The model predicted no novel facts and it failed to solve some serious anomalies like the varying degrees of brightness of the planets. After this system of rotating spheres was abandoned, every single move in the geostatic programme ran counter to the Platonic heuristic. The eccentric displaced the earth from the centre of the circle; the Appollonian and Hipparchan epicycles meant that the real paths of the planets about the earth were not circular; and finally the Ptolemaic equants entailed that even the motion of the epicycle's empty centre was not simultaneously uniform and circular – it was uniform but not circular when viewed from the equant point; it was circular but not uniform when viewed from the centre of the deferent: uniform circularity was replaced by quasi-uniform quasi-circularity.

The use of the equant was tantamount to the abandonment of the Platonic heuristic. No wonder then that at an early stage of this development astronomers like Heracleides and Aristarchus started to experiment with partially or completely heliocentric systems. Each move in the geocentric programme had dealt with certain anomalies but had done so in an *ad hoc* way. No novel predictions were produced, anomalies still abounded and certainly each move had deviated from the original Platonic heuristic.[2]

Copernicus recognized the heuristic degeneration of the Platonic programme at the hands of Ptolemy and his successors. He assumed that the periodicity of planetary motion was connected with – and,

---

[1] The demarcation between 'hard core' and 'heuristics' is frequently a matter of convention as can be seen from the arguments proposed by Popper and Watkins concerning the inter-translatability of what they called 'metaphysics' and 'heuristic' respectively. (Cf. especially Watkins [1958].)

[2] Kuhn claims that 'there were no good reasons for taking Aristarchus seriously' (Kuhn [1962], p. 76). But it is clear that there were – the geocentric programme had already heuristically degenerated.

indeed, exhausted by – combinations of uniform circular motions.[1] Copernicus levelled three charges of *ad hocness* against Ptolemy.

(*a*) The introduction of equants violated the heuristic of Ptolemy's own programme. It was heuristically *ad hoc* (*ad hoc*$_3$)[2]. This objection occurs in the third paragraph of the *Commentariolus*. In the second paragraph Copernicus mentions Callippus's and Eudoxus's vain effort to save the phenomena by a system of concentric spheres.

(*b*) Because of the difference between the solar and the sidereal years, Ptolemy gave two distinct motions to the stellar sphere: the daily rotation and a rotation about the axis of the ecliptic. This was already a major defect of the Ptolemaic system, since the stars, being the most perfect bodies, ought to have a single uniform motion.

In his *Commentariolus*, Copernicus pointed out that the sidereal year provides a more accurate unit of time than the solar year. According to Ravetz,[3] Copernicus must have started from erroneous data and concluded that the difference between the solar and the sidereal years varies irregularly; the stellar sphere must therefore rotate irregularly about the axis of the ecliptic. Thus the Sun moves non-uniformly about the earth. This is yet another violation of the Platonic heuristic, and constitutes further heuristic degeneration.[4]

(*c*) Despite all these violations of the Platonic heuristic, the geostatic programme remained empirically *ad hoc*, that is, it always lagged behind the facts.

Copernicus did not create a completely new programme; he revived the Aristarchan version of the Platonic programme. The hard core of this programme is the proposition that the stars provide the primary frame of reference for physics. Copernicus did not invent a new heuristic but attempted to restore and rejuvenate the Platonic one.[5]

Did Copernicus succeed in creating a more truly Platonic theory than Ptolemy? He did. According to the Platonic heuristic, the stars, being the most perfect bodies, ought ideally to have the most perfect motion, namely a single uniform rotation about an axis. Note that uniform circular motion is perfect because it can be assimilated to a state of rest: all points of the circle being equivalent, uniform circular motion is indistinguishable from rest or absence of change. We have seen that, in Copernicus's time, Ptolemaic astronomers imparted to

---

[1] In view of what we know about Fourier expansions of periodic functions, this is a remarkable mathematical conjecture. Cf. e.g. Kamlah [1971].

[2] Cf. chapter 1, p. 88, nn. 2 and 4.

[3] Ravetz [1966a]. But cf. Gingerich's remark in his [1973], n. 19.

[4] This 'incoherence', in Ravetz's view, suggested to Copernicus that the stars rather than the Earth determine the primary frame of reference for physics. Of course from the point of view of our *present* problem, it does not matter at all what actually triggered off Copernicus's imagination. We are not now concerned with the psychological causes of Copernicus's achievement, but with its appraisal.

[5] It was Kepler who framed the heuristic of the 'new' astronomy, namely the principle that the motion of the planets ought to be explained in terms of heliocentric forces.

the stellar sphere (at least) two distinct motions: a daily rotation and a rotation about the axis of the ecliptic. Also, due to erroneous data, they made this second rotation irregular.

Copernicus, however, fixed the stars, thus making them genuinely immutable. It is true that he had to transfer their motion to the Earth; but in his system the Earth is a planet and planets are anyway less perfect than stars, if only on account of their multiple epicyclic motions. (These multiple epicyclic motions were accepted both by Ptolemaists and by Copernicans.)

Copernicus got rid of the equant and produced a system which, despite the elimination of the equant, contained only about as many circles as the Ptolemaic system.[1]

In addition to its heuristic superiority over the *Almagest*, Copernican astronomy was no worse off in saving the phenomena than Ptolemaic astronomy. Indeed, Copernicus's lunar theory was a definite empirical advance over Ptolemy's. Using the Earth as an equant point, Ptolemy had succeeded in describing the angular motion of the moon; however, the moon would have had, at certain points of its path, twice its (observable) apparent diameter. Copernicus not only dispensed with equants, but also, through replacing equants by epicycles, he happened to improve on the fit between theory and observation.[2]

Copernicus's programme was certainly theoretically progressive. It anticipated novel facts never observed before. It anticipated the phases of Venus. It also predicted stellar parallax, though this was very much a qualitative prediction, because Copernicus had no idea of the size of the planetary system. It was *not*, as Neugebauer put it, 'a step in the wrong direction' from Ptolemy.[3]

---

[1] This mutual replaceability was already known to Islamic astronomers like Ibn-ash-Shatir. As Neugebauer pointed out (cf. Neugebauer [1958] and [1968]), Copernicus used a few equants but since these equants can be replaced by secondary epicycles, they are irrelevant. Copernicus considered uniform circular motions as the only permissible motions in astronomy; this need not prevent him from using equants as computational devices.

[2] In Neugebauer's view, this empirical success might have led Copernicus to believe that the elimination of the equant, apart from restoring the Platonic heuristic to its original purity, would also improve the predictive power of the new theory. But the Copernican system remained anomaly-ridden even in its most highly developed version. One of the most important anomalies in the Copernican programme was the comets whose motion could not be explained in terms of circular motions. This was one of Tycho's most important arguments against Copernicus, and it was one that Galileo found difficult to counter.

[3] Neugebauer [1968], p. 103. He claims: 'Modern historians, making ample use of the advantage of hindsight, stress the revolutionary significance of the heliocentric system and the simplifications it had introduced. Had it not been for Tycho Brahe and Kepler, the Copernican system would have contributed to the perpetuation of the Ptolemaic system in a slightly modified form, but *more pleasing to philosophical minds.*' Which philosophical minds? One wonders how a man of Neugebauer's stature can end a paper on such a casually inaccurate note. But, alas, even the most professional historians who are *in principle* against the philosophy of science, end up with *philosophically motivated blunders.*

But the phases of Venus prediction was not corroborated until 1616. Thus the methodology of scientific research programmes agrees with falsificationism to the extent that Copernicus's system was not *fully* progressive until Galileo, or even until Newton, when its hard core was incorporated in the completely different Newtonian research programme which was immensely progressive. The Copernican system may have constituted heuristic progress within the Platonic tradition, it may have been theoretically progressive but it had no novel *facts* to its credit until 1616.[1] *It seems then that the Copernican Revolution only became a fully fledged scientific revolution in 1616, when it was almost immediately abandoned for the new dynamics-oriented physics.*

From the point of view of the methodology of scientific research programmes the Copernican programme was not further developed but rather abandoned by Kepler, Galileo and Newton. This is a direct consequence of the shift of emphasis from 'hard core' hypotheses to heuristic.[2]

This, rather unwelcome, conclusion seems to be inevitable so long as we regard the prediction of only temporally novel facts as the criterion of progress. However, Zahar was led by considerations quite independent of the history of the Copernican Revolution to propose a new criterion of scientific progress – a criterion which is a very important amendment to that provided by the methodology of scientific research programmes.[3]

5   THE COPERNICAN REVOLUTION IN THE LIGHT OF ZAHAR'S
NEW VERSION OF THE METHODOLOGY OF SCIENTIFIC
RESEARCH PROGRAMMES

I originally defined a prediction as 'novel', 'stunning', or 'dramatic' if it was inconsistent with previous expectations, unchallenged background knowledge and, in particular, if the predicted fact was forbidden by the rival programme. The best novel facts were those which might never have been observed if not for the theory which anticipated it. My favourite examples of such predictions which were corroborated (and hence dramatically supported the theory on the basis of which they were made) were the return of Halley's comet, the discovery of Neptune, the Einsteinian bending of light rays, the Davisson–Germer experiment.[4] But on this view, the Copernican programme became

---

[1] According to Kuhn, the phases of Venus for the heliostatic system was 'not proof, but...propaganda' [1957], p. 224. Of course, it was not proof but it was, in the light of most empirical appraisals, including that of the methodology of scientific research programmes, an *objective* sign of progress. Kuhn seems to agree two pages later: 'Though the telescope *argued much*, it proved nothing' (*op. cit.*, p. 226).

[2] It is then wrong to say that 'The Copernican system of the world had developed into Newton's theory of gravity' (Popper [1963a], p. 98).

[3] Cf. Zahar [1973].

[4] Later I wanted to turn old empirical observations like the Balmer formula into novel facts with respect to Bohr's programme: cf. chapter 1, p. 69. But Zahar solved the problem in a superior way.

empirically progressive only in 1616! If this is so, one can well understand why its early proponents, in want of corroborated excess content, emphasized so much its superior 'simplicity'.

Interestingly, Elie Zahar's modified methodology of scientific research programmes gives a very different picture. Zahar's modification lies primarily in his new conception of 'novel fact'. In his view the explanation of Mercury's perihelion gave crucial empirical support, 'dramatic corroboration', to Einstein's theory, even though, as a low-level empirical proposition, it had been known for almost a hundred years.[1] This was no novel fact in my original sense, yet it was 'dramatic'. But in what sense 'dramatic'? 'Dramatic' in the sense that in Einstein's original design Mercury's anomalous perihelion played no role whatsoever. Its exact solution, was, as it were, an unexpected present from Schwarzschild, a result which was an unintended by-product of Einstein's programme. The same holds for the role of the Balmer formula in Bohr's programme. Bohr's original problem was not to discover the secrets of the hydrogen spectrum but to solve the problem of the stability of the nuclear atom; therefore, Balmer's formula gave 'dramatic' evidential support for Bohr's theory even though it was, temporally speaking, no novel fact.

Let us now look at the situation in 1543 and see if Copernicus's programme had *immediate* support from facts which were novel in Zahar's sense.

Copernicus's proto-hypothesis was that the planets move uniformly on concentric circles enclosing the Sun; the moon moves on an epicycle centred on the Earth.[2] Zahar's claim is that several important facts concerning planetary motions are straightforward consequences of the original Copernican assumptions and that, although these facts were previously known, they lend much more support to Copernicus than to Ptolemy within whose system they were dealt with only in an *ad hoc* manner, by parameter adjustment.

From the basic Copernican model and the assumption that inferior planets have a shorter period while superior planets have a longer period than that of the Earth,[3] the following facts can be predicted *prior to any observation*:

(*i*) *Planets have stations and retrogressions.*

Let us remember that Eudoxus's 26 concentric orbs were already doctored to account for the carefully observed stations and retrogressions. In Copernicus's programme stations and retrogressions are simple logical consequences of the rough model. Moreover, in Copernicus's programme this explains the previously puzzling and unresolved variations in the brightness of planets.

[1] Cf. Zahar [1973].
[2] Cf. the figure drawn by Copernicus on p. 10 of his *De Revolutionibus*.
[3] In the first chapter of *De Revolutionibus* Copernicus explains that this assumption is part of accepted background knowledge, common both to Ptolemy and Copernicus.

(ii) *The periods of the superior planets, as seen from the Earth are not constant.*

For Ptolemy this observational premise is very difficult to explain; for Copernicus it is a theoretical triviality.

(iii) *If an astronomer takes the Earth as the origin of his fixed frame, he will ascribe to each planet a complex motion one of whose components is the motion of the Sun.*

This is an immediate consequence of the Copernican model: a change of origin leads to the addition of the Sun's apparent motion to the motion of every other mobile.

For Ptolemy this is a cosmic accident which one has to accept *after* a careful study of the facts. Thus Copernicus *explains* what for Ptolemy is a fortuitous result, in the same way that Einstein explains the equality of inertial and gravitational masses, which was an accident in Newtonian theory.[1]

(iv) *The elongation of the inferior planets is bounded and the (calculated) periods of the planets strictly increase with their (calculable) distances from the Sun.*

In order to account for the fact that Venus's elongation from the Sun is bounded, Ptolemy resorted to the arbitrary assumption that the Earth, the Sun and the centre of Venus's epicycle remain collinear. It follows by Zahar's criterion of empirical support that Venus's bounded elongation lends little or no support to the Ptolemaic system. Copernicus for his part needs no *ad hoc* assumptions. His theory implies that a planet is inferior if and only if its elongation is bounded. Hence Venus is an inferior planet. Similarly Mars is a superior planet because its elongation is unbounded. *This hypothesis is independently testable* as follows. Let P denote any planet – superior or inferior – and let T_P be the period of P, T_E the period of the earth (i.e. one year), and t_P the time interval between two successive retrogressions of P. Then a simple calculation shows that, since retrogression occurs when the planet passes the Earth, the following relations between T_P, T_E and t_P hold good:

$$\frac{1}{T_P} - \frac{1}{T_E} = \frac{1}{t_P}$$

if P is an inferior planet; and

$$\frac{1}{T_E} - \frac{1}{T_P} = \frac{1}{t_P}$$

if P is a superior planet.

Note that t_P is measurable and that T_E is known and equal to one year. Thus the above equations enable us to calculate T_P.

In the case of a superior planet it follows from the second equation that $1/T_E > 1/t_P$; *i.e.* $T_E < t_P$. Hence we can predict that if a planet's elongation is unbounded, then the interval between two successive

[1] Zahar [1973], pp. 226-7.

retrograde motions of the planet is greater than one year. This is a novel – though well known – fact predicted, and therefore 'explained', by the Copernican programme. It gives support to Copernicus's programme without giving support to Ptolemy's. Neugebauer has a point in claiming that 'the main contribution of Copernicus to astronomy [was] the determination of the absolute dimensions of our planetary system'.[1]

Once he has obtained the periods of the planets, Copernicus goes on to calculate their distances from the Sun. One such method of calculation is described by Kuhn.[2] The period of a planet strictly increases with its distance from the Sun, *i.e.* from the origin of the Copernican frame of reference. This is consistent with accepted background knowledge. The Ptolemaic programme, as such, has no place for planetary distances but only for the angular motions of the planets. *Hence the determination of planetary distances represents excess content of Copernicus's theory over Ptolemy's.*

Ptolemaic astronomy too may be made to yield planetary distances by laying down arbitrarily that

$$\frac{r}{R} = \frac{\text{radius of epicycle}}{\text{radius of deferent}} = \text{distance of an inferior}$$
planet from the Sun (the Earth's distance being taken as unit)

$$\frac{R}{r} = \text{distance of a superior planet.}[3]$$

One can then use these equations to calculate the mean distances of the planets from the Earth. But these equations are grafted in an *ad hoc* way onto the Ptolemaic programme. And it is found, that, although Mercury, Venus and the Sun have approximately the same period, their distance from the Ptolemaic origin, *i.e.* from the Earth, differ widely; and this contradicted the background hypothesis commonly held at the time that the period grows with the distance from the fixed centre to which the motion is referred.

A historical thought-experiment may illuminate the corroborating strength of these facts. Let us imagine that in 1520 – or before – all we knew about the heavens was that the Sun and the planets move periodically relatively to the earth; but our records, because of, say, the cloudy Polish skies, were so scanty that stations and retrogressions have never been experimentally ascertained. Because of his Sun-worship and his belief in the Platonic heuristic, astronomer X proposes the basic Copernican model. Astronomer Y who adheres to the

[1] Neugebauer [1968].
[2] Kuhn [1957], p. 176.
[3] Neugebauer [1968]. One may use also the Aristotelian 'doctrine of plenitude' to arrive at distances; but this doctrine is again heuristically *ad hoc*, besides being both false *and*, within Ptolemy's programme, unfalsifiable.

Platonic heuristic but also to Aristotelian dynamics puts forward the corresponding geocentric model: the Sun and the planets move uniformly on circles centred around the Earth. If so, then X's theory would have been dramatically confirmed by observations carried out later on the coasts of the Mediterranean. The same observations would have refuted Y's hypothesis and compelled Y to resort to a series of *ad hoc* manouevres (assuming that Y was not so disheartened as to abandon his programme instantly).

Zahar's account thus explains Copernicus's achievement as constituting genuine progress compared with Ptolemy. The Copernican Revolution became a great *scientific revolution not because it changed the* European *Weltanschauung*, not – as Paul Feyerabend would have it – *because* it became also a revolutionary change in man's vision of his place in the Universe, but simply because it was *scientifically* superior. It also shows that there were good objective reasons for Kepler and Galileo to adopt the heliostatic assumption, for already Copernicus's (and indeed, Aristarchus's) *rough model* had excess predictive power over its Ptolemaic rival.[1]

Why was then Copernicus not content with his *Commentariolus*? Why did he work for decades to complete his system before having it printed? Because he was not content with mere progress for his programme, but wanted actually to *supersede* Ptolemy's; *i.e.* instead of merely predicting 'novel' facts which Ptolemy's system had not 'predicted', he wanted to explain *all* the true consequences of the Ptolemaic theory. This is why he had to write *De revolutionibus*. But it turned out that apart from his initial successes, Copernicus could save all the Ptolemaic phenomena only in an *ad hoc* and, in its dynamical aspects, very unsatisfactory, way.[2] So Kepler and Galileo took off from the *Commentariolus* rather than from *De revolutionibus*. They took off from the point where the steam ran out of the Copernican programme. Because of the initial success of the rough model and the degeneration of the full programme Kepler discarded the old heuristic and introduced a revolutionary new one, based on the idea of heliocentric dynamics.[3]

Let me end with a trivial consequence of this exposition, which I hope at least *some* of you will find outrageous. Our account is a narrowly internalist one. No place in this account for the Renaissance spirit so dear to Kuhn's heart; for the turmoil of Reformation and

---

[1] Note that this statement does not say whether and why Kepler and Galileo actually became 'Copernicans'.

[2] Zahar's concept of heuristic progress can, of course, be regarded as an *objective* (and 'positivist') explication of 'simplicity' without running into the inconsistencies of naive simplicists like the ones discussed above, in section 2.

[3] This pattern is not unique: after all, Bohr's old quantum theory was abandoned soon after it was accepted and de Broglie's new quantum theory took off from his first crude model rather than from Sommerfeld's and others' sophisticated calculations.

Counter-reformation, no impact of the Churchman; no sign of any effect from the alleged or real rise of capitalism in the sixteenth century; no motivation from the needs of navigation so much cherished by Bernal. The whole development is narrowly internal; its progressive part could have taken place at any time, given a Copernican genius, between Aristotle and Ptolemy or in any year after, say, the 1175 translation of the *Almagest* into Latin, or for that matter, by an Arab astronomer in the ninth century. External history *in this case* is not only secondary; it is nearly redundant.[1] Of course, the system of patronage of astronomy through Church sinecures played a role; but studying it will contribute nothing to our understanding of the Copernican scientific revolution.

6   A POSTSCRIPT ON HISTORY OF SCIENCE AND ITS RATIONAL RECONSTRUCTIONS*

In the previous sections a new solution was proposed to the question why Copernicus's programme (objectively) superseded Ptolemy's. It was superior to it on all three standard criteria for appraising research programmes: on the criteria of theoretical, empirical and heuristic progress. It predicted a wider range of phenomena, it was corroborated by novel facts and, in spite of the degenerative elements of *De Revolutionibus* it had more heuristic unity than the *Almagest*. It was also shown that Galileo and Kepler rejected the Copernican Programme but accepted its Aristarchian hard core. Rather than *initiating* a revolution, Copernicus acted as a midwife to the birth of a programme of which he never dreamt, namely of an anti-Ptolemaic programme, which took astronomy *back* to Aristarchus and at the same time *forward* to a new dynamics.

Having offered an *objective appraisal of Copernicus's achievement*, the historian can proceed to a second class of questions. Why did Kepler and Galileo *accept* Copernicus's hard core and why did they *reject* his Platonic heuristic? Why did people receive his theories as they did? And also, what was Copernicus's problem situation and his motive in starting a new programme?

This question of the *motives and the reception of Copernicus's achievement* is an important one and cannot be answered in strictly 'internal' terms. This present paper does not concern itself with the answer.

---

[1] Of course, our analysis implies that there is a very important, but purely 'external', question to be answered in socio-psychological terms: why did the Copernican Revolution take place when it did, rather than at any other time since Ptolemy? But the answer, if indeed it is possible to give one, will not affect the *appraisal* that we have attempted here. This is a good example of how internal (methodological) history can define what are the important external problems, and why it is therefore of primary importance!

* This section of the paper was written by Lakatos alone, shortly after the completion of the rest of the paper. It is published here for the first time. (*Eds.*)

What I shall, however, try to argue is that (1) the first question can be fully answered without going into the second one and independently from it, and (2) the second question can *only* be answered by assuming explicitly or implicitly, a reply to the first. This implies that philosophy of science is primary, and that sociology and psychology are secondary in writing history of science. Any reply to the first, philosophical question constitutes the backbone of an 'internal' 'rational reconstruction' of history, without which the full history cannot be written.[1]

I argued for this thesis already in my 'History of science and its rational reconstructions' but I shall now try further to clarify some of its points.

The very *problems* of the historian are determined by his methodology (i.e. theory of appraisal). The inductivist will look for the factual basis of Copernican theory and after having invented one in despair, his main external problem will be why people observed certain types of events in Europe rather than in China, in the sixteenth rather than in the tenth century. The falsificationist will look for crucial experiments between Copernicus, Ptolemy and Tycho, and will have to explain (by external myths) why scientists – no doubt, 'irrationally' – accepted Copernican theory *before* the discovery of the parallax and even of the aberration of light. The simplicist will conceal at least some of the complications of *De Revolutionibus* and then has to explain why this overwhelming simplicity did not satisfy Tycho, who, after all, destroyed some of the simplicity in an 'irrational' way. The Kuhnian will cook up a story of the monopoly of Ptolemaic theory until the early sixteenth century and concoct a 'crisis' followed by 'instant conversion'.[2] Neither can those who adopt the methodology of scientific research programmes explain a theory's acceptance or its rejection without adducing further psychological hypotheses. Appraisal alone does not logically imply acceptance or rejection. But the adduced psychological auxiliary hypotheses will vary according to the normative theory of appraisal; and this is the rationale of my relativization of the internal/external distinction to methodology.[3]

Let me show in some pedantic detail why an appraisal criterion alone cannot possibly explain the actual history of science. Let us take the proposition P_3: 'The theory (or research programme) T_1 was superior at time t to T_2'. From this proposition it does not follow that 'All (or some) scientists accepted at time t that T_1 was superior to T_2'. I shall call this proposition $P_{2.1}$. The first proposition may well be true while the second false. But let us add to P_3 a *psychological* premise like $P_{2.2}$: '(All) scientists will – *ceteris paribus* – accept T_1 over T_2 at time t if T_1

---

[1] For a definition of 'rational reconstructions' see *below*, p. 191.
[2] Kuhn does not demarcate (normative) objective appraisal from (descriptive) acceptance and rejection.
[3] Cf. chapter 2 *above*, p. 102.

is superior to T_2 at time t'. From P_3 and $P_{2\cdot2}$, given some weak further psychological assumptions,[1] $P_{2\cdot1}$ does follow. If T_1 and T_2 are research programmes, from the *acceptance* of T_1 as superior ($P_{2\cdot1}$) the decision *to work on* T_1 rather than on T_2 follows only if further substantial psychological assumptions are added.[2]

We find that in this deductive schema for the explanation of scientific change there are both 'third-world' and psychological premises. Moreover, the psychological premises are bound to be different according to the differences in the 'third-world' premises. We need one type of psychological theory in order to explain why scientists accepted Copernicanism as opposed to Tycho's theory before observing the parallax if we are (or assume them to be) falsificationists. But we need another type of psychological theory to explain why they behaved in the same way if we are (or assume them to be) inductivists. If we hold that rational decisions concerning acceptance and rejection of research programmes are based on subconscious or semiconscious applications of Lakatos's or Zahar's methodology, but accompanied by phenomena of false awareness, we may need complex socio-psychological armour to explain a switch from one programme to another.

Our crucial ('internal') 'third-world' premise, as a matter of fact, defines the problem situation for the 'externalist'. The internal skeleton of rational history defines the external problems. For instance, as I have pointed out, for an inductivist all priority problems will seem dysfunctional; for a follower of the methodology of scientific research programmes some may be perfectly functional.[3] The respective psychological/sociological explanatory schemas of some priority disputes may be vastly different. Also, if a theory is rejected because of a single anomaly, falsificationists need only a weak psychological premise (a sort of Falsificationist Rationality Principle) to explain it as *rational* rejection. Those who hold that the operative principle is the methodology of scientific research programmes have to devise a possibly very sophisticated theory of false awareness to explain – in the same case – the rejection as *rational*.

All historians of science who distinguish between progress and degeneration, science and pseudoscience, are bound to use a 'third-world' premise of appraisal in explaining scientific change. *It is the use of such a premise in explanatory schemas describing scientific change that I called 'rational reconstruction of the history of science'*. There are different

---

[1] These assumptions will spell out that the *ceteris paribus* clause is satisfied. For instance that the scientist did not misread the rival theories; or that the books containing T_1 and T_2 were materially available to them; that the hard core of the superior programme is consistent with their religion or ideology.

[2] The indices are not entirely arbitrary. P_3 is a proposition about Frege's and Popper's 'third world' of objective knowledge; $P_{2\cdot i}$ are propositions about the second world of beliefs and mental decisions and actions. (Cf. e.g. Popper [1972].)

[3] Cf. chapter 2, p. 116.

rival rational reconstructions for any historical change and one reconstruction is better than another if it explains more of the actual history of science; that is, rational reconstructions of history are research programmes, with a normative appraisal as hard core and psychological hypotheses (and initial conditions) in the protective belt. These historiographical research programmes are to be appraised as any other research programme for progress and degeneration. Which historiographical research programme is superior may be tested by seeing how successfully they explain scientific progress. In the case of the Copernican Revolution this talk was only programmatic: the real test will come only when the appraisal is supplemented by a full explanation.

Finally, I wish to clarify a few points arising from earlier discussions of my theory.

First, it is not the case that I *propose* a rational reconstruction of history of science *as opposed to* describing and explaining it. Rather I maintain that all historians of science *who hold that the progress of science is progress in objective knowledge*, use, willy-nilly, some rational reconstruction.

Secondly, in my particular programme of rational reconstruction (for which I now accept Zahar's important amendment), there is no 'attempt to protect [myself] from real history'.[1] This Kuhnian charge stems probably from a rather unsuccessful joke of mine. Some years ago I wrote that 'One way to indicate discrepancies between history and its rational reconstruction is to relate the internal history *in the text*, and indicate *in the footnotes* how actual history "misbehaved" in the light of its rational reconstruction'.[2] Of course, such parodies may be written, and may even be instructive; but I never said that this is the way in which history actually ought to be written and, indeed, I never wrote history in this way except for one occasion.[3]

Kuhn's charge that my conception of history 'is not history at all but philosophy fabricating examples', is misconceived. I hold that all histories of science are *always* philosophies fabricating examples. Philosophy of science determines to a large extent historical explanation; and Kuhn provided us with probably the most colourful one. But, equally, all physics or any kind of empirical assertion (i.e. theory) is 'philosophy fabricating examples'. Surely since Kant and Bergson this is a commonplace. But, of course, some fabrications in physics are better than others and some fabrications in history are better than others. And I offer sharp criteria using which one can compare rival fabrications both in physics *and* in its history – and I claim that my fabrications contain more truth than Kuhn's.

[1] Kuhn [1971], p. 143.
[2] Chapter 2 *above*, p. 120; quoted and criticized in Kuhn [1971], p. 142.
[3] I used this style extensively in my *Proofs and Refutations*, but there my purpose was to distill a methodological message from the history, rather than to write history itself.

5

Newton's effect on scientific standards*

(a) Justificationism and its two poles: dogmatism and scepticism

Schools in the theory of knowledge draw a demarcation between two vastly different sorts of knowledge: *episteme*, that is, proven knowledge, and *doxa*, that is, mere opinion. The most influential schools – the 'justificationist' schools[1] – rank *episteme* exceedingly high and *doxa* exceedingly low; indeed, according to their extreme canons only the former deserves the name 'knowledge'. To quote a leading seventeenth century justificationist: 'For with me, to know and be certain is the same thing; what I know, that I am certain of; and what I am certain of, that I know. What reaches to knowledge, I think may be called certainty; and what comes short of certainty, I think cannot be called knowledge.'[2] Or, as a twentieth century justificationist put it: 'We cannot know a proposition unless it is in fact true.'[3] According to this school then, knowledge is proven knowledge, the growth of knowledge is the growth of proven knowledge, which, of course, is *eo ipso* cumulative. The dominance of justificationism in the theory of knowledge cannot be characterized better than by the fact that the theory of knowledge came to be called 'epistemology', the theory of *episteme*. Mere *doxa* was not deemed worthy of serious investigation: growth of *doxa* was regarded as a particularly absurd idea, since in the orthodox justificationist view,[4] the hallmark of progress was the

---

* Early drafts of this paper were written in 1963–4. Lakatos returned to it several times, but he still regarded it as in need of substantial revision. It is published here for the first time. We have at various points slightly modified the text of Lakatos's final type-script. We have supplied titles both to the whole paper and to section 2(a). Many quotations were incomplete and the references to them omitted; these have been completed where possible. (*Eds.*)

[1] The recognition of 'justificationism' as one of the most influential traditions in modern European thought, and its first analysis and criticism, are due to Karl Popper; cf. his classic [1960a], pp. 39–71. I discussed some aspects of the empiricist versions of justificationism in my [1968b] (volume 2, chapter 8).

[2] Locke [1697], p. 145.   [3] Keynes [1921], p. 11.

[4] I call those justificationists 'orthodox' who hold that *doxa* is absolutely valueless; as some contemporary justificationists – like Schlick – put it, 'meaningless'. I call those justificationists who allow some heuristic value to *doxa*, 'liberal' justificationists. But both orthodox and liberal agree that *doxa* has no place in the end product.

increase of rational *episteme* and the gradual decrease in irrational doxa.

While justificationists agreed about the value of *episteme* and the valuelessness of *doxa*, they greatly differed about the *limitations* of *episteme*. Almost all of them agreed that *episteme* was possible, but they disagreed about the range of those propositions which may be proved. Pyrrhonian sceptics thought that *no* proposition can be proved, academic sceptics thought that at least one proposition – 'we cannot know' – can be proved.[1] These universal and quasi-universal sceptics were the epistemological pessimists. 'Dogmatists' were more optimistic. Some of them thought that one may have (epistemic) knowledge about religious and moral truth, but about nothing else;[2] others that it can also be extended to logic, mathematics and sublunar reality; epistemological optimists of the seventeenth and eighteenth centuries swept even the 'sublunar' restriction away and hoped that all secrets of nature will, in the end, yield to the powers of rational inquiry. Yet others thought that, while we can achieve *epistemic* knowledge of the laws of nature, religion and perhaps morality are doomed to remain arbitrary *doxa*. Most of the history of epistemology is the story of the infighting among rival schools of justificationism about the demarcation between *episteme* on the one hand, and *doxa*, the sink of uncertainty and error, of futile and inconclusive discussion, on the other.[3] The demarcation line came to be referred to as the '*limitations of human knowledge*' and the term *doxa* came to be replaced by 'metaphysics'.

Another very important problem about which justificationists differ is the problem of what exactly constitutes *episteme*. According to *essentialism*,[4] episteme must be final, ultimate truth (and then 'to prove' means 'to prove that a proposition is final, ultimate truth'). Essentialists thought that description of appearances, however precise, cannot be called 'knowledge', and that an argument for a phenomenal theory cannot be called 'proof'. According to some philosophers Ptolemy's school *did* have a most exact description of celestial phenomena but this description was only a description of shadows in Plato's cave and such description – even if perfect – remains mere *doxa*. The human

---

[1] Pyrrhonian sceptics called academic sceptics 'negative dogmatists'. ('Dogmatist' was, of course, a nickname, given by the Pyrrhonian sceptics to their opponents, who thought that at least *some* propositions are provable.)

[2] It is important to note that the term 'sceptic' was generally used as a swearword in the fight between rival schools of dogmatism. People whose epistemological interest was in religion, morality and politics, termed those who were sceptics with regard to this domain but dogmatists with regard to science, 'sceptics'. People whose epistemological interest lay in science, called their Church opponents 'sceptics'. Because of this it is very important that the terms 'dogmatist' and 'sceptic' should always be relativized to specific domains.

[3] The most dramatic battle was between theological and scientific dogmatism, culminating in Galileo's trial.

[4] The term is Popper's (cf. Popper [1945], volume II, chapter 11, or Popper [1963a], pp. 103 ff.). Its origin is Aristotle's dictum: 'there is knowledge of each thing only when we know its essence' (*Metaphysics*, 1031 b7).

mind has then its limitations: about some things it can achieve (explanatory) certainty, *i.e.* ultimate truth; about other things only (descriptive) certainty, *i.e.* phenomenal truth. It was Newton himself who conducted the great crusade against essentialism; who had the term 'knowledge' extended to proven truths about appearances, proven truths which are not ultimate truths.[1] I shall call this Newtonian position '*defensive positivism*'.[2]

(b) Psychologistic justificationism

Dogmatists, as I have already said, set exceedingly high standards for knowledge. In the frenzy of the religious wars of the sixteenth century dogmatism reached a peak. Religious knowledge was definitely meant to be certain and ultimate. As Luther put it: 'A Christian ought. ... to be certain of what he affirms, or else he is not a Christian'.[3] The slightest doubt brings anathema: 'Anathema to the Christian who will not be certain of what he is supposed to believe and who does not comprehend it.'[4] In order to get to heaven one needs then certain religious knowledge; doubt, let alone error, leads to eternal damnation.

Scientific knowledge in the seventeenth century was regarded by most of its representatives as an integral part of theological knowledge: most scientists, like Descartes, Kepler, Galileo, Newton, and Leibniz, were after God's Blueprint of the Universe.[5] Thus scientific knowledge too was meant to be proven and ultimate. As Maclaurin put it: Natural philosophy leads us

to the knowledge of the Author and Governor of the universe. To study nature is to search into his workmanship. . . . False schemes of natural philosophy may lead us to atheism, or suggest opinions, concerning the Deity and the universe, of most dangerous consequence to mankind; and have been frequently employed to support such opinions.[6]

Whether ultimate truth was contingent or necessary – that is, whether God created the world completely of his free will or not – formed a central problem for seventeenth century scientist–theologians. Newtonians advocated the former, Cartesians the latter.

Justificationist epistemology has then two main problems:[7] how to discover (ultimate) Truth and how to prove that it *is* the Truth. Justificationist epistemology can be characterized by its two main

---

[1] Cf. *below*, p. 204 ff.   [2] See *below*, section 2(*a*).
[3] Cf. Luther [1525], p. 603.   [4] *Ibid.*, p. 605.
[5] For the decline of the rival, instrumentalist–fallibilist tradition of Copernicus and others, cf. *below*, p. 199 ff.
[6] Maclaurin [1748], pp. 3–4.
[7] Justificationism, of course, was not born in the sixteenth century. Its roots go back to antiquity: the arch-dogmatist was, after all, Aristotle, and the arch-sceptic, Pyrrho. But in appraising the intellectual climate in which Newtonian science came into being, one had better concentrate on its modern version.

problems: (1) *the problem of the foundations of knowledge (the logic of justification)* and (2) *the problem of the growth of knowledge (the problem of method, logic of discovery, heuristic)*.

The 'logic of justification' was to solve the problem of how to recognize Truth when one has found it. The only existing paradigm of knowledge, Euclid's Geometry, was deductively organized. This suggested one obvious solution: to establish some anchors of Truth – let us call them *basic propositions* – and a machinery of safe truth-transmission from these basic propositions to other propositions – some sort of infallible *logic*. But where should one look for basic propositions? Should one look for them among powerful propositions with high content? In this case the natural immediate light of intuition must be very strong to establish them. Or should one look for them among weaker, nearly tautologous propositions with low content? In this case the logic must be very strong in order to increase their truth-content in the course of transmission. The main problem of the school of thought which favoured the first approach was to justify their *basic propositions*; the main problem of the school of thought which favoured the second approach was to justify content-increasing inductive *logic*.[1]

But how can one '*prove*' that a basic proposition is *true* (even less ultimately true)? How can one '*prove*' that an inference is *valid*?

Dogmatists were deeply divided on this issue. Some thought that both questions should be decidable by having the *propositions – or inferences – in themselves (as existing in the 'third world'*[2]) inspected by some objective mind which could possibly be represented by a machine. Indeed, as it turned out – three centuries later – an important part of *logic* could be checked, at least in a weaker sense, in this way. But the Leibnizian dream of a universal decision machine which would decide the truth or falsehood of any proposition has never come true. This is how dogmatists fell back on a new, psychological 'second-world'[3] criterion.

To understand this criterion let us remember that the dogmatists have always maintained that there are human faculties – the senses, intellect or the ability to receive divine communication – which, separately or jointly, enable humans to recognize the truth of what we called 'basic propositions'. But it was well known that all of these faculties may be deceivers. Therefore dogmatists set up an *ad hoc* theory: that human faculties do not deceive when they are in a 'healthy', 'right', 'normal' or, as they said later, 'scientific' state. Basic statements are then proven if recognized as true by the 'healthy', 'right', 'normal' or 'scientific' mind. That is, the question of whether

---

[1] This demarcation of the two schools contrasts sharply with the traditional rationalist-empiricist (or rather intellectualist–sensationalist) demarcation. This traditional demarcation is defined in psychologistic terms, while ours is defined in objective, logical Popperian terms, which, of course, were then not available.

[2] Cf. Popper [1972], chapters 2 and 3.   [3] *ibid.*

or not a proposition is genuinely proved is to be decided by an examination of the discoverer's mind: if it is 'scientific', then the proposition is accepted.

All schools of dogmatism agreed that there are certain types of propositions which can be recognized as true by the right mind. But they differed with regard to the *type of possible basic statements* and they also differed with regard to what constitutes a *right mind*. These two problems gave rise to two dogmatist research programmes: to the search for a criterion for basic propositions and to the search for a criterion of the right mind.

There were two major research programmes centred upon the quest for *basic propositions*: one was the empiricists' search for 'pure sensation statements', the other was the rationalists' search for *a priori* first principles.

There were many different theories of the criterion of the *right mind*. Aristotle – and the Stoics – thought that the right mind is simply the medically healthy mind; according to Descartes the right mind is primarily the one [that has been] steeled in the fire of sceptical doubt and [has] then [found] itself – and God's guiding hand – in the final loneliness of pure thought. According to Baconians, the right mind is the *tabula rasa*, devoid of all content, so that it can receive the imprint of nature without distortion, etc. All schools of dogmatism can then be characterized by the particular *psychotherapy* by which they prepare the mind to receive the grace of proven truth in the course of a mystical communion.

Since for justificationism the growth of knowledge is *eo ipso* cumulative, there is no place for a logic of discovery as distinct from the logic of justification; for justificationists 'to discover is to prove'. What some of them call 'logic of discovery' or 'heuristic' is usually nothing but the preliminary psychotherapy that must precede the start of cumulative scientific growth. This approach has two consequences, basic for this justificationist outlook. First, the logic of justification becomes an appraisal of the scientific mind, a check not of the discovery but of the discoverer, whether he has undergone the preliminary psychotherapy properly. Thus a bad heuristic and a false psychology came to serve as logic of justification. Secondly, if one abandons a theory, then the very fact of the abandonment shows that the refuted theory was not really a result of scientific communion with truth, and that the psychotherapy had failed. Each scientific change is then regarded as a conversion from false faith to true faith, as a change from a pseudoscientific state of mind to a scientific state of mind.

(c) Justificationist fallibilism

In the seventeenth century many thinkers held that justificationist standards should be abandoned in important matters. Dogmatism in religion, moral and political affairs had led to cruel wars, massacres and anarchy in the preceding century. The reaction took the form of *tolerant sceptical enlightenment* according to which no one could justify his position so flawlessly as to justify killing his opponent as a heretic: everyone had the right to his beliefs.[1] The ancient sceptics' teaching about fallibility and suspense of judgment was revived and canvassed. All proofs, theological, scientific, and even mathematical, were called into question.

On the other hand, it became increasingly clear that in ordinary human affairs one cannot suspend judgment for lack of *episteme*: 'he that will not stir till he infallibly knows the business he goes about will succeed...will have little else to do, but to sit still and perish'.[2] Why not then agree with the sceptics that there can be no *episteme* but point out – against them – that there can be relevant, verisimilar *doxa* which must not be rejected only because it is not *episteme*? In the seventeenth century many seemed to be prepared to explore this way and develop some sort of *fallibilism*. Martin Clifford, the theologian, wrote in 1675, that 'all the miseries which have followed the variety of opinions since the Reformation have proceeded entirely from these two mistakes, the tying Infallibility to whatsoever we think Truth, and *damnation* to whatsoever we thing *error*';[3] and Glanvill, the house-philosopher of the Royal Society, argues in the same year: 'If I should say, we are to expect no more from our Experiments and Inquiries, than great likelyhood, and such degrees of probability, as might deserve an hopeful assent; yet thus much of diffidence and uncertainty would not make me a *Sceptick*; since *They* taught, that no one thing was *more probable* than an *other*; and so with-held assent from all things.'[4] Locke preserved the term 'knowledge' or 'science' for proven, ultimate truth and thought that 'natural philosophy is not capable of being made a science';[5] but, unlike the sceptics, he claimed that it may have the 'twilight of probability' (by which he meant 'likeliness to be true'[6]). 'Our knowledge being short, we want something else...Judgment supplies the want of knowledge'.[7]

But what should be the *standards* for *doxa*, for mere hypotheses? Here

---

[1] 'Since...it is unavoidable to the greatest part of men, if not all, to have several opinions, without certain and indubitable proofs of their truths; it would, methinks, become all men to maintain peace and the common offices of humanity and friendship in the diversity of opinions.' Locke [1690], IV, 16, section 4.

[2] *Ibid.*, IV, 14, section 1, quoted in Laudan [1967], p. 214, n. 12.

[3] Martin Clifford [1675], p. 14. Quoted in Popkin [1964], p. 16.

[4] Joseph Glanvill: 'Of Scepticism and Certainty' (Essay II in his [1675], p. 45).

[5] Locke [1690], IV, 12, section 10.   [6] *Ibid.*, IV, 15, section 3.

[7] *Ibid.*, VI, 14, sections 1 and 3.

seventeenth century fallibilists did not have to break completely new ground. They inherited from antiquity a theory – Ptolemaic astronomy – which, although not regarded as ultimate truth, was nevertheless respected for its predictive achievement, or, as it was said, for its success in 'saving the phenomena'. According to standards developed to judge such 'hypotheses', a hypothesis is acceptable if it is consistent with the facts. But then further problems arise. What happens, for instance, if *several* hypotheses are equally consistent with the facts? Such a situation had already arisen with different schools *within* the Ptolemaic tradition.[1] Two types of solution were offered. One was Theon's: to adopt the hypothesis which was also consistent with the established first principles of some dogmatist school;[2] the other was Ptolemy's, which suggested choosing the *simpler* alternative.[3] The debate between these two schools went on in the Arabic–Jewish astronomy of the Middle Ages. Averroes and his disciples were on the one side, while Maimonides was the most influential representative on the other side.[4] Indeed, some dogmatists wanted to rule that, whether or not there was an alternative, hypotheses *must* conform with established *episteme*. Thus in the sixteenth century Aristotelian Jesuits proposed the following criteria for acceptability of hypotheses: they should be consistent (1) with the facts, (2) with Aristotelian physics, (3) with the Scriptures.[5] The second requirement was later abandoned and the third circumvented by Bellarmino's suggestion (made in order to accommodate Copernican theory) that no *doxa* concerning phenomena can be inconsistent with any *episteme* concerning ultimate reality. But then consistency with facts (or rather with 'phenomena') and simplicity remain the only criteria for hypotheses.

These criteria, although very unsatisfactory, represented an encouraging beginning for elaborating critical standards for *doxa*. Some thinkers in the seventeenth century were prepared to accept scepticism in religious affairs and fallibilism in scientific and practical affairs, and interest increased in developing some code for appraising *doxa*, not only in natural philosophy but in law, history, etc.[6] But this budding scepticism *cum* fallibilism soon degenerated into a curious sort of quasi-justificationism. The fallibilist pronouncements just quoted were not followed up by an elaboration of interesting *new* critical standards for *doxa*, of rules of acceptance, rejection, and, above all, comparison of fallible theories. The fallibilists soon developed a second-world appraisal of *doxa*: good *doxa* is the one accepted by the good mind, which is exactly the same as the justificationist good mind. But since they thought that the criterion of good *doxa* was the same as that

[1] For an excellent discussion cf. Duhem [1908], esp. chapter 1, pp. 14 ff.
[2] *Ibid.*, pp. 15–16.   [3] *Ibid.*, p. 18.
[4] *Ibid.*, chapter 2, *passim*.
[5] Clavius [1581]; discussed by Duhem, *op. cit.*, chapter 7.
[6] Cf. Popkin [1970].

of *episteme*, the difference between this quasi-justificationist fallibilism and dogmatism remained purely verbal.

In order to understand better the degeneration of early fallibilism into justificationism, one should remember that essentialists divided propositions into two: those propositions which are proven ultimate truths and those propositions which are not. Because of this basic conflation of truth, ultimate truth and proven truth the main problem of fallibilism was not so much to assess the distance of a (normally false) proposition from truth – as Popper's contemporary concept of verisimilitude does[1] – but to assess the distance of a proposition from *ultimate* truth. When Glanvill talks about degrees of 'verisimilitude' or 'probability', he means distance from ultimate truth. All *better* sorts of *doxa* – like Ptolemaic or Copernican astronomy – are assumed to be equally true:[2] the idea that false propositions may have large truth-contents which may be compared, is a Popperian idea alien to that age. The problem of this early fallibilism was how near propositions are, not to truth, but to proven ultimate truth.[3]

What about the residual sceptics? Let us remember that while fallibilists were prepared to value *doxa*, sceptics were not. But then they had somehow to solve the problem of practical action. Few sceptics ever maintained that suspense of action follows necessarily from suspense of judgment. None of them elaborated the separation of theory and practice as dramatically as Hume. According to Hume, we act on beliefs caused by nature. A belief about a matter of fact (or a value judgment) 'is the necessary result of placing the mind in [certain] circumstances'.[4] The sceptic then claims that all statements of matter of fact 'are evidently incapable of *demonstration*',[5] but they are capable of *proof*, in the sense that they can be supported by such 'arguments from experience as leave no room for doubt or opposition'.[6] Thus nature's

---

[1] Popper's 'verisimilitude' is the difference between the truth-content and the falsity-content of a proposition. Cf. chapter 10 of his [1963a]* The concept of verisimilitude has turned out to be a problematic one. See Tichý [1974], Miller [1974] and the subsequent discussion. (*Eds.*)

[2] A characteristic quote: 'For the best principles, excepting Divine and Mathematical, are but Hypotheses; within the circle of which, *we may indeed conclude many things with security from error. But yet the greatest certainty... is still hypothetical. So that we may affirm that things are thus and thus, according to the principles we have espoused: But we strangely forget ourselves, when we plead a necessity of their being so in Nature, and an impossibility of their being otherwise* (Glanvill [1665], p. 145, my italics; quoted in Laudan [1967], p. 220). Thus for Glanvill – and for most of his contemporaries – some 'hypotheses' may be statements which, although not 'necessary' (that is, proven), are still 'secure from error'. Laudan overlooks this possibility in his interesting [1967].

[3] Of course if one wants to preserve the term 'truth' for 'proven and ultimate truth', the Ptolemaic and Copernican astronomies cannot be said to be 'true'; if however one – understandably – does not want to call them 'false', one has to say that they are 'neither true nor false'. This characteristic terminology of the sixteenth and seventeenth century was then due to the conflation of truth and ultimate proven truth.

[4] Hume [1777], v, part I, p. 46.

[5] *Ibid.*, XII, part III, p. 164, my italics.   [6] *Ibid.*, p. 56, n. 1.

proofs make up for reason's lack of *demonstrations* and the sceptic can submit himself to, and act upon, his – *proven* – natural beliefs. Within his study, this sort of Humean mitigated sceptic is a genuine sceptic; outside in the world, he is a practical man, relying on proofs. The criterion of 'proof' is, however, psychologistic and this psychologism is indistinguishable from justificationist psychologism. The Humean separation of theory and practice is in fact a separation of a sceptical theory and a dogmatist practice based on 'morally certain hypotheses'.[1]

This strange unification of fallibilism and scepticism under the *aegis* of justificationism is responsible for many confusions in the history of ideas. What caused it? What made both fallibilism and scepticism yield to dogmatism? The answer is simple: *Newton*. Newtonian success routed both scepticism and fallibilism;[2] gave justificationism another two hundred and fifty years' lease of life; turned tolerant enlightenment into militant enlightenment; and postponed the crucial development of genuine fallibilism until Einstein – and Popper.

2   NEWTONIAN METHODOLOGY VERSUS NEWTONIAN METHOD
(a) *Newton's problem: the clash between standards and achievements*

Great works of art may change aesthetic standards – great scientific achievements may change scientific standards. The history of standards is the history of the critical – and not so critical – interaction between standards and achievements.

Newton's theory, according to the justificationist standards of the day, was non-knowledge. Either Newton's theory had to be rejected or the justificationist critical standards had to yield and be replaced by standards to fit the achievement. In fact, the outcome was a curious compromise: under the name of 'mitigated scepticism' or 'positivism' a new, less rigorous version of justificationism was established; and with a curious *lebenslüge* it was agreed that Newton's theory lived up magnificently to these weaker standards. The lie was upheld for centuries.

In order to understand the Newtonian compromise, let us have a look at the standard seventeenth century forms of scientific criticism.

The (scientific) sceptics harped on the time-honoured method of arguing from the *infinite regress in proofs and definitions*: they liked to point to the allegedly unproven premises of the opponent's argument

---

[1] The use of this term by Descartes and the Cartesians indicated how naturally Hume's ideas grew out of Cartesianism.

[2] As MacLaurin put it in 1748: 'The variety of opinions and perpetual disputes amongst philosophers has induced not a few, of late as well as in former times, to think that it was vain labour to endeavour to acquire certainty in natural knowledge, and to ascribe this to some unavoidable defect in the principles of the science. But it has appeared...from what we learn from Sir Isaac Newton, that the fault has lain in the philosophers themselves, and not in philosophy' (MacLaurin [1748], pp. 95–6).

and demand their proof; to point to the allegedly undefined terms and demand their definition. However, such criticism soon becomes monotonous and unconvincing. They occasionally succeeded in supplementing it by another method: the *sceptical proliferation of theories*. Sceptics liked to show that theories which seem to be strongly supported by the evidence are not the only ones which are supported by that evidence: any fact (or 'phenomenon') can be explained in different ways and the only rational thing one can do in such an epistemological *impasse* is to suspend judgment. Sceptical proliferation of theories does not aim at better conjectures – for the sceptic all conjectures are equally conjectural – it aims at discrediting, at making suspect and thus 'refuting', eliminating all conjectures. But, as a matter of fact, no rival theory was produced which could 'neutralize' Newton's.

More dangerous were the criticisms of those (scientific) sceptics who were (religious) dogmatists. They pointed to the inconsistency of Newton's theory with theology.

But the criticism that worried Newtonians much more than sceptical attacks was criticism from their fellow natural philosophers, themselves dogmatists in this very field. Nothing is ever considered as more menacing for a dogmatist school of thought than criticism from within, for such criticism threatens the very survival of their research programme. The main threat came from Cartesian rationalists. This essay concentrates – as did the Newtonians themselves – primarily on this threat.

(b) Newtonians against metaphysical criticism

The main weapon of Cartesian rationalists was *metaphysical (or 'essentialist') criticism*. This criticism hinged on the assumption that only *ultimate* truths deserve to be accepted in the body of knowledge, and that the general features of the 'essential' structure of the universe are *a priori* recognizable.

Cartesians knew very well that we can 'save the phenomena' in many possible ways by different hypotheses. But this is not science *yet*. Science starts with the first hypothesis which actually can be deduced from *a priori*, clear and distinct, first principles (perhaps with the help of some plausible auxiliary model). Then that hypothesis – the 'mediate cause' – is *proved* (and by this proof made 'intelligible').

Cartesians also thought that mediate causes could be induced from phenomena: and, moreover, that *some* mediate causes can only be inferred from experiments. But, encouraging though such *preliminary* inductions may be, they are not yet Proofs. And as long as there is no Proof the result is only a hypothesis, not science. Cartesians claimed that Newton's theory was not *proved* in their sense, since it was not derivable from Cartesian metaphysics.

Already in 1688 the first French review of Newton's *Principia* points out that his theory of gravitation 'has not been proved; the demonstration that depends on it can therefore only be mechanics'.[1] Huygens, in a letter to Leibniz in 1690 about Newton's gravitational 'principle' writes that he often 'wondered how he [Newton] could have given himself all the trouble of making such a number of investigations and difficult calculations that have no other foundation than this very principle'.[2] Huygens was firmly opposed to Newton's theory. Leibniz, in 1711, in a letter to Hartsoeker, wrote that 'the method of those who say, after M. de Roberval's *Aristarchus*, that all bodies attract one another by a law of nature, which God made in the beginning of things...[maintain] a fiction invented to support an ill-grounded opinion'.[3]

Such criticisms can do terrible damage to the progress of a research programme. Research programmes are fragile affairs, and too severe criticism may deter talented people from working *within* and developing them: they will prefer to work for rival programmes or search for new ones. The problems and techniques for solving them are very different [if one aims to explain the phenomena in terms of] Cartesian vortices [than if one aims to explain them using] Newtonian forces, and Newton and Newtonians were in despair at seeing their approach discredited. Leibniz's letter, written in 1711, and published in the *Memoires of Literature* in 1712, angered Cotes who immediately called Newton's attention to it.[4] Newton decided that something must be done.

What could the Newtonians do? They could, of course, try to provide a Cartesian proof of their theory of gravitation before going on further with their research programme. Indeed Newton himself chose this way. He interrupted his work on his programme and worked very hard, but on his own admission unsuccessfully, on such a Cartesian proof for many years.[5]

Newton's efforts later baffled his successors who, born into a world dominated by the spectacular growth of his research programme – and not Cartesian philosophy – found his principles not only securely proven but also perfectly intelligible and in no need of further explanation. But Newton himself and his personal disciples never considered the theory of gravitation anything but an intermediate solution.

In 1693 he still warned:

---

[1] Cf. Koyré [1965], p. 115.   [2] *Ibid.*, pp. 117–18.

[3] *Ibid.*, p. 141. It may be worth mentioning that metaphysical criticism can be formulated as a demand for intelligibility. Crude Cartesians (like Leibniz) opposed Newtonian gravitation because attraction, as it stood, was unintelligible. *Sophisticated* Cartesians (like Newton himself) thought that it must be made intelligible by an intelligible explanation. Also cf. *below*, p. 204.

[4] Cotes [1712–13].

[5] Cf. e.g. Jourdain [1915].

That Gravity should be innate, inherent and essential to Matter, so that one Body may act upon another at a Distance thro' a Vacuum, without the Mediation of any thing else, by and through which their Action and Force may be conveyed from one to another, is to me so great an Absurdity, that I believe no Man who has in philosophical Matters a competent Faculty of thinking, can ever fall into it.[1]

Newton went to immense trouble to convince his admirers not to ignore Cartesian criticism. Indeed, the very last sentence in Pemberton's exposition of his theory is this: 'To acquiesce in the explanation of any appearance by asserting it to be a general power of attraction, is not to improve our knowledge in philosophy, but rather to put a stop to our farther search.'[2] But, after Newtonians failed in their repeated efforts, they became convinced that the task of 'explaining' gravity (that is, explaining it 'intelligibly') must be left for later generations and that their research programme could go on regardless. Metaphysical criticism as a ground for rejecting a theory, or, better, for holding up or stopping a research programme, must then be ignored. Therefore, while agreeing that his law could and should be further explained, he proposed to weaken the Cartesian concept of 'proof' (that is, criterion of scientific acceptability) by demanding for propositions only an empirical–experimental but not a rational–metaphysical proof. This was the crucial motivation of Newton's methodological concern which increased sharply between the first and the second edition of his Principia: to modify (indeed, lower) the critical standards of his day just enough to save his research programme – not an inch more. This was the reason why he made the famous changes and insertions in the second and third editions of the Principia;[3] this was why Cotes wrote his brilliant polemic preface to the second edition.

The most important new methodological rule which Newton inserted into the second edition expresses this modification very succinctly. According to his famous Rule IV, metaphysical criticism must not be allowed to make us reject inductive proofs:

In Philosophia experimentali, Propositiones ex Phaenomenis per Inductionem collectae, non obstantibus Hypothesibus [contrariis], pro veris aut accurate quam proxime haberi debent, donec alia occurrerint Phaenomena per quae aut accuratoires reddantur aut exceptionibus obnoxiae. Hoc fieri debet ne argumentum Inductionis tollatur per Hypotheses.[4]

[1] Letter to Bentley, 25 February 1693; cf. Cohen [1958], pp. 302–3.
[2] Pemberton [1728], p. 407.
[3] For a discussion of these changes and insertions, cf. Koyré [1965], *passim*.
[4] Newton hesitated at length over the question of whether or not to include the word 'contrariis'. Finally he decided to cross it out (cf. Koyré [1965], pp. 271 ff.). The editors, Bentley and Halley have reinserted it, but Koyré misunderstood Newton probably much more than the editors. Cf. *below*, p. 206, n. 2. * An English translation of Rule IV is: 'In experimental philosophy we are to look upon propositions inferred by general induction from phenomena as accurately or very nearly true, notwithstanding any contrary hypotheses that may be imagined, till such time

This rule amounts to a truncation of the Cartesian model[1] of explanation. The ladder of science is [now considered to be] open-ended at the top [*i.e.* we may not quite be able to deduce the causes of phenomena from first principles], but we may still have science, the lower rungs may still be scientific, even if the whole ladder does not yet exist. The only necessary requirement is that the 'phenomena' should be true and the induction correct: as Newton put it:

And therefore I could wish all objections were suspended, taken from *Hypotheses* or any other heads than these two; Of showing the insufficiency of Experiments to determine these *Queries* or prove any other parts of my Theory, by assigning the flaws and defects in my conclusions drawn from them; Or of producing other Experiments which directly contradict me, if any such may seem to occur.[2]

The *first* type of criticism Newton would permit is criticism of his inductive *premises* (which he later called 'phenomena') and criticism of his inductive *argument*; the *second* type is the production of a counterexample. (This is puzzling. Is not the second equivalent to criticism of the inductive argument? To this problem we shall soon return.)

But Newton's Rule IV rejects two kinds of justificationist criticism widely accepted in his age. First, he rejects any criticism that his theory is unproven because it has no self-evident Cartesian premises and that therefore 'he built without foundations'.[3] Secondly, he rejects any criticism which argues that his theory is not only unproven from, but, in fact, is in contradiction with, *a priori* first principles.

The two kinds of criticism [which Newton rejects] might well have been confused in the seventeenth and eighteenth centuries. If one interprets Newton's theory of gravitation as an *ultimate* theory, as one which attributes [real existence to the] gravitational attractive power, then this *contradicts* Cartesian metaphysics according to which there is only push and pull and no action at a distance. However, if one interprets it as an intermediate theory, with attractive force yet to be explained in Cartesian terms, then this *does not contradict* Cartesian philosophy; in this interpretation 'attraction' is only a *word* that 'allude[s] to what is real, though [its] signification is confused',[4] and which was only chosen as a 'commodious term...to avoid a useless and tedious circumlocution'.[5]

However, the essentialist interpretation of definitions[6] was so dominant at the time that the latter way out was difficult to grasp. This

as other phenomena occur, by which they may either be made more accurate, or liable to exceptions. This rule we must follow, that the argument of induction may not be evaded by hypotheses.' (*Eds.*)
[1] Cf. *above*, p. 202.
[2] Letter to Oldenburg, 8 July 1672, reprinted in Cohen [1959], p. 94.
[3] Descartes' disparaging remark on Galileo; cf. his [1638], p. 380.
[4] Pemberton [1728], p. 10.   [5] MacLaurin [1748], pp. 110 ff.
[6] For 'essentialism in definitions' cf. Popper [1945], volume 2, chapter 11.

is why Newton hesitated over the inclusion of *contrariis* in his Rule IV;[1] his final decision to cross it out was [based upon the fact that] by crossing it out he gave the Rule a form which speaks out clearly against *both* kinds of criticism in a succinct form.[2]

But Newton's Rule IV has a further important implication. This is against a sceptical proliferation of theories; [the fact that] somebody [can] put forward an alternative hypothesis which fits all phenomena but is not *proved* inductively has to be rejected as an argument for suspending judgment. [Also,] if somebody puts forward [such] a hypothesis without inductive proof, even if it is true, it is not yet a discovery and has no place in the history of science (for to discover is to prove). Rule IV also implies the rejection of hypothetico-deductive (or 'eliminative') induction, that is, the method of proving a hypothesis by disproving its alternatives: Newton thought this method was powerless.

I cannot think it effectual for determining truth, to examine the several ways by which Phaenomena may be explained, unless where there can be a perfect enumeration of all those ways... You know, the proper Method for *inquiring* after the properties of things is... not by deducing it only from a confutation of contrary suppositions, but by deriving it from Experiments concluding positively and directly.[3]

This methodology makes us understand Newton's anger in his priority dispute with Hooke. The dispute started with Halley's letter to Newton, of 22 May 1686, about the reception of his *Principia* in the Royal Society. Halley let Newton know that according to Hooke's comment, Hooke 'had some pretensions'[4] to the discovery of the inverse square law of gravitation. Hooke claimed that Newton acquired the law from him and only 'the demonstration of the curves generated thereby belonged wholly to Newton',[5] and that he expected at least a mention in Newton's preface. The letter left no doubt that Halley sympathized with Hooke. Newton's reply is well-known, as well as Halley's second letter appealing to Newton 'not to let his resentment run so high';[6] it is also well-known that Newton finally agreed that the first edition should include a sentence: 'The inverse square law of gravity holds in all the celestial motions, as was discovered also independently by my countrymen, Wren, Hooke, and Halley'.[7] The story up to this point is told by Brewster.[8] But Newton's acknowledgment was half-hearted; he felt that his precursors did not deserve it: he was

[1] Cf. *above*, p. 204, n. 4.
[2] Koyré thought the Rule was 'probably an allusion to the principles of conservation of Descartes, and of Leibniz' (*op. cit.*, p. 271). That is, Koyré missed the point completely.
[3] Letter of 8 July 1672; reprinted in Cohen [1958], p. 93.
[4] Brewster [1855], volume I, p. 308.   [5] *Ibid.*
[6] Brewster, *op. cit.*, p. 310.
[7] Newton omitted foreigners like Borelli, Ballialdus and Huyghens.
[8] Brewster, *op. cit.*, pp. 307 ff.

the real discoverer. The methodological rules of the second edition, by implication withdraw the earlier acknowledgment [which had been] forced upon him: a mere hypothesis, without inductive experimental proof, is not a discovery.[1]

This discussion shows how much methodology is condensed within Newton's Rule IV. The Rule practically demands the prohibition of almost all possible criticism and by implication is a plea for concentration instead on the development of his research programme. Kuhn is very near to the historical truth (*if* restricted to this particular but important context) when he says that 'science starts when criticism stops';[2] Newtonian science certainly started by gagging both metaphysical criticism and any attempt at proliferation of research programmes. Feyerabend rightly blames Newton's Rule IV for 'theoretical monism'.[3]

To sum up: Newtonian methodologists were, on the whole, concerned with discrediting and eliminating the most dangerous patterns of criticism directed against their research programme. Newton himself foresaw the difficulties of having his theories accepted: 'For I see a man must either resolve to put out nothing new, or to become a slave to defend it'.[4] The main purpose of Newtonian methodological polemic was to persuade the Cartesians to 'act fairly, and not deny the same liberty to us which they demand for themselves. Since the *Newtonian* philosophy appears true to us, let us have the liberty to embrace and retain it.'[5] *This is why they were forced, almost against their will to oppose the tyranny of self-evident, a priori first principles and thus to change standards of scientific proof and criticism and indeed, the very concept of knowledge.*

For obvious reasons, I first meant to call this Newtonian conception of knowledge '*anti-essentialist justificationism*'. This term would have had all the correct – Popperian – connotations, but it sounded too German and so I opted for '*defensive positivism*'. This had the disadvantage that the 'justificationist' character of the position does not appear in its name, but it also had the advantage of stressing the difference between the defensive positivism of the late seventeenth and of the eighteenth centuries and the aggressive positivism of the nineteenth and early twentieth centuries. The former was an attempt to eliminate the *pressure* of 'first principles', and defend low-level

---

[1] True, Hooke never had the mathematics to *deduce* the inverse square law from an *ellipse*, although he could have deduced it from a circle. (Cf. Bonnar and Phillips [1957], p. 85.) But since the full formula anyway cannot be proved from Kepler's three laws, we have to reject this Newtonian explanation of his claim to the law of gravitation. This superiority can be explained only with the help of the concept of 'research programme'; cf. chapter 1.

[2] Cf. his [1970a], p. 6: 'it is precisely the abandonment of critical discourse that marks the transition to a science'.

[3] Cf. his [1970c].

[4] Newton [1676].

[5] Cotes [1717], p. xxvii.

empirically motivated inquiry; the latter wanted to eliminate 'first-principles' altogether and destroy any high-level metaphysically motivated inquiry. (It seemed to me important to stress this demarcation because of the well-known efforts of aggressive positivists to conflate the two trends and turn Newton into their forerunner.)

(c) Newton's idea of experimental proof and its credo quid absurdum

Newton's defensive positivism was directed against a single, though important, aspect of the justificationism of his age. But while he refused to admit that a theory is proved *only* if it is proved from – to use Pope's jocular expression – 'high priori'[1] premises, he still thought that only proven propositions had a place in science. But his standards for scientific proofs – 'experimental proofs' – were *weaker* than the Aristotelian–Cartesian standards. Let us now have a closer look at these standards.

The most interesting point that strikes one's eye is that Newton did not think that his 'experimental proofs' were as conclusive as Cartesian proofs. Indeed, according to him 'arguing from Experiments and Observations by Induction be no Demonstration of general Conclusions'[2] although experimental–inductive proof 'is the highest evidence that a Proposition can have in [my] philosophy'.[3]

After these statements it should be no surprise that Newton warns that there might well be 'exceptions' to his valid proofs; or, as we have already mentioned, that he invites two *different* kinds of criticism of his inductive proofs: either by examining their premises and the validity of their inferences or by offering counterexamples.[4] What is the explanation of this strange dichotomy and of the implied assumption that an experimental proof may be valid but still lead to false conclusions?

The most obvious solution seems to be the one offered later by Hume.[5] We may describe Hume's position as suggesting that a valid experimental–inductive proof is not a third-world relation between two propositions, a true *A* and a correctly derived *B*, but a psychologistic relation between a 'certain' belief in *A* and a 'certain' belief in *B*; the latter being established by an 'operation of the mind' leading from *A* to *B* which compels the absolute consent of a *scientific* mind.[6]

---

[1] *The Dunciad*, book IV. 1, 471.
[2] Query 31 in his [1717], p. 404.
[3] Newton [1713], p. 155.
[4] Cf. e.g. Newton [1672], p. 94.
[5] Cf. *above*, p. 200.
[6] I want to distinguish between *psychological*, plainly second-world concepts, like 'belief', and *psychologistic* concepts like 'rational belief' in the sense of 'belief of a clear mind'. While *psychology* may be defined as the theory of mind, *psychologism* is the theory of a 'healthy', 'normal', 'clear', 'ideal', 'empty', 'purged', 'unbiased', 'objective', 'rational', or 'scientific' mind.

But then it is clear that there is no guarantee that valid arguments, determined by the laws of the scientific mind yield true propositions. In this case we may have two inconsistent propositions *A* and *B* where both *A* and *B* are supported by valid experimental proofs. According to Newton, phenomena are always 'certain' and the worst thing that can happen to an 'inductive generalization' is that the 'domain of validity' of an otherwise 'certain' inductive assertion has later to be restricted by exceptions.[1]

The inherent psychologism of Newton's concept of experimental proof puts him into the category of justificationist fallibilism:[2] Newtonian standards are those of justificationist fallibilism. They are not third-world standards but psychologistic standards. The proof of the phenomena is guaranteed by the 'lack of speculative bias', 'carefulness', and 'experimental skill'; the proof of the inductive generalization is guaranteed by the 'caution' and 'sagacity' of the theoretician:[3] one could well call them 'proofs by pedigree'.[4] Kepler's laws were proved by Kepler's 'reliability' as an observer; Newton's laws by Newton's 'sagacity' at making inductive inferences.

But Kepler's addiction to speculation, and therefore his unreliability as an observer, was too well known at the time [for this account to be believed]; it was also well known that Kepler had many other 'laws' (e.g.: 'the comets move on straight lines', not to mention his musical laws of the heavens[5]) which Newtonians never cared even to mention. But then they had to explain how three of Kepler's many laws could still be taken as 'the solid and unshakable foundation of modern astronomy'.[6] They did this with the theory of Kepler's (of course, strictly temporary) conversion from a speculative Saul into an inductivist Paul under the influence of a letter from Tycho de Brahe. In this letter Brahe enjoined Kepler to drop speculation for the sake of observations.[7] [Kepler, so the story went, did precisely this and hence

---

[1] Newton claimed certainty for his law of gravitation. In his anonymous review of the *Commercium epistolicum* he writes about himself: 'One would wonder that Mr. Newton should be reflected upon for not explaining the Causes of Gravity by Hypotheses; as if it were a Crime to content himself with Certainties and let Uncertainties alone.' I discussed this in detail in my [1963–4], especially Part II.

[2] Cf. *above*, p. 198 ff.

[3] Incidentally, the association of *theory* and *fallibility* is a quite recent one: originally 'theory' was a synonym of 'theorem'.

[4] Cf. Popper [1963a], pp. 25 ff., and Agassi [1963], pp. 12 ff., for Popper's classic exposition of this concept and for Agassi's excellent elaboration and historical illustrations.

[5] Kepler [1619].   [6] Argo's tribute to Kepler, cf. W. C. Rufus [1931], p. 34.

[7] In this letter Tycho, as Kepler himself tells us, advised Kepler 'first to lay a solid foundation for his views by actual observation, and by ascending from these to strive to reach the causes of things'. According to Brewster, 'by the magic of the whole Baconian philosophy thus compressed by anticipation into a nutshell, Kepler abandoned for a while his visionary speculations'; 'for a while', that is, for long enough to discover his three laws of planetary motion (Brewster [1855], I, p. 265).

his] three laws 'were not based on theory of any kind, but were intended to sum up facts of observations'.[1]

But, of course, Kepler's three laws were false. Moreover, by 1686 it was generally *known* that they were false; that planets did not move in nice ellipses,[2] that changes in the speeds of Jupiter and Saturn were not in accordance with Kepler's 'second law'[3] and that the Moon's motion too was very different from a simple Keplerian model.[4] Newton's compartmentalized mind cannot be better characterized than by contrasting Newton, the methodologist, who claimed that he *derived* his laws from Kepler's 'phaenomena', with Newton, the scientist, who knew very well that his laws *directly contradicted* these phenomena. Let us remember his clearly conditional 'as if' statement which indicated that his derivation only applied to a crude model of a planetary system with a *fixed* sun and planets which do not attract each other: '*if* the sun *were* at rest, and the other planets *did not* act one upon another, their orbits *would be* ellipses, having the sun in their common focus; and they would describe areas proportional to the time'.[5] The conditionals are dramatically counterfactual: Newton's third law of dynamics *forbids* the existence of a planetary system with a sun at rest (i.e. with only heliocentric forces and no forces from the planets to the sun) and Newton's theory of universal gravitation *forbids* the existence of a planetary system with only heliocentric and no interplanetary forces. But then Feyerabend's statement that Newton, taking the Phenomena as the foundation of his theory, 'turns part of the new theory into its own foundation',[6] is false: Newton turns the *negation* of his new theory into its own foundation.

Incidentally, one of the reasons why Newtonians were not struck by this absurdity was that they were more worried by the *unprovenness* of their Phenomena than by their *falsity*. Again, let us remember that in the seventeenth century these two counted as *equally* serious charges. Kepler's laws concerning the planets implied that the Earth and the planets orbit the Sun: the unprovenness of this assertion was argued very seriously by the Churchmen. This, I am convinced, is the reason why Newton starts his Phenomena with the 'circumjovial' and 'circumsaturnal' planets (Phenomena I and II) and leaves the 'primary' planets for Phenomenon III: the statements about the first two are much better proved by Galileo's telescopic observations. Indeed, if one looks carefully at his Phenomena II, IV and V, they do *not* state *Kepler's*

---

[1] Lamb [1923], p. 231.
[2] Kepler himself realized in 1625 that Saturn and Jupiter do not move in ellipses.
[3] Halley mentions in 1676 that Saturn had slowed down and Jupiter had become faster since Kepler's time.
[4] Again Halley, who had a *penchant* for checking ancient observations, noticed that the moon is accelerating over the centuries.
[5] This is the second sentence following prop. XIII, theorem XIII of book III of the *Principia* (my italics) p. 421. *It is ironic that the theorem itself is the very same sentence, repeated in categorical mood, with the conditional clauses removed.*
[6] See his fascinating [1970c]. For this paper cf. also *below*, p. 213, n. 5.

laws for the primary planets. Kepler was a devoted 'Copernican' in the sense that he assumed a *fixed* Sun and a heliocentric universe. Newton's starting point is weaker: his Phenomenon III states that the moon-like 'horns' of the primary planets prove that they 'encompass the Sun', whether the Sun itself moves around the Earth or not; his Phenomena IV and V *explicitly* state that the statements about periodic times and the areas swept out by radii from the planets to the Sun are independent of whether the Sun moves around the Earth or the Earth around the Sun.[1] In order to *prove* (approximately) Kepler's laws in their original heliocentric interpretations, he found that he needed at least one unproven 'hypothesis': 'that the centre of the system of the world is immovable'. He adds to this famous *Hypothesis I*: 'This is acknowledged by all, while some contend that the Earth, others that the Sun, is fixed in that centre'. But if one agrees to this, admittedly hypothetical, proposition, the heliocentricity of the universe, as Newton shows, follows logically: with this masterpiece of diplomacy, Newton admitted that heliocentricity *is* hypothetical, but it rests on a very, very weak – *i.e.* very plausible – hypothesis.[2]

This much then for Newton's 'basic statements': the *Phenomena*. But even if the *Phenomena* were true statements, would Newton's theory follow logically from them? From the model consisting of a fixed Sun and one single planet (or several planets which do not interact) the inverse square law really follows. But there is much more to Newton's law of gravitation than the inverse square relation: Newton's law involves gravitational masses.[3] How can a validly drawn conclusion, whether deductive or inductive, contain [essential] terms not contained in the premises? How can it even command compelling assent?

But if no theory can be proved from phenomena, then Newton's demarcation criterion between proven and unproven theories falls down and any unrefuted theory is equally good. If so, was he rational in rejecting his own Cartesian explanation of gravitation? Why was that less proved than his lower level law of gravitation?

[1] Let us quote the two: *Phenomenon IV*: 'That the fixed stars being at rest, the periodic times of the five primary planets, and (*whether of the sun about the earth, or*) of the earth about the sun, are as the 3/2th power of their mean distances from the sun' (p. 404, my emphasis); *Phenomenon V*: 'Then the primary planets, *by radii drawn to the earth, describe areas in no wise proportional to the times; but the areas which they describe by radii drawn to the sun are proportional to the times of description*' (p. 405, my italics).

[2] Koyré explains why Newton called the thesis about the immobility of the centre of the universe a 'hypothesis' by claiming that 'Newton, *no doubt*, was well aware that it could, after all, be utterly false' (Koyré [1965], p. 40; my italics). Was he? Koyré uses here the term 'no doubt' as an abbreviation for the clumsy expression 'although I have no argument or evidence for this totally arbitrary assumption'.

[3] Leibniz seems to have seen this: 'I am strongly in favour of experimental philosophy, but M. Newton is departing very far from it when he claims that all matter is heavy (or that every part of matter attracts every other part) which is certainly not proved by experiment' (letter to Abbé Conti at the end of 1715, quoted in Koyré [1965], p. 144).

The schizophrenic combination of the mad Newtonian methodology, resting on the *credo quid absurdum* of 'experimental proof' and the wonderful Newtonian *method* strikes one now as a joke. But from the rout of the Cartesians until 1905 nobody laughed. Most textbooks solemnly claimed that first Kepler '*deduced*' his laws 'from the accurate observations of planetary motion by Tycho Brahe', then Newton '*deduced*' his law from 'Kepler's laws and the law of motion' but also 'added' perturbation theory as a crowning achievement.[1] The philosophical bric-a-brac, hurled by Newtonians at their contemporary critics to defend their 'proofs' by hook or by crook, were taken as pieces of eternal wisdom instead of being recognized for the worthless rubbish that they really were. Newtonians falsified the history of thought freely in order to be able to appeal to alleged authorities: Newtonians invented the myth of the Great Conflict between Bacon and Descartes and falsely claimed to have strictly followed the method of 'analysis–synthesis',[2] that is, the time-honoured method of Euclidean Geometry, of 'the only Science that it hath pleased God hitherto [before Newton] to bestow on mankind'.[3] Indeed, this method was attributed by some to Newton himself:

> In order to proceed with perfect security, and to put an end for ever to disputes, he proposed that, in our enquiries into nature, the methods of *analysis* and *synthesis* should be both employed in a proper order; that we should begin with the phaenomena, or effects, and from them investigate the powers or causes that operate in nature; that, from particular causes we should proceed to the more general ones, till the argument end in the most general: this is the method of *analysis*. Being once possessed of these causes, we should then descend in a contrary order; and from them, as established principles, explain all the phaenomena that are their consequences, and prove our explications: and this is the *synthesis*.[4]

(Newtonians insinuated that Cartesians ignore *analysis*; but in fact analysis–synthesis played exactly the same role in the Cartesian model.)

Before concluding this section it should be mentioned that Newtonians claimed that once they proved their theories from facts by 'analysis' they could predict, new, undreamt of facts, far beyond the original experimental premises in the 'synthesis'. They stressed that their theory 'led to the knowledge of such things, that it would have been reputed no less than madness for any one, before they had been discovered, even to have conjectured that our faculties ever have

---

[1] This phraseology has survived until today among physicists; cf. e.g. Symon [1960], pp. 132–3. But for many of them this is much more than phraseology: Max Born claims that 'the fact that Newton's law is a logical consequence of Kepler's laws' is the basis of his whole philosophy of science (Born [1949], p. 129).

[2] For a discussion of the Euclidean method of 'analysis–synthesis' cf. my [1963–4], p. 10, n. 2; p. 243, n. 1; and p. 308, n. 3. The flexible but fallible form of analysis–synthesis in informal mathematics which I discuss in Part IV of my [1963–4] is similar to Newton's 'analysis–synthesis'.

[3] Hobbes [1651], part I, chapter IV, p. 22.

[4] MacLaurin [1748], pp. 8–9.

reached so far'.[1] Thus the Newtonian model of proof/explanation/discovery[2] has to be modified.

But Newtonians did not explicitly *require* a theory to entail *new* phenomena in order to be satisfactory; curiously, it was Huygens and Leibniz, their adversaries, who first made this requirement explicit. And no Newtonian ever seems to have been concerned about the curious feature of this logic that the facts which constituted the starting point for analysis (say, Kepler's laws) were inconsistent with some of the facts proved from them at the end of the synthesis. What was perfectly acceptable in the 'analysis', was in fact rejected in the 'synthesis'.

The first to break the myth of Newtonian foundations and inductive 'logic' was Duhem. His two chapters on the 'Criticism of the Newtonian Method' in his classic *The Aim and Structure of Physical Theory*, published in 1905, contains a brilliant and crushing criticism, which reveals some of the skeletons in the Newtonian cupboard. It is amazing how this criticism was ignored until resuscitated by Popper and his school. Popper, in his crusade against inductivism, revived and improved Duhem's arguments in two papers, published in 1948 and in 1957.[3] His papers were ignored just as were Duhem's; they were finally given wider circulation by Feyerabend, who took up their main point in 1962.[4]

What is the explanation of the strange fact that nobody before Duhem took up Newton's challenge to his critics to find fault with his premises and with the validity of his argument? I think the explanation of this in the seventeenth and eighteenth centuries is that people were accustomed to approach a theory primarily with *metaphysical* criticism. Therefore the battle *then* was about the very standards of criticism: they criticized Newton for the 'unintelligibility' of his theory and did not care whether it was valid or not on Newton's proposed criteria. Newton won the battle about the criterion. But by that time the unprecedented and truly miraculous success of his research programme built up such a religious atmosphere,[5] or, if you wish, a bandwagon effect, that the hypocrisy of *credo quid absurdum* came easily and naturally.

[1] Pemberton [1728].
[2] Cf. *above*, p. 205. I there point out that this model is a truncation of the original Cartesian model.
[3] Popper [1948] and [1957a].
[4] Feyerabend [1962]. Unfortunately this exposition, for all its merits, is, in many respects, not as good (and definitely not as clear) as Duhem's and Popper's original expositions. Still it was acclaimed by some philosophers of science of repute (who obviously were ignorant of Duhem and Popper and who curiously also missed Feyerabend's clear references to both) as the 'major epistemological breakthrough' of 1962. (Cf. e.g. Hesse [1963], p. 108.)
[5] This atmosphere is impressively described by Feyerabend in his already quoted [1970c].

(d) Newtonians and factual criticism

Newtonians thus claimed to use 'phenomena' – propositions allegedly established by experiment – as premises of their experimental proofs, as verifying evidence of the strongest sort. But were they prepared to accept them as falsifying evidence? Under what conditions would they have been prepared to abandon their research programme?

The relation between theory and counterexample is one of the most obscure parts of Newtonian methodology: they were more confused in this respect than in any other.

When they criticized the theories of their adversaries, they were naive, and indeed, very aggressive falsificationists. Thus Newton claimed that 'the hypothesis of vortices is utterly irreconcilable with astronomical phenomena'.[1] Cartesian argument was ridiculed by Cotes in the Preface to the second edition of the *Principia*. But Newton's 'refutation' was fishy: Descartes' original theory of vortices was so vague, so woolly that it was, strictly speaking, irrefutable. So Newton first improved it, made it precise by developing a sharp hydrodynamical version of it solely to show that it was inconsistent with Kepler's laws (which, incidentally, he knew to be false), and therefore demanded that it be rejected. But Huygens in 1688 – and Leibniz in 1689 – correctly observed that Newton refuted only a particular version of the vortex hypothesis; one can easily put forward a new version, which is immune to Newton's criticism.[2] Moreover, John Bernoulli, in 1730, won the prize of the French Academy with a treatise explaining celestial phenomena with the help of Cartesian vortices. The Abbé de Molières even claimed to disprove Newton's mathematical theory of vortices, and he scored some points. Indeed, de Molières set out to prove Newton's theory of gravitation from a version of Cartesian vortex theory.[3] Voltaire had reason to complain in his 1738 that 'there are yet some Philosophers attached to their Vortices of subtile Matter, who would willingly reconcile these imaginary vortices with [Newton's] demonstrable Truths'[4] – the more so since he himself went on to reintroduce 'non-Cartesian' vortices into celestial mechanics![5] John Stuart Mill believed Newton's assertion that his theory of gravitation refuted Cartesian vortex theory: Whewell mentions this – correctly – as an example of Mill's illiteracy.[6]

---

[1] Cf. the end of Book I of *Principia*. Of course, the same 'phenomena', i.e. Kepler's laws, were equally irreconcilable with Newton's theory. Yet for a century they served to prove Newton's theory and refute Descartes's!

[2] Cf. Koyré [1965], especially pp. 117 and 136.

[3] Cf. Brunet [1931], volume I, chapter III.

[4] Voltaire [1738], p. 235.

[5] *Ibid.*, pp. 320–3.

[6] Whewell [1856], p. 261. Whewell puts it very politely; he quotes Mill without mentioning his name, and states, reservedly, that the great resilience of theories to refutation is 'unfamiliar to those who have only slightly attended to the history of science'. Also cf. *below*, p. 221.

However, Newtonians, when their own theories came under the fire of factual criticism, assumed sophisticated nonchalance and rarely seemed to be worried. According to Newton's ruling: 'If no Exception occur from Phenomena, the Conclusion may be pronounced generally. But if at any time afterwards any Exception shall occur from Experiments, it may then begin to be pronounced with such exceptions as occur.'[1]

But Newtonians had little doubt that their programme would finally digest all the 'exceptions'; and this required a great deal of self-confidence, for 'exceptions', or 'anomalies', 'recalcitrant instances', abounded. It is characteristic, for instance, that nobody thought that the well-known fact that the comets' tails seem repulsed rather than attracted by the Sun, was a refutation of Newton's theory, although it was acknowledged as a *problem* – or 'puzzle', as Kuhn would call it – within Newton's research programme. Halley hoped its solution would be inserted in the first edition of the *Principia*. While it was still in the press, he wrote to Newton: 'I doubt not but this may follow from your principles with the like ease as all the other phenomena; but a proposition or two concerning these will add much to the beauty and perfection of your Theory of Comets.'[2] Although Newton did not reply, no Newtonian was unduly worried.

The same tranquility was displayed at the many divergences between Newton's theory of the Moon and the observations. These divergences were regarded as problems but few thought there was anything wrong with the research programme: it was rather the researchers who were at fault. Newton's 'theory of the Moon' was in fact first published many years after the first edition of the *Principia*, in 1702, in David Gregory's *Astronomiae Physicae et Geometricae Elementa*. It calmly states that Newton's theory 'agrees *very nearly* with the phenomena as he had proved by very *many places* of the Moon observed by the celebrated Mr. Flamsteed'.[3] But we have to remember that Newtonians never let the authority of observations prevail against their research programme; with the help of their positive heuristic they produced one theory after the other to accommodate counterexamples;[4] but frequently they ignored observational counterevidence altogether: they knew not only that theories had to be constantly tested by observations but also observations by *their* theories. The 'best observations' – a term very frequent in Newtonian literature – were those which corroborated their research programmes.[5] This transpires beautifully from Newton's and Flamsteed's correspondence. Flamsteed, the first Astronomer

---

[1] Newton [1717], p. 404. For this 'exception-barring' tradition cf. my [1963-4], especially p. 124.

[2] Halley [1687], p. 474.   [3] Gregory [1702], p. 332.

[4] For the difference between 'theory' and 'research programme' and for the idea of 'positive heuristic' cf. chapter 1.

[5] For instance, MacLaurin writes that Kepler 'gave himself the liberty to imagine several...analogies, that have no foundation in nature, and are overthrown by the *best* observations' (MacLaurin [1748], p. 51, my italics).

Royal, was a real, unschizophrenic inductivist; he slowed down Newton's and his associates' work more than anybody else had done, by refusing to let them have the results of observations he made of the Moon. At first, Newton and Flamsteed corresponded frequently, but Flamsteed soon became annoyed by Newton's use of Flamsteed's data as touchstones of his, Newton's, lunar theories, the first dozen of which ended up in Newton's wastepaper-basket.[1] He complained to his friend Lowthorp in 1700:

> [Newton] had made lunar tables once to answer his conceived laws, but when he came to compare them with the heavens, (that is, the moon's observed places) he found he had mistook, and was forced to throw them all aside: that I had imparted above 200 of her observed places to him, which one would think would be sufficient to limit any theory by; and since he has altered and suited his theory till it fitted these observations, 'tis no wonder that it represents them: but still he is more beholden to them for it than he is to his speculations about gravity, which had misled him.[2]

But he does not mention to Lowthorp that some of *his observations* too ended up in the wastepaper-basket. For instance, Newton visited him on 1 September 1694 when working full time on his lunar theory and told him to reinterpret some of his data since they contradicted his theory and explained to him exactly how to do it. Flamsteed obeyed Newton and wrote to him on 7 October 1694: 'Since you went home, I examined the observations I employed for determining the greatest equations of the earth's orbit, and considering the moon's places at the times...I find that (*if, as you intimate, the earth inclines on that side the moon then is*) you may abate about 20″ from it.'[3] Thus Newton constantly criticized and corrected Flamsteed's observational, touchstone theories. Newton taught Flamsteed for instance a better theory of the refractive power of the atmosphere which Flamsteed accepted and which corrected his original 'data'. One can understand the constant humiliation and slowly increasing fury of this great observer, having his data criticized and improved by a man, who, on his own confession, made no observations himself.[4]

By 1700 Newton and Flamsteed were not on corresponding terms any more, but earlier, when Newton still went to great lengths to get Flamsteed's data, he tried to explain to him patiently, that his (Newton's) 'theory will be a demonstration of their exactness... Without such a theory to recommend them, they will only be thrown into the heap of the observations of former astronomers, till somebody shall arise that, by perfecting the theory of the moon, shall discover

---

[1] 'Wastepaper-baskets' were containers used in the seventeenth century for the disposal of some first versions of manuscripts which self-criticism – or private criticism of learned friends – ruled out on the first reading. In our age of publication explosion most people have no time to read their manuscripts, and the function of wastepaper-baskets has now been taken over by scientific journals.

[2] Baily [1835], p. 176.   [3] Cf. Brewster [1855], volume 2, p. 168, my italics.

[4] Cf. Newton [1694].

your observations to be exacter than the rest'. But, warned Newton, this can be achieved only by someone 'who understands the theory of gravity as well, or better than I do'.[1] It is clear from this letter that Newtonians measured the exactness of observations with the help of their theories; and when they claim that 'certain observations, and plain facts, perpetually appear in contradiction to [the Cartesians'] boasted speculations',[2] then by 'certain observations and plain facts' they mean well-corroborated consequences of *their* programme. When the facts seemed to contradict their theories, they did their best to 'supply the defects of sense by a well regulated imagination'.[3] [The following is an example of the Newtonians' respect for facts (when it suited them). MacLaurin] writes that as the philosopher's

knowledge of nature is founded on the observation of sensible things, he must begin with these, and must often return to them, to examine his progress by them. Here is his secure hold; and as he sets out from thence, so if he likewise trace not often his steps backwards with caution, he will be in hazard of losing his way in the labyrinths of nature.[4]

It is also interesting that Flamsteed's charge that Newton fiddled with his theory to adapt it to the data is similar to Cotes's charge against the Cartesians who, after Newton 'has abundantly proved from the clearest reasons' that 'phenomena can by no means be accounted for by vortices', still 'can spend their time so idly as in patching up a ridiculous figment and setting it off with new comments of their own'.[5]

In the light of this discussion it may then seem rather ironical that Newtonians complained that Cartesian first principles had been regarded as 'of so great authority as not to be overturned by contradictory observations, or by extravagant consequences that arose from them'.[6] Incidentally, let us not forget, concerning 'extravagant consequences', that according to the original Newtonian theory the perturbations of our planetary system were leading to catastrophe with dramatic speed; a problem which Newtonians solved by claiming that God occasionally restores the balance of the system. Cartesians shouted that this maligns God's art; Newtonians retorted that their God is active, not dead, as the Cartesians would have it. Laplace finally showed that the occasional restoration of balance can be explained within the Newtonian research programme, without the *ad hoc* hypothesis of God. (But as Poincaré showed, even if Laplace's solution was not *ad hoc*, it was not final either.)

But, returning to the Newton–Flamsteed controversy, let us ask, who was right, who was wrong? Was Flamsteed wrong against Newton, but Cotes right against Leibniz? This is certainly the consensus *now*; but

---

[1] *Ibid.*, pp. 151–2.   [2] MacLaurin [1748], p. 90.
[3] MacLaurin, *op. cit.*, p. 17.
[4] *Ibid.*, p. 18. Incidentally if the Newtonian analysis–synthesis is a model both of proof and of discovery, why this zig-zag?
[5] Cotes [1717], p. xxviii.   [6] MacLaurin [1748], p. 94.

why? Why do we accept now that Newtonian progress was 'real', while Cartesian progress was 'fiction grafted on fiction'?[1]

Newton's confused methodology does not give any answer. Was there any rational reason to accept Kepler's laws as 'foundations' of Newtonian theory and as the 'refutation' of Descartes' theory? What was the rationale, if any, of not letting the tide of anomalies sweep away Newton's programme altogether?[2] Or was this only religious faith? Was the replacement of Cartesian vortices by the Newtonian void a religious conversion? Or a change in intellectual fashion?

Newtonians did not take all refutations easily. According to Cajori, the main reasons why Newton did not publish his theory of gravitation from 1666 – when he 'deduced' the inverse square relation from Kepler's laws – to 1687 was that a false 'observation report' about the length of an arc on the Earth made him think that his theory was false and, indeed, he abandoned it.[3] Even if the story is incorrect, it *might* have been true. But what is then the difference between a serious *refutation* and dozens of 'harmless' anomalies?[4] One looks vainly for guidance in Newton's writings. In a letter of 8 July 1672 Newton wrote to Oldenburg that he would accept 'experiments which directly contradict [him]' as valid objections. But if *his* basic experiments were correct and his inductive conclusions from these were flawless, no such thing could, of course, occur; the escape clause 'directly' turns his statement into an empty phrase. Newtonians had no criteria for the rejection of a theory in face of counterevidence, just as they had no criteria for the rejection of a theory without counterevidence.[5] But they *did* reject *some* theories in both kinds of situation. Was their decision completely irrational, or was there a method behind the mad methodology?

When, and under what circumstances, would Newtonians give up their position and try a new research programme? When does the host of open problems add up to a 'crisis' during which alternatives can be explored? Voltaire concluded his celebrated *Elements of Newton's Philosophy* in 1738 by 'confessing' to the existence of a series of open

[1] *Ibid.*

[2] For a long list of anomalies cf. e.g. Whewell [1837], where the long chapter 'Sequel to the Epoch of Newton' is nothing but the fascinating story of the war of Newtonians with the anomalies. We can also read about the weak hearted who, like Euler and Clairaut, faltered, tried alternative approaches and finally were humiliated for their disloyalty by the victories of orthodox Newtonians.

[3] Voltaire tells us this 'curious Anecdote' which, he says, shows 'with what Sincerity Sir Isaac proceeded in his Search after Truth' (Voltaire [1738], p. 197 ff).

[4] Incidentally, many people believe the myth that the anomalous behaviour of Mercury's perihelion was the last anomaly in Newtonian theory and that it has been solved by Einstein's theory. But the perihelion of Mercury is only 'less' anomalous in the light of Einstein's than in the light of Newton's theory. Moreover, Einstein's theory *inherited* quite a few of Newtonian anomalies. Let us take for instance 'Chandler's wobble'. Both Newton's and Einstein's theory predict a wobble of the spinning earth in each 300 days or so; but, *alas*, it wobbles every 428 days.

[5] Cf. *above*, section 2(a).

problems; but this did not shake his faith, stated in the Introduction, that 'there is but one Way that leads to Truth' and that is Newton's. Following this way 'From Truth to Truth the human Mind rises...'[1] MacLaurin, in 1748, did not hesitate to claim that '[Newton's] philosophy, being founded on experiment and demonstration, cannot fail till reason or the nature of things are changed'[2] and that Newton 'left to posterity little more to do, but to observe the heavens, and compute after his models'.

But two years before MacLaurin wrote, Clairaut found that the progress of the Moon's apogee is in reality twice what would follow from Newton's theory, and he proposed an additional term to Newton's formula involving the inverse fourth power of the distance. (MacLaurin seems not to have known of this, or perhaps he just ignored it – for he never mentioned open problems.) But as it turned out, Clairaut's mathematics was wrong, and in fact later a correct calculation was found among Newton's unpublished manuscripts. But even so, a small discrepancy remained: a 'secular acceleration'. In 1770 the Paris Academy put up a prize for the solution of this problem. Euler won this prize with an essay in which he first concluded that 'it appears to be established, by indisputable evidence, that the secular inequality of the moon's motion cannot be produced by the [Newtonian] forces of gravitation', and he proposed a rival formula again involving an additional term, which, in a sequel published a year later, he tried to explain from the resistance of Cartesian ether. However, Laplace in 1787 showed that the problem can be solved *better* within the Newtonian research programme. He scathingly pointed out that the 'brilliance' of Newton's programme is exactly that it turns each difficulty into a new victory. This, says Laplace, is 'the surest sign of truth'.[3]

Did Clairaut and Euler make a methodological blunder – as Kuhn would surely say – when they tried alternative research programmes to solve Newtonian puzzles and only wasted time, energy and talent? Or was it rather Poincaré who made a mistake – as Popper and Feyerabend would surely say – by sticking to Newton's theory and by not daring to proceed to special relativity theory when it was within his reach?

[1] Voltaire [1738], p. 241.
[2] MacLaurin [1748], p. 8. Incidentally, note the psychologistic clause of the statement. Why should the true laws of heaven change because human reason changes?
[3] Laplace [1824], p. 39. It should be here mentioned that Newton's authority strangled the development of Newtonian philosophy in Britain. Open problems were freely and aggressively discussed in France, where there was a rival research programme, but not in Britain, where there was none; also the better Leibnizian notation in the infinitesimal calculus equipped the continental scientists with better means to solve the vast mathematical problems. (The difference between Voltaire's and MacLaurin's textbooks is characteristic: the former ends up with the anomalies, the latter never mentions them.)

Newton's methodology provides no answer to any of these questions.

(e) Newton's double legacy

Newton left to the world his scientific research programme and his critical standards for judging such programmes. The extent of the impact of this schizophrenic achievement on the history of thought was tremendous. Newton set off the first major scientific research programme in human history; he and his brilliant followers established, *in practice*, the basic features of scientific methodology. *In this sense one may say that Newton's method created modern science.*[1]

On the other hand, Newton inherited his epistemology from a theology-dominated era, from 'justificationism'; and even though he modified its dominant Aristotelian–Cartesian version, he still remained its prisoner. The Newtonians' main methodological *problem*, in Pemberton's classic formulation was how '*to steer a just course between the conjectural method of proceeding...and demanding so rigorous a proof, as will reduce all philosophy to mere scepticism, and exclude all prospect of making any progress in the knowledge of nature*'.[2] But the Newtonian solution of this problem, although better than the Cartesian one, was still very weak. The confusion, the poverty of Newton's *theory of scientific achievement* contrasts dramatically with the clarity, the richness of his *scientific achievement*. His theory about why he rejected Cartesian vortex theory and why he accepted his own theory of gravitation was utterly absurd. But the incredible *success* of his research programme presented his philosophical admirers with the problem of defending his *theory of his success* and of the defeat of his rivals. The first Newtonians were confused and inconsistent in their methodology. The later vulgarizers however who could not follow the Newtonian research programme but could only scan its slogans, selected the crudest of them and arranged them into a colourful consistent subset. These crude slogans then gave rise to many – occasionally rival – philosophical projects and, especially, to two major philosophical research programmes: first, to find a rock bottom, an indubitably certain *empirical basis* of science in the form of 'pure sensations', or, failing that, of theory-impregnated or conventional basic propositions of some sort; and secondly, to solve the problem of how one can validly deduce/induce the laws of nature from this empirical basis. The first research programme gave rise to *justificationist philosophical psychology* and to the programmes of (linguistic) 'reductionism' and of the establishment of a 'theory-free, neutral, observational language' in logical positivism.[3] The second research programme gave rise to

---

[1] Of course, this is not to deny that he stood on the shoulders of Galileo.

[2] Pemberton [1728], p. 23.

[3] Incidentally, it may be worth mentioning that the third-world character of the Carnapian programme originates with Popper. Originally Carnap, a typical sceptico-

inductive logic.[1] *In this sense one may say that, while Newton's method created modern science, Newton's theory of method created modern philosophy of science.*

Moreover, the worst part of Newton's theory of method was set up as a rulebook for the underdeveloped disciplines and especially for the social sciences. Newtonianism, preached by semiliterates, like John Stuart Mill, who never *read* Newton, exerted a powerful influence in keeping underdeveloped disciplines underdeveloped.[2]

The influence of Newtonian success reached even political thought. It created a veritable euphoria among the dogmatists: before Newton the problem was whether it is possible at all to arrive at *episteme*; after Newton the problem became *how* it was possible to arrive at *episteme*, and how one can extend it to other spheres of knowledge. Without appreciating this problemshift one cannot understand eighteenth century thought. The struggle over the recognition of Newton's celestial mechanics as *episteme* took some time; but once it was recognized, the whole intellectual climate underwent a tremendous change. Much of eighteenth century thinking was determined by two major seventeenth century events conflicting in their effect. One was the tremendous suffering and chaos created by catholic–protestant warfare. The other was Newton's discoveries. The reaction to the first was *tolerant sceptical enlightenment*: there was no way to obtain proven truth about the most essential matters, therefore everyone should have the right to his beliefs. The best known exponent of this position was Bayle. The reaction to the second was *intolerant dogmatist enlightenment*: the light of science – to be extended to all domains of human knowledge – was to dispel pre-Newtonian darkness and also the darkness of the Church.[3] The leader of this movement was the Newtonian Voltaire.[4] The influence of intolerant dogmatist enlightenment soon superseded that of its tolerant sceptical counterpart and bred the ideas of totalitarian democracy. Scientific scepticism, defeated by Newton, degenerated into Humean psychologism and joined forces with dogmatism: human *reason* may not give assent to Newton, but human *nature* must. But then the study of (unchanging, external, universal) human nature will lead us to a theory of (monolithic) 'healthy' belief.

The influence of Newtonian success was then possibly the most powerful influence on modern thought. But it is not the aim of this essay to map out the whole story; our attention is focussed on, if not

dogmatist, started from the position of 'methodological solipsism' and wanted to establish basic propositions at second-world level, in the form of Neurathian 'protocol statements': 'At 9 o'clock *I saw*...' It was Popper who, in 1932, persuaded him to replace second-world 'protocol-statements' by third-world 'basic statements'.

[1] For the degenerating problemshifts in inductive logic cf. volume 2, chapter 8.
[2] One is tempted to say that Newton created *two cultures*; one which developed his method, another which 'developed' his methodology.
[3] The discovery of distant lands – a third important factor – worked both ways.
[4] This analysis, if correct, makes nonsense of the Marxist approach to the history of the eighteenth century.

necessarily limited to, the problem it has presented to the philosophy of knowledge.

One may say that philosophy of science from 1687 to 1934 consisted, in the main, of two schools, best characterized by their evaluation of Newton's theory of gravitation. One school, the successor of the 'dogmatists' (whether of the empiricist or of the rationalist brand) claimed that they had proved or could prove that Newton's facts *were* facts and that Newton's argument from fact to theory *was*, in some objective, third-world sense, valid. The other school, the successors of the 'sceptics', claimed that Newton's theory cannot (or may not) be objectively *proved*, but its final success (a hard fact!) can be *explained* in psychological, second-world terms. The dogmatists tried to *prove* too much, the sceptics tried to *explain* too much: for Newton's theory was false. But the fact that Newton's theory was false – and later recognized to be false – does not turn the problem of proving it or explaining the inevitable assent to it into a 'pseudoproblem'. Such investigations do not *necessarily* lead to degenerating problemshifts. Just as a heuristically generated sequence of false propositions may imply an increasing number of interesting true propositions, a heuristically generated sequence of incorrectly stated problems may contain the solution of an increasing number of correctly stated problems. Some of those very few who could follow, to some degree, Newton's actual method and not just his methodology, could, in trying to solve these problems, make a few steps toward narrowing the gap between professed methodology and the Newtonians' actual method, even though they did not realize that the problem itself must be shifted. The three philosophers who contributed most to this process were Adam Smith, Whewell, and LeRoy.

The crucial problemshift however came only after Einstein's theory in fact had superseded Newton's: the problem was now to explain the success not of Newton's victorious but of his defeated theory; and also to explain its defeat. Popper was the first to look at the problem in this way; and thereby he ushered in a new epoch in philosophy.

References

Achinstein, P. [1970]: 'Inference to Scientific Laws', in R. Stuewer (*ed.*): *Historical and Philosophical Perspectives in Science, Minnesota Studies in the Philosophy of Science*, **5**, pp. 87–111. University of Minnesota Press.

Agassi, J. [1959]: 'How are Facts Discovered?', *Impulse*, **3**, No. 10, pp. 2–4.

Agassi, J. [1963]: *Towards an Historiography of Science*. Wesleyan University Press.

Agassi, J. [1964*a*]: 'The Confusion between Physics and Metaphysics in the Standard Histories of Sciences', in the *Proceedings of the Tenth International Congress of the History of Science*, 1964, **1**, pp. 231–8.

Agassi, J. [1964*b*]: 'Scientific Problems and Their Roots in Metaphysics', in M. Bunge (ed.): [1964], pp. 189–211.

Agassi, J. [1966]: 'Sensationalism', *Mind*, N.S. **75**, pp. 1–24.

Agassi, J. [1968]: 'The Novelty of Popper's Philosophy of Science', *International Philosophical Quarterly*, **8**, pp. 442–63.

Agassi, J. [1969]: 'Popper on Learning from Experience', in N. Rescher (*ed.*): *Studies in the Philosophy of Science*, pp. 162–71. American Philosophical Quarterly Monograph Series.

Ayer, A. J. [1936]: *Language, Truth and Logic*. London: Victor Gollancz. (2nd edition, 1946.)

Baily, F. [1835]: *An Account of the Revd John Flamsteed, the First Astronomer-Royal*. London: Order of the Lords Commissioners of the Admiralty.

Bartley, W. W. [1968]: 'Theories of Demarcation between Science and Metaphysics', in I. Lakatos and A. E. Musgrave (*eds.*): [1968], pp. 40–64.

Beck, G. and Sitte, K. [1933]: 'Zur Theorie des β-Zerfalls', *Zeitschrift für Physik*, **86**, pp. 105–19.

Bernal, J. D. [1954]: *Science in History*, 1st edition. London: Watts.

Bernal, J. D. [1965]: *Science in History*, 3rd edition. London: Watts.

Bernstein, J. [1967]: *A Comprehensible World: On Modern Science and its Origins*. New York: Random House.

Bethe, H. and Peierls, R. E. [1934]: 'The "Neutrino"', *Nature*, **133**, p. 532.

Beveridge, W. [1937]: 'The Place of the Social Sciences in Human Knowledge', *Politica*, **2**, pp. 459–79.

Black, M. [1967]: 'Induction', in P. Edwards (*ed.*): *The Encyclopedia of Philosophy*, vol. 4, p. 169. New York: Macmillan.

Bohr, N. [1913*a*]: 'On the Constitution of Atoms and Molecules', *Philosophical Magazine*, **26**, pp. 1–25, 476–502 and 857–75.

Bohr, N. [1913*b*]: Letter to Rutherford, 6 March, published in N. Bohr [1963], pp. xxxviii–ix.

Bohr, N. [1913*c*]: 'The Spectra of Helium and Hydrogen', *Nature*, **92**, pp. 231–2.

Bohr, N. [1922]: 'The Structure of the Atom', *Nobel Lectures*, vol. 2. Amsterdam: Elsevier, 1965.

Bohr, N. [1926]: Letter to *Nature*, **117**, p. 264.

Bohr, N. [1932]: 'Chemistry and the Quantum Theory of Atomic Constitution', Faraday Lecture, 1930, *Journal of the Chemical Society*, 1932/1, pp. 349–84.

Bohr, N. [1933]: 'Light and Life', *Nature*, **131**, pp. 421–3 and 457–9.

Bohr, N. [1936]: 'Conservation Laws in Quantum Theory', *Nature*, **138**, pp. 25–6.

Bohr, N. [1949]: 'Discussion with Einstein on Epistemological Problems in Atomic Physics', in P. A. Schilpp (*ed.*): *Albert Einstein, Philosopher–Scientist*, vol. 1, pp. 201–41. La Salle: Open Court.

Bohr, N. [1963]: *On the Constitution of Atoms and Molecules*, New York: Benjamin.

Bonnar, F. T. and Phillips, M. [1957]: *Principles of Physical Science*. Reading, Massachusetts: Addison–Wesley.

Born, M. [1948]: 'Max Karl Ernst Ludwig Planck', *Obituary Notices of Fellows of the Royal Society*, **6**, pp. 161–80.

Born, M. [1949]: *Natural Philosophy of Cause and Chance*. Oxford University Press.

Born, M. [1954]: 'The Statistical Interpretation of Quantum Mechanics'. *Nobel Lectures*, vol. 3. Amsterdam: Elsevier, 1964.

Braithwaite, R. B. [1938]: 'The Relevance of Psychology to Logic', *Aristotelian Society Supplementary Volumes*, **17**, pp. 19–41.

Braithwaite, R. B. [1953]: *Scientific Explanation*. Cambridge University Press.

Brewster, D. [1855]: *Memoirs of the Life, Writings and Discoveries of Sir Isaac Newton*. Two volumes. Edinburgh: Thomas Constable. Reprinted with a new introduction by R. S. Westfall in *Sources of Science*, **14**. New York and London: Johnson Reprint Corporation.

Brunet, P. [1931]: *L'introduction des Théories de Newton en France au XVIIᵉ Siècle*. Two volumes. Paris: Blanchard.

Bunge, M. (*ed.*) [1964]: *The Critical Approach to Science and Philosophy*. New York: The Free Press.

Callendar, M. L. [1914]: 'The Pressure of Radiation and Carnot's Principle', *Nature*, **92**, p. 553.

Canfield, J. and Lehrer, K. [1961]: 'A Note on Prediction and Deduction', *Philosophy of Science*, **28**, pp. 204–8.

Cantor, G. [1971]: 'Henry Brougham and the Scottish Methodological Tradition', *Studies in the History and Philosophy of Science*, **2**, pp. 69–89.

Carnap, R. [1932–3]: 'Über Protokollsätze', *Erkenntnis*, **3**, pp. 215–28.

Carnap, R. [1935]: Review of Popper's [1934], *Erkenntnis*, **5**, pp. 290–4.

Clavius, C. [1581]: *In Sphaeram Ioannis de Sacro Bosco commentarius nunc iterum ab ipso Anctore recognitus, et multis ac variis locis locupletatus*. Rome: ex officina Dominici Basae.

Clifford, M. [1675]: *A Treatise of Human Reason*. London: Henry Brome.

Coffa, A. [1968]: 'Deductive Predictions', *Philosophy of Science*, **35**, pp. 279–83.

Cohen, I. B. (*ed.*) [1958]: *Isaac Newton's Papers and Letters on Natural Philosophy*. Cambridge University Press.

Cohen, I. B. [1960]: *The Birth of a New Physics*. London: Heinemann.

Compton, A. H. [1919]: 'The Size and Shape of the Electron', *Physical Review*, **14**, pp. 20–43.

Cotes, R. [1712–13]: 'Letter to Newton', in Edleston, J. (*ed.*): [1850], pp. 181–4.

Cotes, R. [1717]: Preface to 2nd edition of *Principia*, pp. xx–xxxiii.

Crookes, W. [1886]: Presidential Address to the Chemistry Section of the British Association, *Report of British Association*, pp. 558–76.

Crookes, W. [1888]: Report at the Annual General Meeting, *Journal of the Chemical Society*, **53**, pp. 487–504.

Davisson, C. J. [1937]: 'The Discovery of Electron Waves', *Nobel Lectures*, vol. 2. Amsterdam: Elsevier, 1965.

Descartes, R. [1638]: 'Letter to Mersenne, 11 October', in C. Adam and P. Tanner, (*eds.*): *Oeuvres de Descartes*, vol. II, pp. 379–405. Paris: Librairie Philosophique J. Vrin, 1969.

Dirac, P. A. M. [1936]: 'Does Conservation of Energy Hold in Atomic Processes?', *Nature*, **137**, pp. 298–9.

Dirac, P. A. M. [1951]: 'Is there an Aether?', *Nature*, **168**, pp. 906–7.

Dorling, J. [1968]: 'Length Contraction and Clock Synchronisation: The Empirical Equivalence of the Einsteinian and Lorentzian Theories', *The British Journal for the Philosophy of Science*, **19**, pp. 67–9.

Dorling, J. [1971]: 'Einstein's Introduction of Photons: Argument by Analogy or Deduction from the Phenomena?', *The British Journal for the Philosophy of Science*, **22**, pp. 1–8.

Dreyer, J. L. E. [1906]: *History of the Planetary Systems from Thales to Kepler*. Republished as *A History of Astronomy from Thales to Kepler*. New York: Dover, 1953.

Duhem, P. [1906]: *La théorie physique, son objet et sa structure.* (English translation of 2nd (1914) edition: *The Aim and Structure of Physical Theory*. Princeton University Press, 1954.)

Duhem, P. [1908]: ΣΩΖΕΙΝ ΤΑ ΦΑΙΝΟΜΕΝΑ, *Annales de Philosophie Chrétienne*, **6**. Reprinted in book form as *To Save the Phenomena*. Translated by E. Doland and C. Maschler. Chicago University Press, 1969.

Eccles, J. C. [1964]: 'The Neurophysiological Basis of Experience', in M. Bunge (*ed.*): [1964], pp. 266–79.

Edleston, J. [1850]: *Correspondence of Sir Isaac Newton and Professor Cotes*. Cambridge University Press.

Ehrenfest, P. [1911]: 'Welche Züge der Lichtquantenhypothese spielen in der Theorie der Wärmestrahlung eine wesentliche Rolle?', *Annalen der Physik*, **36**, pp. 91–118.

Ehrenfest, P. [1913]: *Zur Krise der Lichtäther-Hypothese*. Berlin: Springer.

Einstein, A. [1909]: 'Über die Entwicklung unserer Anschauungen über das Wesen und die Konstitution der Strahlung', *Physikalische Zeitschrift*, **10**, pp. 817–26.

Einstein, A. [1927]: 'Neue Experimente über den Einfluss der Erdbewegung auf die Lichtgeschwindigkeit relativ zur Erde', *Forschungen und Fortschritte*, **3**, p. 36.

Einstein, A. [1928]: Letter to Schrödinger, 31.5.1928, in K. Przibram (*ed.*): *Briefe Zur Wellenmechanik*. Vienna: Springer, 1963.

Einstein, A. [1931]: 'Gedenkworte auf Albert A. Michelson', *Zeitschrift für angewandte Chemie*, **44**, p. 658.

Einstein, A. [1949]: 'Autobiographical Notes', in P. A. Schilpp (*ed.*): *Albert Einstein, Philosopher–Scientist*, vol. I, pp. 2–95. La Salle: Open Court.

Elkana, Y. [1971]: 'The Conservation of Energy: a Case of Simultaneous Discovery?', *Archives Internationales d'Histoire des Sciences*, **24**, pp. 31–60.

Ellis, C. D. and Mott, N. F. [1933]: 'Energy Relations in the β-Ray Type of Radioactive Disintegration', *Proceedings of the Royal Society*, Series A, **141**, pp. 502–11.

Ellis, C. D. and Wooster, W. A. [1927]: 'The Average Energy of Disintegration of Radium E', *Proceedings of the Royal Society*, Series A, **117**, pp. 109–23.

Evans, E. J. [1913]: 'The Spectra of Helium and Hydrogen', *Nature*, **92**, p. 5.

Ewald, P. [1969]: 'The Myth of Myths', *Archive for History of Exact Sciences*, **6**, pp. 72–81.

Feigl, H. [1964]: 'What Hume Might Have Said to Kant', in M. Bunge (*ed.*): [1964], pp. 45–51.

Fermi, E. [1933]: 'Tentativo di una teoria dell emissione dei raggi "beta"', *Ricerci Scientifica*, **4**(2), pp. 491–5.

Fermi, E. [1934]: 'Versuch einer Theorie der β-Strahlen. I', *Zeitschrift für Physik*, **88**, pp. 161–77.

Feyerabend, P. K. [1961]: 'Comments on Grünbaum's "Law and Convention in Physical Theory"', in H. Feigl and G. Maxwell (*eds.*): *Current Issues in the Philosophy of Science*, p. 155–61. University of Minnesota Press.

Feyerabend, P. K. [1962]: 'Explanation, Reduction and Empiricism', in H. Feigl and G. Maxwell (*eds.*): *Minnesota Studies in the Philosophy of Science*, **3**, pp. 28–97. University of Minnesota Press.

Feyerabend, P. K. [1963]: 'Review of Kraft's *Erkenntnislehre*', *British Journal for the Philosophy of Science*, **13**, pp. 319–23.

Feyerabend, P. K. [1964]: 'Realism and Instrumentalism: Comments on the Logic of Factual Support', in M. Bunge (*ed.*): [1964], pp. 280–308.

Feyerabend, P. K. [1965]: 'Reply to Criticism', in R. S. Cohen and M. Wartofsky (*eds.*): *Boston Studies in the Philosophy of Science*, **2**, pp. 223–61. Dordrecht: Reidel.

Feyerabend, P. K. [1968–9]: 'On a Recent Critique of Complementarity', *Philosophy of Science*, **35**, pp. 309–31 and **36**, pp. 82–105.

Feyerabend, P. K. [1969a]: 'Problems of Empiricism II', in R. G. Colodny (*ed.*): *The Nature and Function of Scientific Theory*. University of Pittsburgh Press.

Feyerabend, P. K. [1969b]: 'A Note on Two "Problems" of Induction, *British Journal for the Philosophy of Science*, **19**, pp. 251–3.

Feyerabend, P. K. [1970a]: 'Consolations for the Specialist', in I. Lakatos and A. Musgrave (*eds.*): [1970], pp. 197–230.

Feyerabend, P. K. [1970b]: 'Against Method', in *Minnesota Studies for the Philosophy of Science*, **4**. University of Minnesota Press.

Feyerabend, P. K. [1970c]: 'Classical Empiricism', in R. E. Butts and J. W. Davis (*eds.*): *The Methodological Heritage of Newton*, pp. 150–70. Oxford: Basil Blackwell.

Feyerabend, P. K. [1972]: 'Von der beschränkten Gültigheit methodologischer Regeln', in R. Bubner, K. Cramer and R. Wiehl (*eds.*): *Dialog als Methods*, pp. 124–71.

Feyerabend, P. K. [1974]: *Against Method*. London: New Left Books.

Forman, P. [1969]: 'The Discovery of the Diffraction of X-Rays by Crystals: A Critique of the Critique of the Myths', *Archive for History of Exact Sciences*, **6**, pp. 38–71.

Fowler, W. A. [1912]: 'Observations of the Principal and Other Series of lines in the Spectrum of Hydrogen', *Monthly Notices of the Royal Astronomical Society*, **73**, pp. 62–71.

Fowler, W. A. [1913a]: 'The Spectra of Helium and Hydrogen', *Nature*, **92**, p. 95.

Fowler, W. A. [1913b]: 'The Spectra of Helium and Hydrogen', *Nature*, **92**, p. 232.

Fowler, W. A. [1914]: 'Series Lines in Spark Spectra', *Proceedings of the Royal Society of London (A)*, **90**, pp. 426–30.

Fresnel, A. [1818]: 'Lettre à Francois Arago sur L'Influence du Mouvement Terrestre dans quelques Phénomènes Optiques', *Annales de Chimie et de Physique*, **9**, pp. 57 ff.

Galileo [1615]: 'Letter to the Grand Duchess', in S. Drake (*ed.*): *Discoveries and Opinions of Galileo*, pp. 173–216. Garden City: Doubleday, 1957.

Galileo [1632]: *Dialogue on the Great World Systems*. University of Chicago Press.

Gamow, G. A. [1966]: *Thirty Years that Shook Physics*. Garden City: Doubleday.

Gingerich, O. [1973]: 'The Copernican Celebration', *Science Year*, 1973, pp. 266–7.

Gingerich, O. [1975]: '"Crisis" versus Aesthetic in the Copernican Revolution', in A. Beer (*ed.*): *Vistas in Astronomy*, **17**.

Glanvill, J. [1665]: *Scepsis Scientifica*. London: E. Coates.

Glanvill, J. [1675]: *Essays on Several Important Subjects in Philosophy and Religion*. London: Thomas Tomkins.

Gregory, D. [1702]: *Astronomiae Physicae et Geometricae Elementa*.

Grünbaum, A. [1959a]: 'The Falsifiability of the Lorentz–Fitzgerald Contraction Hypothesis', *British Journal for the Philosophy of Science*, **10**, pp. 48–50.

Grünbaum, A. [1959b]: ' Law and Convention in Physical Theory', in H. Feigl and G. Maxwell (*eds.*): *Current Issues in the Philosophy of Science*, pp. 40–155. University of Minnesota Press.

Grünbaum, A. [1960]: 'The Duhemian Argument', *Philosophy of Science*, **11**, pp. 75–87.

Grünbaum, A. [1966]: 'The Falsifiability of a Component of a Theoretical System', in P. K. Feyerabend and G. Maxwell (*eds.*): *Mind, Matter and Method: Essays in Philosophy and Science in Honor of Herbert Feigl*, pp. 273–305. University of Minnesota Press.

Grünbaum, A. [1969]: 'Can we Ascertain the Falsity of a Scientific Hypothesis?', *Studium Generale*, **22**, pp. 1061–93.

Hall, R. J. [1970]: 'Kuhn and the Copernican Revolution', *British Journal for the Philosophy of Science*, **21**, pp. 196–7.

Halley, E. [1687]: 'Letter to Newton, 5 April', in H. W. Turnbull (*ed.*). [1960], vol. ii, 1676–87, pp. 473–4.

Hanson, N. R. [1973]: *Constellations and Conjectures*. Dordrecht: D. Reidel.

Heisenberg, W. von. [1955]: 'The Development of the Interpretation of Quantum Theory', in W. Pauli (*ed.*): *Nils Bohr and the Development of Physics*. London: Pergamon.

Hempel, C. G. [1937]: Review of Popper's [1934], *Deutsche Literaturzeitung*, pp. 309–14.

Hempel, C. G. [1952]: 'Some Theses on Empirical Certainty', *The Review of Metaphysics*, **5**, pp. 620–1.

Henderson, W. J. [1934]: 'The Upper Limits of the Continuous β-ray Spectra of Thorium C and C^{11}', *Proceedings of the Royal Society*, Series A, **147**, pp. 572–82.

Hesse, M. [1963]: 'A New Look at Scientific Explanation', *The Review of Metaphysics*, **17**, pp. 98–108.

Hesse, M. [1968]: Review of Grünbaum's [1966], *The British Journal for the Philosophy of Science*, **18**, pp. 333–5.

Hevesy, G. von [1913]: 'Letter to Rutherford, 14 October', quoted in N. Bohr [1963], p. xlii.

Hobbes, T. [1651]: *Leviathan*. Oxford: James Thornton, 1881.

Holton, G. [1969]: 'Einstein, Michelson, and the "Crucial" Experiment', *Isis*, **6**, pp. 133–97.

Hume, D. [1777]: *Enquiries Concerning the Human Understanding and Concerning the Principles of Morals*. L. A. Selby-Bigge (*ed.*): 2nd edition. Oxford: Clarendon Press, 1966.

Hund, F. [1961]: 'Göttingen, Copenhagen, Leipzig im Rückblick', in F. Bopp (*ed.*): *Werner Heisenberg und Die Physik unserer Zeit*. Braunschweig: Vieweg.

Jaffe, B. [1960]: *Michelson and the Speed of Light*. London: Heinemann.

Jammer, M. [1966]: *The Conceptual Development of Quantum Mechanics*. New York: McGraw-Hill.

Jeans, J. [1948]: *The Growth of Physical Science*. Cambridge University Press.

Joffé, A. [1911]: 'Zur Theorie der Strahlenden Energie', *Annalen der Physik*, **35**, p. 474.

Johnson, F. R. [1959]: 'Commentary on Derek J. de S. Price,' in M. Clagett (*ed.*): *Critical Problems in the History of Science*, pp. 219–21. University of Wisconsin Press.

Jourdain, P. E. B. [1915]: 'Newton's Hypotheses of Ether and of Gravitation from 1672 to 1679', *The Monist*, **25**, pp. 79–106.

Juhos, B. [1966]: 'Über die empirische Induktion', *Studium Generale*, **19**, pp. 259–72.

Kamlah, A. [1971]: 'Kepler im Licht der modernen Wissenschaftstheorie', in H. Lenk (*ed.*): *Neue Aspekte der Wissenschaftstheorie*, pp. 205–20. Braunschweig: Vieweg.

Kepler, J. [1604]: *Ad Vitellionem Paralipomena*, in M. Caspar (*ed.*): *Gesammelte Werke*, **2**. Munich: C. H. Beck.

Kepler, J. [1619]: *Harmonice Mundi*, in *Gesammelte Werke*, **6**, Munich: C. H. Beck, 1940.

Keynes, J. M. [1921]: *A Treatise on Probability*. Cambridge University Press.

Koestler, A. [1959]: *The Sleepwalkers*. London: Hutchinson.

Konopinski, E. J. and Uhlenbeck, G. [1935]: 'On the Fermi theory of β-radioactivity', *Physical Review*, **48**, pp. 7–12.

Koyré, A. [1965]: *Newtonian Studies*. London: Chapman and Hall.

Kraft, V. [1925]: *Die Grundformen der wissenschaftlichen Methoden*. Vienna and Leipzig: Hölder–Pichler–Tempsky.

Kraft, V. [1966]: 'The Problem of Induction', in P. Feyerabend and G. Maxwell (*eds.*): *Mind, Matter and Method*, pp. 306–17. University of Minnesota Press.

Kramers, H. A. [1923]: 'Das Korrespondenzprinzip und der Schalenbau des Atoms', *Die Naturwissenschaften*, **11**, pp. 550–9.

Kudar, J. [1929–30]: 'Der wellenmechanische Charakter des β-Zerfalls, I–II–III', *Zeitschrift für Physik*, **57**, pp. 257–60, **60**, pp. 168–75 and 176–83.

Kuhn, T. S. [1957]: *The Copernican Revolution*. Chicago University Press.

Kuhn, T. S. [1962]: *The Structure of Scientific Revolutions*. Princeton University Press. (Second edition, 1970.)

Kuhn, T. S. [1963]: 'The Function of Dogma in Scientific Research', in A. C. Crombie (*ed.*): *Scientific Change*, pp. 347–69. London: Heinemann.

Kuhn, T. S. [1968]: 'Science: The History of Science', in D. L. Sills (*ed.*): *International Encyclopedia of the Social Sciences*, vol. **14**, pp. 74–83. New York: Macmillan.

Kuhn, T. S. [1970a]: 'Logic of Discovery or Psychology of Research?', in I. Lakatos and A. E. Musgrave (*eds.*): [1970], pp. 1–24.

Kuhn, T. S. [1970b]: 'Reflections on my Critics', in I. Lakatos and A. Musgrave (eds.): [1970], pp. 237–78.

Kuhn, T. S. [1971]: 'Notes on Lakatos', in R. C. Buck and R. S. Cohen (eds.): Boston Studies in the Philosophy of Science, **8**, pp. 137–46. Dordrecht: Reidel.

Lakatos, I.: see Lakatos bibliography, pp. 237–9.

Lamb, H. [1923]: Dynamics. Second edition. Cambridge University Press.

Laplace, M. [1824]: Exposition du Système du Monde. Fifth edition. Paris: Bachelier.

Larmor, L. [1904]: 'On the Ascertained Absence of Effects of Motion through the Aether, in Relation to the Constitution of Matter, and on the Fitzgerald–Lorentz Hypothesis', Philosophical Magazine, Series 6, **7**, pp. 621–5.

Laudan, L. L. [1965]: 'Grünbaum on "The Duhemian Argument"', Philosophy of Science, **32**, pp. 295–9.

Laudan, L. L. [1967]: 'The Nature and Sources of Locke's Views on Hypotheses', Journal of the History of Ideas, **28**, pp. 211–23.

Leibniz, G. W. [1677]: 'Towards a Universal Characteristic', in P. P. Wiener (ed.): Leibniz Selections, pp. 17–25. New York: Scribner.

Leibniz, G. W. [1678]: Letter to Conring, 19 March, in L. Loemker (ed.): Leibniz's Philosophical Papers and Letters, pp. 186–91. Dordrecht: Reidel, 1967.

LeRoy, E. [1899]: 'Science et Philosophie', Revue de Metaphysique et de Morale, **7**, pp. 375–425, 503–62, 706–31.

LeRoy, E. [1901]: 'Un Positivisme Nouveau', Revue de Metaphysique et de Morale, **9**, pp. 138–53.

Locke, J. [1690]: Essay Concerning Human Understanding (A. S. Pringle-Pattison (ed.)). Oxford: Clarendon Press, 1924.

Locke, J. [1697]: 'Second Letter to Stillingfleet, 29 June' as Mr Locke's Reply to the Bishop of Worcester's Answer to his Letter, in The Works of John Locke, vol. 6. London: Thomas Tegg, 1823.

Lorentz, H. A. [1886]: 'De l'influence du Mouvement de la Terre sur les Phénomènes Lumineux', Versl. Kon. Akad. Wetensch. Amsterdam, **2**, pp. 297–358. Reprinted in H. A. Lorentz: Collected Papers, **4**, pp. 153–218. The Hague: Nijhoff, 1937.

Lorentz, H. A. [1892a]: 'The Relative Motion of the Earth and the Ether', Versl. Kon. Akad. Wetensch. Amsterdam. **1**, pp. 74–7. Reprinted in H. A. Lorentz: Collected Papers, **4**, pp. 219–23.

Lorentz, H. A. [1892b]: 'Stokes' Theory of Aberration', Versl. Kon. Akad. Wetensch. Amsterdam, **1**, pp. 97–103. Reprinted in H. A. Lorentz: Collected Papers, **4**, pp. 224–31.

Lorentz, H. A. [1895]: Versuch einer Theorie der electrischen und optischen Erscheinungen in bewegten Körpern. Sections 89–92. Leipzig: Teubner.

Lorentz, H. A. [1897]: 'Concerning the Problem of the Dragging Along of the Ether by the Earth', Versl. Kon. Akad. Wetensch. Amsterdam, **6**, pp. 266–72. Reprinted in H. A. Lorentz: Collected Papers, **4**, pp. 237–44.

Lorentz, H. A. [1923]: 'The Rotation of the Earth and its Influence on Optical Phenomena', Nature, **112**, pp. 103–4.

Luther, M. [1525]: De Servo Arbitrio, in D. Martin Luther's Werke, **18**. Weimar: H. Böhlau, 1883–1948.

Lykken, D. T. [1968]: 'Statistical Significance in Psychological Research', Psychological Bulletin, **70**, pp. 151–9.

McCulloch, J. R. [1825]: The Principles of Political Economy: With a Sketch of the Rise and Progress of the Science. Edinburgh: William and Charles Tait.

MacLaurin, C. [1748]: *An Account of Sir Isaac Newton's Philosophy*. Reprinted in L. L. Laudan (*ed.*): *The Sources of Science Series*, **74**. London and New York: Johnson Reprint Corporation, 1968.

McMullin, E. [1971]: 'The History and Philosophy of Science: a Taxonomy', *Minnesota Studies in the Philosophy of Science*, **5**, pp. 12–67. University of Minnesota Press.

Margenau, H. [1950]: *The Nature of Physical Reality*. New York: McGraw–Hill.

Marignac, C. [1860]: 'Commentary on Stas' Researches on the Mutual Relations of Atomic Weights'. Reprinted in *Prout's Hypothesis*, Alembic Club Reprints, **20**, pp. 48–58.

Maxwell, J. C. [1871]: *Theory of Heat*. London: Longmans.

Medawar, P. B. [1967]: *The Art of the Soluble*. London: Methuen.

Medawar, P. B. [1969]: *Induction and Intuition in Scientific Thought*. London: Methuen.

Meehl, P. [1967]: 'Theory Testing in Psychology and Physics: A Methodological Paradox', *Philosophy of Science*, **34**, pp. 103–15.

Meitner, L. [1933]: 'Kernstruktur', in H. Geiger and J. Scheel (*eds.*): *Handbuch der Physik*, Zweite Auflage, **25**/1, pp. 118–62. Berlin: Springer.

Meitner, L. and Orthmann, W. [1930]: 'Uber eine absolute Bestimmung der Energie der primären β-Strahlen von Radium E', *Zeitschrift für Physik*, **60**, pp. 143–55.

Merton, R. [1957]: 'Priorities in Scientific Discovery', *American Sociological Review*, **22**, pp. 635–59.

Merton, R. [1963]: 'Resistance to the Systematic Study of Multiple Discoveries in Science', *European Journal of Sociology*, **4**, pp. 237–82.

Merton, R. [1969]: 'Behaviour Patterns of Scientists', *American Scholar*, **38**, pp. 197–225.

Michelson, A. [1881]: 'The Relative Motion of the Earth and the Luminiferous Ether', *American Journal of Science*, Series 3, **22**, pp. 120–9.

Michelson, A. [1891–2]: 'On the Application of Interference Methods to Spectroscopic Measurements, I–II', *Philosophical Magazine*, Series 3, **31**, pp. 338–46 and **34**, pp. 280–99.

Michelson, A. [1897]: 'On the Relative Motion of the Earth and the Ether', *American Journal of Science*, Series 4, **3**, pp. 475–8.

Michelson, A. and Gale, H. G. [1925]: 'The Effect of the Earth's Rotation on the Velocity of Light', *Astrophysical Journal*, **61**, pp. 137–45.

Michelson, A. and Morley, E. W. [1887]: 'On the Relative Motion of the Earth and the Luminiferous Ether', *American Journal of Science*, Series 3, **34**, pp. 333–45.

Milhaud, G. [1896]: 'La Science Rationnelle', *Revue de Metaphysique et de Morale*, **4**, pp. 280–302.

Mill, J. S. [1843]: *A System of Logic*. London: Longmans, 1967.

Miller, D. C. [1925]: 'Ether-Drift Experiments at Mount Wilson', *Science*, **61**, pp. 617–21.

Miller, D. W. [1974]: 'Popper's Qualitative Theory of Verisimilitude', *British Journal for the Philosophy of Science*, **25**, pp. 166–77.

Morley, E. W. and Miller, D. C. [1904]: 'Letter to Kelvin', published in *Philosophical Magazine*, Series 6, **8**, pp. 753–4.

Moseley, H. G. J. [1914]: 'Letter to Nature', *Nature*, **92**, p. 554.

Mott, N. F. [1933]: 'Wellenmechanik und Kernphysik', in H. Geiger and J. Scheel (*eds.*): *Handbuch der Physik*, Zweite Auflage, **24**/1, pp. 785–841.

Musgrave, A. E. [1968]: 'On a Demarcation Dispute', in I. Lakatos and A. E. Musgrave (eds.): [1968], pp. 78–85.

Musgrave, A. E. [1969a]: Impersonal Knowledge, PhD Thesis, University of London.

Musgrave, A. E. [1969b]: Review of Ziman's 'Public Knowledge: An Essay Concerning the Social Dimensions of Science', in The British Journal for the Philosophy of Science, 20, pp. 92–4.

Musgrave, A. E. [1971]: 'Kuhn's Second Thoughts', British Journal for the Philosophy of Science, 22, pp. 287–97.

Musgrave, A. E. [1974]: 'The Objectivism of Popper's Epistemology', in P. A. Schilpp (ed.): The Philosophy of Sir Karl Popper, pp. 560–96. La Salle, Illinois: Open Court.

Naess, A. [1964]: 'Reflections About Total Views', Philosophy and Phenomenological Research, 25, pp. 16–29.

Nagel, E. [1961]: The Structure of Science. New York: Harcourt, Brace and World.

Nagel, E. [1967]: 'What is True and False in Science: Medawar and the Anatomy of Research', Encounter, 29, no. 3, pp. 68–70.

Nature, [1913–14]: 'Physics at the British Association', Nature, 92, pp. 305–9.

Neugebauer, O. [1958]: The Exact Sciences in Antiquity. New York: Dover, 1969.

Neugebauer, O. [1968]: 'On the Planetary Theory of Copernicus', Vistas in Astronomy. 10, pp. 89–103.

Neurath, O. [1935]: 'Pseudorationalismus der Falsifikation', Erkenntnis, 5, pp. 353–65.

Newton, I. [1672]: 'Letter to the Editor of the Philosophical Transactions of the Royal Society, 8 July', Reprinted in I. B. Cohen (ed.): [1958].

Newton, I. [1676]: 'Letter to Oldenburg, 18 November', in H. W. Turnbull (ed.) [1960], vol. II.

Newton, I. [1686]: Principia Mathematica. Translated by F. Cajori. Berkeley: University of California Press, 1960.

Newton, I. [1694]: 'Letter to Flamsteed, 16 February', in F. Baily [1835], p. 151.

Newton, I. [1713]: 'Letter to Roger Cotes, 28 March', in J. Edleston (ed.): [1850], pp. 154–6.

Newton, I. [1717]: Opticks. 4th edition. Dover, 1952.

Nicholson, J. W. [1913]: 'A Possible Extension of the Spectrum of Hydrogen', Monthly Notices of the Royal Astronomical Society, 73, pp. 382–5.

Pannekoek, A. [1961]: A History of Astronomy. New York: Interscience Publishers.

Pauli, W. [1961]: 'Zur älteren und neuren Geschichte des Neutrinos', in W. Pauli: Aufsätze und Vorträge über Physik und Erkenntnistheorie, pp. 156–80.

Pearce Williams, L. [1968]: Relativity Theory: Its Origins and Impact on Modern Thought.

Pearce Williams, L. [1970]: 'Normal Science and its Dangers,' in I. Lakatos and A. Musgrave (eds.): [1970], pp. 49–50.

Peierls, R. E. [1936]: 'Interpretation of Shankland's Experiment', Nature, 137, p. 904.

Pemberton, H. [1728]: A View of Sir Isaac Newton's Philosophy. London: S. Palmer.

Planck, M. [1900a]: 'Über eine Verbesserung der Wienschen Spektralgleichung', Verhandlungen der Deutschen Physikalischen Gesellschaft, 2, pp. 202–4. English translation in Ter Haar [1967].

Planck, M. [1900b]: 'Zur Theorie des Gesetzes der Energieverteilung im Normalspektrum', *Verhandlungen der Deutschen Physikalischen Gesellschaft*, **2**, pp. 237–45. English translation in Ter Haar [1967].

Planck, M. [1929]: 'Zwanzig Jahre Arbeit am Physikalischen Weltbild', *Physica*, **9**, pp. 193–222.

Planck, M. [1948]: *Scientific Autobiography*. London: Williams and Norgate, 1950.

Poincaré, H. [1891]: 'Les géométries non euclidiennes', *Revue des Sciences Pures et Appliquées*, **2**, pp. 769–74.

Poincaré, H. [1902]: *La Science et l'Hypothèse*. Translated into English as *Science and Hypothesis*. New York: Dover.

Polanyi, M. [1951]: *The Logic of Liberty*. London: Routledge and Kegan Paul.

Polanyi, M. [1958]: *Personal Knowledge. Towards a Post-Critical Philosophy*. London: Routledge and Kegan Paul.

Polanyi, M. [1966]: *The Tacit Dimension*. London: Routledge and Kegan Paul.

Popkin, R. [1967]: 'Skepticism', in P. Edwards (*ed.*): *The Encyclopedia of Philosophy*, vol. **7**, pp. 449–60. New York: Macmillan.

Popkin, R. [1968]: 'Scepticism, Theology and the Scientific Revolution in the Seventeenth Century', in I. Lakatos and A. Musgrave (*eds.*): [1968], pp. 1– 28.

Popkin, R. [1970]: 'Scepticism and the Study of History', in A. D. Beck and W. Yourgrau (*eds.*): *Physics, Logic and History*, pp. 209–30. New York and London: Plenum.

Popper, K. R. [1933]: 'Ein Kriterium des empirischen Charakters theoretischer Systeme', *Erkenntnis*, **3**, pp. 426–7.

Popper, K. R. [1934]: *Logik der Forschung*. Vienna: Springer. Expanded English edition: Popper [1959a].

Popper, K. R. [1935]: 'Induktionslogik und Hypothesenwahrscheinlichkeit', *Erkenntnis*, **5**, pp. 170–2; published in English in his [1959a], pp. 315– 17.

Popper, K. R. [1940]: 'What is Dialectic?', *Mind*, N.S. **49**, pp. 403–26; reprinted in Popper [1963a], pp. 312–35.

Popper, K. R. [1945]: *The Open Society and Its Enemies*. Two volumes. London: Routledge and Kegan Paul.

Popper, K. R. [1948]: 'Naturgesetze und theoretische Systeme', in S. Moser (*ed.*): *Gesetz und Wirklichkeit*, pp. 65–84. Innsbruch and Vienna: Tyrolia Verlag.

Popper, K. R. [1957a]: 'The Aim of Science', *Ratio*, **1**, pp. 24–35. Reprinted in his [1972], pp. 191–205.

Popper, K. R. [1957b]: *The Poverty of Historicism*. London: Routledge and Kegan Paul.

Popper, K. R. [1957c]: 'Three Views Concerning Human Knowledge', in H. D. Lewis (*ed.*): *Contemporary British Philosophy*, pp. 355–88. Reprinted in Popper [1963a], pp. 97–119.

Popper, K. R. [1958]: 'On the Status of Science and of Metaphysics', *Ratio*, **1**, pp. 97–115. Reprinted in Popper [1963a].

Popper, K. R. [1959a]: *The Logic of Scientific Discovery*. London: Hutchinson.

Popper, K. R. [1959b]: 'Testability and "ad-Hocness" of the Contraction Hypothesis', *British Journal of the Philosophy of Science*, **10**, p. 50.

Popper, K. R. [1960a]: 'On the Sources of Knowledge and Ignorance', *Proceedings of the British Academy*, **46**, pp. 39–71. Reprinted in Popper [1963a].

Popper, K. R. [1960b]: 'Philosophy and Physics', published in *Atti del XII Congresso Internazionale di Filosofia*, vol. **2**, pp. 363–74.

Popper, K. R. [1962]: 'Facts, Standards, and Truth: A further Criticism of Relativism', *Addendum* to the Fourth Edition of Popper [1945].

Popper, K. R. [1963a]: *Conjectures and Refutations*. London: Routledge and Kegan Paul.

Popper, K. R. [1963b]: 'Science: Problems, Aims, Responsibilities', *Federation Proceedings*, **22**, pp. 961–72.

Popper, K. R. [1967]: 'Quantum Mechanics without "the Observer"', in M. Bunge (*ed.*): *Quantum Theory and Reality*. Berlin: Springer.

Popper, K. R. [1968a]: 'Epistemology without a Knowing Subject', in B. Rootselaar and J. Staal (*eds.*): *Proceedings of the Third International Congress for Logic, Methodology and Philosophy of Science*, pp. 333–73. Amsterdam: North Holland. Reprinted as Popper [1972], chapter 3.

Popper, K. R. [1968b]: 'On the Theory of the Objective Mind', in *Proceedings of the XIV International Congress of Philosophy*, **1**, pp. 25–53. Reprinted as Popper [1972], chapter 4.

Popper, K. R. [1968c]; 'Remarks on the Problems of Demarcation and Rationality', in I. Lakatos and A. Musgrave (*eds.*): [1968], pp. 88–102.

Popper, K. R. [1969a]: 'A Realist View of Logic, Physics and History', in W. Yourgrau and A. D. Breck (*eds.*): *Physics, Logic and History*. New York and London: Plenum Press.

Popper, K. R. [1969b]: *Logik der Forschung*. 3rd edition.

Popper, K. R. [1970]: 'Normal Science and its Dangers', in I. Lakatos and A. Musgrave (*eds.*): [1970], pp. 51–8.

Popper, K. R. [1971]: 'Conjectural Knowledge: My Solution of the Problem of Induction', *Revue Internationale de Philosophie*, **95–96**, pp. 167–97. Reprinted as Popper [1972], chapter 1.

Popper, K. R. [1972]: *Objective Knowledge*. Oxford: Clarendon Press.

Popper, K. R. [1974]: 'Replies to my Critics', in P. A. Schilpp (*ed.*): *The Philosophy of Karl Popper*, pp. 961–1197. La Salle: Open Court.

Power, E. A. [1964]: *Introductory Quantum Electrodynamics*. London: Longmans.

Price, D. J. de S. [1959]: 'Contra-Copernicus: a Critical Re-estimation of the Mathematical Planetary Theory of Ptolemy, Copernicus, and Kepler', in M. Clagett (*ed.*): *Critical Problems in the History of Science*, pp. 197–218. University of Wisconsin Press.

Prokhovnik, S. J. [1967]: *The Logic of Special Relativity*. Cambridge University Press.

Prout, W. [1815]: 'On the Relation between the Specific Gravities of Bodies in their Gaseous State and the Weights of their Atoms', *Annals of Philosophy*, **6**, pp. 321–30. Reprinted in *Prout's Hypothesis*, Alembic Club Reprints, **20**, 1932.

Quine, W. V. O. [1953]: *From a Logical Point of View*. Harvard University Press.

Rabi, I. I. [1956]: 'Atomic Structure', in G. M. Murphy and M. H. Shamos (*eds.*): *Recent Advances in Science, Physics and Applied Mathematics*, pp. 27–46. Science editions, New York: Wiley.

Ravetz, J. [1966a]: *Astronomy and Cosmology in the Achievement of Nicolaus Copernicus*. Warsaw: Polish Academy of Sciences.

Ravetz, J. [1966b]: 'The Origins of the Copernican Revolution'. *Scientific American*, **215**, pp. 88–98.

Reichenbach, H. [1951]: *The Rise of Scientific Philosophy*. Los Angeles: University of California Press.

Rufus, W. C. [1931]: 'Kepler as an Astronomer', in *Johan Kepler, 1571–1639, A Tercentenary Commemoration of His Life and Work*, pp. 1–38. Baltimore: The Williams and Wilkins Company.

Runge, C. [1925]: 'Äther und Relativitätstheorie', *Die Naturwissenschaften*, **13**, p. 440.

Russell, B. A. W. [1914]: *The Philosophy of Bergson*. Cambridge: Bowes and Bowes.

Russell, B. A. W. [1919]: *Introduction to Mathematical Philosophy*. London: George Allen and Unwin.

Russell, B. A. W. [1943]: 'Reply to Critics', in P. A. Schilpp (*ed.*): *The Philosophy of Bertrand Russell*, pp. 681–741. La Salle: Open Court.

Russell, B. A. W. [1946]: *History of Western Philosophy*. London: George Allen and Unwin.

Rutherford, E., Chadwick, J. and Ellis, C. D. [1930]: *Radiations from Radioactive Substances*. Cambridge University Press.

Santillana, G. de [1953]: 'Historical Introduction' to Galileo [1632].

Scheffler, I. [1967]: *Science and Subjectivity*. New York: Bobbs–Merrill.

Schlick, M. [1934]: 'Über das Fundament der Erkenntnis', *Erkenntnis*, **4**, pp. 79–99. Published in English in A. J. Ayer (*ed.*): *Logical Positivism*, pp. 209–27. New York: The Free Press, 1959.

Schrödinger, E. [1958]: 'Might perhaps Energy be merely a Statistical Concept?', *Il Nuovo Cimento*, **9**, pp. 162–70.

Shankland, R. S. [1936]: 'An Apparent Failure of the Photon Theory of Scattering', *Physical Review*, **49**, pp. 8–13.

Shankland, R. S. [1964]: 'Michelson–Morley Experiment', *American Journal of Physics*, **32**, pp. 16–35.

Shapere, D. [1964]: 'The Structure of Scientific Revolutions', *Philosophical Review*, **63**, pp. 383–4.

Shapere, D. [1967]: 'Meaning and Scientific Change', in R. G. Colodny (*ed.*): *Mind and Cosmos*, pp. 41–85. University of Pittsburgh Press.

Smith, A. [1773]: 'The Principles which Lead and Direct Philosophical Inquiries Illustrated by the History of Astronomy', in D. Stewart (*ed.*): *Adam Smith: Essays on Philosophical Subjects*. 1799.

Soddy, F. [1932]: *The Interpretation of the Atom*. London: Murray.

Sommerfeld, A. [1916]: 'Zur Quantentheorie der Spektrallinien', *Annalen der Physik*, **51**, pp. 1–94 and 125–67.

Stebbing, L. S. [1914]: *Pragmatism and French Voluntarism*. Girton College Studies, **6**.

Stegmüller, W. [1966]: 'Explanation, Prediction, Scientific Systematization and Non-Explanatory Information', *Ratio*, **8**, pp. 1–24.

Stokes, G. G. [1845]: 'On the Aberration of Light', *Philosophical Magazine*, Third Series, **27**, pp. 9–15.

Stokes, G. G. [1846]: 'On Fresnel's Theory of the Aberration of Light', *Philosophical Magazine*, Third Series, **28**, pp. 76–81.

Symon, K. R. [1963]: *Mechanics*. Second Edition. Reading, Massachusetts: Addison–Wesley.

Synge, J. [1952–4]: 'Effects of Acceleration in the Michelson–Morley Experiment', *The Scientific Proceedings of the Royal Dublin Society*, New Series, **26**, pp. 45–54.

Ter Haar, D. [1967]: *The Old Quantum Theory*. Oxford: Pergamon.

Thomson, J. J. [1929]: 'On the Waves associated with β-rays, and the Relation between Free Electrons and their Waves', *Philosophical Magazine*, Seventh Series, **7**, pp. 405–17.

Tichý, P. [1974]: 'On Popper's Definitions of Verisimilitude', *British Journal for the Philosophy of Science*, **25**, pp. 155–60.

Toulmin, S. [1967]: 'The Evolutionary Development of Natural Science', *American Scientist*, **55**, pp. 456–71.

Toulmin, S. [1972]: *Human Understanding*. Oxford: Clarendon Press.

Treiman, S. B. [1959]: 'The Weak Interactions', *Scientific American*, **200**, March, pp. 72–84.

Truesdell, C. [1960]: 'The Program toward Rediscovering the Rational Mechanics in the Age of Reason', *Archive of the History of Exact Sciences*, **1**, pp. 3–36.

Turnbull, H. W. (*ed.*) [1960]: *The Correspondence of Isaac Newton*. Cambridge University Press.

Uhlenbeck, G. R. and Goudsmit, S. [1925]: 'Ersetzung der Hypothese vom unmechanischen Zwang durch eine Forderung bezüglich des inneren Verhaltens jedes einzelnen Electrons', *Die Naturwissenschaften*, **13**, pp. 953–4.

Urbach, P. [1974]: 'Progress and Degeneration in the "IQ Debate"', *The British Journal for the Philosophy of Science*, **25**, pp. 99–135 and 235–59.

Voltaire, F. M. A. [1738]: *The Elements of Sir Isaac Newton's Philosophy*. Translated by J. Hanna. London: Frank Cass and Company, 1967.

van der Waerden, B. L. [1967]: *Sources of Quantum Mechanics*. Amsterdam: North Holland.

Watkins, J. W. N. [1952]: 'Political Tradition and Political Theory: an Examination of Professor Oakeshott's Political Philosophy', *Philosophical Quarterly*, **2**, pp. 323–37.

Watkins, J. W. N. [1957]: 'Between Analytic and Empirical', *Philosophy*, **32**, pp. 112–31.

Watkins, J. W. N. [1958]: 'Influential and Confirmable Metaphysics', *Mind*, **67**, pp. 344–65.

Watkins, J. W. N. [1960]: 'When are Statements Empirical?', *British Journal for the Philosophy of Science*, **10**, pp. 287–82.

Watkins, J. W. N. [1963]: 'Negative Utilitarianism', *Aristotelian Society Supplementary Volume*, **37**, pp. 95–114.

Watkins, J. W. N. [1964]: 'Confirmation, the Paradoxes and Positivism', in M. Bunge (*ed.*): [1964], pp. 92–115.

Watkins, J. W. N. [1967]: 'Decision and Belief', in R. Hughes (*ed.*): *Decision Making*. London: British Broadcasting Corporation.

Watkins, J. W. N. [1968]: 'Hume, Carnap and Popper', in I. Lakatos (*ed.*): [1968], pp. 271–82.

Watkins, J. W. N. [1970]: 'Against Normal Science', in I. Lakatos and A. Musgrave (*eds.*): [1970]: pp. 25–37.

Watkins, J. W. N. [1971]: 'CCR: A Refutation', *Philosophy*, **47**, pp. 56–61.

Westman, R. S. [1972]: 'Kepler's Theory of Hypothesis and the "Realist Dilemma"', *Studies in History and Philosophy of Science*, **3**, pp. 233–64.

Whewell, W. [1837]: *History of the Inductive Sciences, from the Earliest to the Present Time*. Three volumes. (Frank Cass, 1967.)

Whewell, W. [1840]: *Philosophy of the Inductive Sciences, 'Founded upon their History*. Two volumes. London: Frank Cass, 1967.

Whewell, W. [1851]: 'On the Transformation of Hypotheses in the History of Science', *Cambridge Philosophical Transactions*, **9**, pp. 139–47. Reprinted in R. E. Butts (*ed.*): *William Whewell's Theory of Scientific Method*. University of Pittsburgh Press, 1968.

Whewell, W. [1858]: *Novum Organon Renovatum*. Being the second part of the Philosophy of the Inductive Sciences. Third edition.

Whewell, W. [1860]: *On the Philosophy of Discovery, Chapters Historical and Critical.* Being the third part of the Philosophy of the Inductive Sciences. Third edition.

Whittaker, E. T. [1947]: *From Euclid to Eddington.* Cambridge University Press.

Whittaker, E. T. [1953]: *History of the Theories of Aether and Electricity,* vol. 2. London: Longmans.

Wisdom, J. O. [1963]: 'The Refutability of "Irrefutable" Laws', *The British Journal for the Philosophy of Science,* **13**, pp. 303–6.

Worrall, J. [1976a]: 'Thomas Young and the "Refutation" of Newtonian Optics', in C. Howson (*ed.*): *Method and Appraisal in the Physical Sciences,* pp. 102–79. Cambridge University Press.

Worrall, J. [1976b]: 'The Nineteenth Century Revolution in Optics: a Case Study in the Interaction between Philosophy of Science and History of Science', University of London PhD. Thesis, unpublished.

Wu, C. S. [1966]: 'Beta Decay', in *Rediconti della Scuola Internazionale di Fisico "Enrico Fermi",* XXXII Corso.

Wu, C. S. and Moskowski, S. A. [1966]: *Beta Decay.* New York: Interscience.

Zahar, E. [1973]: 'Why did Einstein's Research Programme Supersede Lorentz's?', *The British Journal for the Philosophy of Science,* **24**, pp. 95–123 and 223–62. Reprinted in C. Howson (*ed.*): *Method and Appraisal in the Physical Sciences.* Cambridge University Press, 1976.

Lakatos bibliography[1]

[1946a]: 'Citoyen és Munkásosztály', *Valóság*, **1**, pp. 77–88.
[1946b]: 'A Fizikai Idealizmus Bírálata', *Athenaeum*, **1**, pp. 28–33.
[1947a]: 'Huszadik Szársad: Társadalomtudományi és politikai szemle', *Forum*, **1**, pp. 316–20.
[1947b]: 'Eötvos Collegium – Györffy Kollégium', *Valóság*, **2**, pp. 107–24.
[1947c]: Review of K. Jeges: 'Megtanulom a Fizikat', *Társadalmi Szemle*, **1**, p. 472.
[1947d]: Review of J. Hersy; 'Hirosima', *Társadalmi Szemle*, **1**.
[1947e]: 'Vigilia, Szerkeszti Juhász Vilmos és Sik Sandor', *Forum*, **1**, pp. 733–6.
[1961]: 'Essays in the Logic of Mathematical Discovery'. Unpublished PhD dissertation. Cambridge.
[1962]: 'Infinite Regress and Foundations of Mathematics', *Aristotelian Society Supplementary Volume*, **36**, pp. 155–94. Republished as chapter 1 of volume 2.
[1963]: Discussion of 'History of Science as an Academic Discipline' by A. C. Crombie and M. A. Hoskin, in A. C. Crombie (*ed.*): *Scientific Change*, pp. 781–5. London: Heinemann. Republished as chapter 13 of volume 2.
[1963–4]: 'Proofs and Refutations', *British Journal for the Philosophy of Science*, **14**, pp. 1–25, 120–39, 221–43, 296, 342. Republished in revised form as part of Lakatos [1976c].
[1967a]: *Problems in the Philosophy of Mathematics*. Edited by Lakatos. Amsterdam: North Holland.
[1967b]: 'A Renaissance of Empiricism in the Recent Philosophy of Mathematics?' in I. Lakatos (*ed.*): [1967a], pp. 199–202. Republished in much expanded form as Lakatos [1976b].
[1967c]: *Dokatatelstva i Oprovershenia*. Russian translation of [1963–4] by I. N. Veselovski. Moscow: Publishing House of the Soviet Academy of Sciences.
[1968a]: *The Problem of Inductive Logic*. Edited by Lakatos. Amsterdam: North Holland.
[1968b]: 'Changes in the Problem of Inductive Logic', in I. Lakatos (*ed.*): [1968a], pp. 315–417. Republished as chapter 8 of volume 2.
[1968c]: 'Criticism and the Methodology of Scientific Research Programmes', *Proceedings of the Aristotelian Society*, **69**, pp. 149–86.
[1968d]: 'A Letter to the Director of the London School of Economics', in C. B. Cox and A. E. Dyson (*eds.*): *Fight for Education, A Black Paper*, pp. 28–31. London: Critical Quarterly Society. Republished as chapter 12 of volume 2.
[1969]: 'Sophisticated versus Naive Methodological Falsificationism', *Architectural Design*, **9**, pp. 482–3. Reprint of part of [1968c].

[1] References to 'volume 2' are to Lakatos [1977b]. We have included as many of Lakatos's Hungarian writings as we have been able to trace.

[1970a]: 'Falsification and the Methodology of Scientific Research Programmes', in Lakatos and A. Musgrave (eds.): [1970], pp. 91–196. Republished as chapter 1 of this volume.

[1970b]: Discussion of 'Knowledge and Physical Reality' by A. Mercier, in A. D. Breck and W. Yourgrau (eds.): Physics, Logic and History, pp. 53–4. New York: Plenum Press.

[1970c]: Discussion of 'Scepticism and the Study of History' by R. H. Popkin, in A. D. Breck and W. Yourgrau (eds.): Physics, Logic and History, pp. 220–3. New York: Plenum Press.

[1971a]: 'Popper zum Abgrenzungs– und Inductionsproblem', in H. Lenk (ed.): Neue Aspekte der Wissenschaftstheorie, pp. 75–110. Braunschweig: Vieweg. German translation of [1974c] by H. F. Fischer. Republished as chapter 3 of this volume.

[1971b]: History of Science and its Rational Reconstructions', in R. C. Buck and R. S. Cohen (eds.): P.S.A. 1970 Boston Studies in the Philosophy of Science, 8, pp. 91–135. Dordrecht: Reidel. Republished as chapter 2 of this volume.

[1971c]: 'Replies to Critics', in R. C. Buck and R. S. Cohen (eds.): P.S.A. 1970, Boston Studies in the Philosophy of Science, 8, pp. 174–82. Dordrecht: Reidel.

[1974a]: 'History of Science and its Rational Reconstructions', in Y. Elkana (ed.): The Interaction Between Science and Philosophy, pp. 195–241. Atlantic Highlands, New Jersey: Humanities Press. Reprint of [1971b].

[1974b]: Discussion Remarks on Papers by Ne'eman, Yahil, Beckler, Sambursky, Elkana, Agassi, Mendelsohn, In Y. Elkana (ed.): The Interaction Between Science and Philosophy, pp. 41, 155–6, 159–60, 163, 165, 167, 280–3, 285–6, 288–9, 292, 294–6, 427–8, 430–1, 435. Atlantic Highlands, New Jersey: Humanities Press.

[1974c]: 'Popper on Demarcation and Induction', in P. A. Schilpp (ed.): The Philosophy of Karl Popper, pp. 241–73. La Salle: Open Court. Republished as chapter 3 of this volume.

[1974d]: 'The Role of Crucial Experiments in Science', Studies in the History and Philosophy of Science, 4, pp. 309–25.

[1974e]: 'Falsifikation und die Methodologie Wissenschaftlicher Forschungsprogramme', in I. Lakatos and A. Musgrave (eds.): Kritisismus und Erkenntnisfortschrift. German translation of [1970a] by Á. Szabó.

[1974f]: 'Die Geschichte der Wissenschaft und Ihre Rationalen Reconstruktionen', in I. Lakatos and A. Musgrave (eds.): Kritisismus und Erkenntnisfortschrift. German translation of [1971b] by P. K. Feyerabend.

[1974g]: Wetenschapsfilosofie en Wetenschapsgeschiedenis. Boom: Mepple. Dutch translation of [1970a] by Karel van der Lenn.

[1974h]: 'Science and Pseudoscience', in G. Vesey (ed.): Philosophy in the Open. Open University Press. Republished as the introduction to this volume.

[1976a]: 'Understanding Toulmin', Minerva, 14, pp. 126–43. Republished as chapter 11 of volume 2.

[1976b]: 'A Renaissance of Empiricism in the Recent Philosophy of Mathematics?', British Journal for the Philosophy of Science, 27, pp. 201–23. Republished as chapter 2 of volume 2.

[1976c]: Proofs and Refutations: The Logic of Mathematical Discovery. Edited by J. Worrall and E. G. Zahar. Cambridge University Press.

[1977a]: The Methodology of Scientific Research Programmes: Philosophical Papers, volume 1. Edited by J. Worrall and G. P. Currie. Cambridge University Press.

[1977b]: Mathematics, Science and Epistemology: Philosophical Papers, volume 2. Edited by J. Worrall and G. P. Currie. Cambridge University Press.

With Other Authors

[1968]: *Problems in the Philosophy of Science.* Edited by I. Lakatos and A. Musgrave. Amsterdam: North Holland.

[1970]: *Criticism and the Growth of Knowledge.* Edited by I. Lakatos and A. Musgrave. Cambridge University Press.

[1976]: 'Why Did Copernicus's Programme Supersede Ptolemy's?', by I. Lakatos and E. G. Zahar, in R. Westman (*ed.*): *The Copernican Achievement*, pp. 354–83. Los Angeles: University of California Press. Republished as chapter 5 of this volume.

Index of names

(Indexes compiled by Alex Bellamy)

Subject Index